THE BUSINESS OF BIOTECHNOLOGY

"If you cannot—in the long run—tell everyone what you have been doing, your doing has been worthless."

Erwin Schrödinger
Austrian Physicist (1887–1961)

THE BUSINESS OF BIOTECHNOLOGY

From the Bench to the Street

R. DANA ONO

With 23 contributing authors

Butterworth–Heinemann
Boston London Oxford Singapore Sydney Toronto Wellington

Recognizing the importance of preserving what has been written, it is the policy of Butterworth–Heinemann to have the books it publishes printed on acid-free paper, and we exert our best efforts to that end.

Library of Congress Cataloging-in-Publication Data

The Business of biotechnology : from the bench to the street / edited by R. Dana Ono.
p. cm.
Includes bibliographical references and index.
ISBN (invalid) 0-7506-9119-0 (case bound)
1. Biotechnology industries. I. Ono, R. Dana (Richard Dana)
HD9999.B442B88 1991
338.4'76606—dc20 91-16857
CIP

British Library Cataloguing in Publication Data

The business of biotechnology : from the bench
to the street.
I. Title II. Ono, R. Dana
338.476606

ISBN 0-7506-9119-0

Butterworth–Heinemann
80 Montvale Avenue
Stoneham, MA 02180

10 9 8 7 6 5 4 3 2 1

Printed in the United States of America

This book is dedicated to Dr. Karel F. Liem, mentor and friend, who gave me that window of opportunity at Harvard.

CONTENTS

CONTRIBUTING AUTHORS

Lisa E. Burns is President of Burns McClellan, Inc., a public relations firm specializing in health care and biotechnology. Previously, Ms. Burns was Senior Vice President, Director of Health Care and Biotechnology Marketing at Fleishman-Hillard, Inc., New York, where she founded the company's health care and biotechnology division. Prior to joining Fleishman-Hillard, she was Senior Vice President, Director, National Media Relations at Edelman Public Relations, New York, and Director of Health Care Media Services at Burson-Marsteller, New York. She has spent the last eight years specializing in health care and biotechnology. Ms. Burns planned the public relations launch of Armour's Factor VIII: C (Monoclate), the media activity for the launch of Schering-Plough's Interferon (Intron A), and Sandoz's cyclosporine.

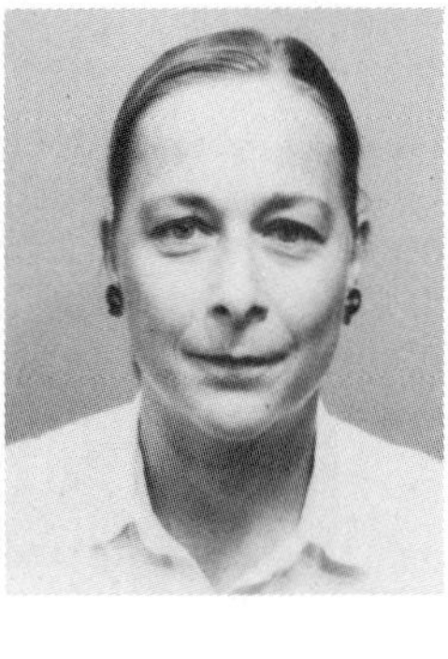

Susan K. Clymer is Managing Director of NichiBei Bio, Inc., a business development and management consulting firm specializing in U.S.–Japan transactions in the Life Sciences. Prior to the formation of NichiBei Bio, Inc., Ms. Clymer spent a year in Japan as a Fulbright Scholar examining corporate biotechnology research and investment strategies. While in Japan, she was affiliated with the Japan Bioindustry Association, a MITI-sponsored biotechnology trade association. Ms. Clymer was the Manager of Business Development at California Biotechnology, Inc. prior to her Fulbright and is a co-author of The Congressional Office of Technology Assessment (OTA) report on U.S. competitiveness in biotechnology. The author is an internationally recognized expert on Japanese biotechnology and a graduate of Georgetown University.

Allan R. Ferguson is President at 3i Ventures, Boston, Massachusetts, a venture capital firm. Mr. Ferguson was most recently Senior Vice President at 3i Venture. He has held senior operating management positions at Johnson & Johnson and at Damon Biotech, where he worked in manufacturing, engineering, development, and planning. He has extensive experience in the start-up and development of medical products. Mr. Ferguson holds a B.S. in chemical engineering from the University of Delaware.

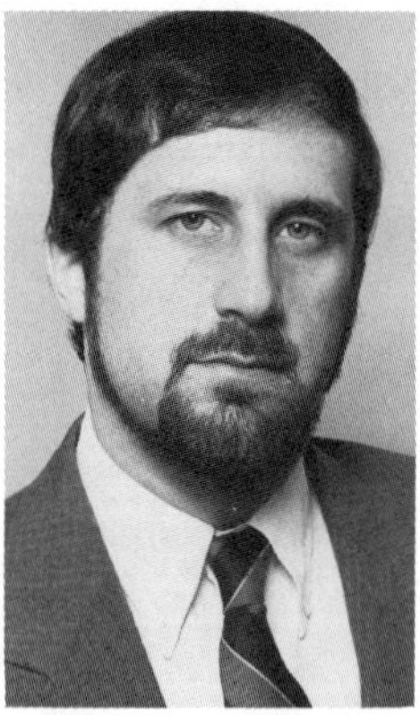

David J. Glass is currently President of D. Glass Associates, Inc., an environmental and agricultural biotechnology consulting firm. Dr. Glass was formerly Vice President of Government and Regulatory Affairs for BioTechnica Agriculture (a wholly owned subsidiary of BioTechnica International), which commercializes improved crop varieties and seed treatments arising from biotechnology. Dr. Glass is a spokesperson on biotechnology, its regulation, and its impact on agriculture. He has appeared before Congressional and state legislature committees. He is a founder and former president of the Massachusetts Biotechnology Council, Inc., and is a member of the Executive Committee of the International Food Biotechnology Council. He holds a B.S. in biology from Cornell University and a Ph.D. in biochemistry from Princeton University.

Richard D. Godown is President of the Industrial Biotechnology Association (IBA). He has worked with the public relations firm of Burson-Marsteller in Washington, D.C., and is a former general counsel of the National Association of Manufacturers and Associate Director of the Business Roundtable. As an attorney and government relations and communications professional, Mr. Godown has appeared on television, on radio, and in the print media as an industry spokesperson. He has appeared on behalf of the biotechnology industry before House and Senate committees and helped lead IBA to two significant legislative victories: the adoption of the Drug Export Act and the enactment of Process Patent legislation.

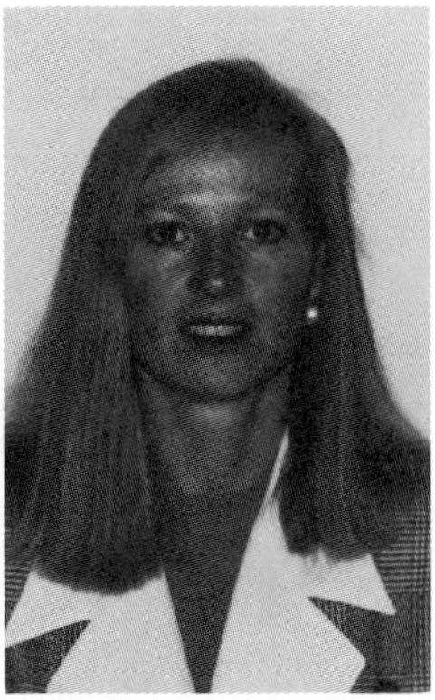

Suzanne S. Groet is a Program Manager in Corporate Development at Genzyme Corporation, a diversified health care products company headquartered in Cambridge, Massachusetts. She has directed the development of the company's grants and contracts program and focuses on the development of new business opportunities. Previously, at Battelle Columbus Laboratories, Ms. Groet was involved in a broad spectrum of bioenvironmental and biomedical research programs. She holds a B.S. from the University of Rhode Island and an M.S. in botany from the University of Connecticut.

Carol V. Hall is a Principal in BioVenture Consultants, a consulting firm that specializes in the assessment of emerging bioscience technologies. Dr. Hall has served as a biotechnology analyst at Shearson, Lehman Brothers and as a member of Sutro & Company's corporate finance department. Most recently, she was with Jesse & Co. a private investment banking group. Dr. Hall received her Ph.D in molecular biology from Stanford University and completed postdoctural work at DNAX Research Institute, a subsidiary of Schering-Plough Corporation. She earned a certificate of business administration from the University of Pennsylvania's Wharton School and is a chartered financial analyst.

Mark A. Hofer is Vice President, General Counsel, and Assistant Secretary for Genzyme Corporation, a diversified human health care products company with its headquarters in Cambridge, Massachusetts. Prior to joining Genzyme, he was Chief Patent Counsel and Assistant Secretary for Integrated Genetics, Inc., until its merger into Genzyme. He served for a number of years as Patent Counsel for Johnson & Johnson, specializing in biotechnology. He has also worked with the New York patent and trademark firm of Watson Leavenworth Kelton & Taggart.

Robert Kupor joined Kidder, Peabody as a biotechnology analyst and is now Vice President. Prior to joining Kidder, Peabody, he worked at Cable, Howse & Ragan in Seattle, Washington. He has done postdoctoral research at the University of California, San Francisco, and has worked as an assistant professor of biology at the University of Tennessee at Chattanooga. Over the past several years he has been included in the Institutional Investor All-American Research Team in biotechnology. Dr. Kupor received his B.A. in biology from Brooklyn College, a Ph.D. in microbiology from Harvard University, and an M.B.A. in finance from the University of Washington.

Donald A. Lotufo is the Managing Director of M/P Lotufo Associates, Inc., an executive search firm. Mr. Lotufo was formerly Executive Vice President and Chief Operating Officer with Halbrecht Associates, Inc., Stamford, Connecticut. He has been Senior Vice President and a Principal with Management Advisors of Princeton, Inc. (a national executive search organization), and has been associated with The Bristol Myers and Gillette companies as Corporate Director of Personnel. His industry knowledge includes consumer package goods, pharmaceutical and health care, biotechnology, information and communications, financial services, and a wide range of industrial organizations. Mr. Lotufo has lectured internationally and written on organizational behavior, executive selection, manpower planning, and other vital management issues.

Christina L. Lowell is a former partner in the San Francisco Bay area practice of KPMG Peat Marwick, and served as the U.S. coordinator of the firm's Biotechnology Services. She specializes in early-stage high technology companies and has worked with industries such as medical technology, computer and software technology, leasing, venture capital and biotechnology. Currently, Ms. Lowell has left the world of public accounting to join a *start-up* company as the Chief Financial Officer and experience "growth" first hand.

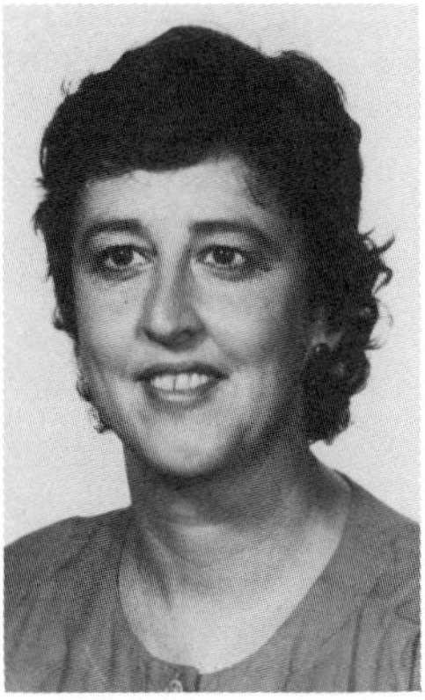

Lita L. Nelsen is the Associate Director of the Massachusetts Institute of Technology's Technology Licensing Office, specializing in licensing of biotechnology-related inventions. In this position she has had primary responsibility for over 100 license agreements and for the start-up of over a dozen new biotechnology companies founded around MIT technology. She has worked in chemical engineering research and marketing, primarily in the fields of membrane processing, medical devices, and biotechnology. Ms. Nelsen received B.S. and M.S. degrees in chemical engineering from MIT and an M.B.A. as a Sloan Fellow, also at MIT.

Melvin S. Oka is Assistant Director in the Biological Process Sciences at SmithKline Beecham Pharmaceuticals. A noted authority in cell biology and biopharmaceutical production technologies, he is the author of several fundamental papers in the field. Dr. Oka received his B.S. in zoology from the University of California, Berkeley, and a Ph.D. in cell biology from the University of Texas. After postdoctoral studies at Baylor College of Medicine, he began a career in industry working with Damon Biotech, Inc.

R. Dana Ono is President and Chief Executive Officer of ARCTURUS Pharmaceutical Corporation. Dr. Ono was formerly Vice President of Business Development at Enzytech, Inc., a biotechnology company in Cambridge, Massachusetts. Dr. Ono has held senior management positions at Integrated Genetics, Inc., Regis McKenna, Inc., and Damon Biotech, Inc. His fields of expertise are strategic planning, product management, technology development, licensing, and commercial development. A founding director of the Massachusetts Biotechnology Council, Inc., Dr. Ono is a frequent spokesperson for the biotechnology industry. Dr. Ono received his A.B. from The Johns Hopkins University in earth and planetary sciences and his A.M. and Ph.D. in biology from Harvard University, where he also completed a program in business administration.

Debra A. Peattie is a Founding Staff Scientist at Vertex Pharmaceuticals, Inc., located in Cambridge, Massachusetts. Before joining Vertex, a rational drug design company, Dr. Peattie was an Assistant Professor in the Department of Tropical Public Health at the Harvard School of Public Health. After graduating from Hollins College, Virginia, with an undergraduate degree in chemistry, Dr. Peattie enrolled as a graduate student at Harvard University in the Department of Biochemistry and Molecular Biology, where she began researching chemical DNA sequencing techniques and molecular biology under the tutelage of Nobel laureate, Dr. Walter Gilbert. Following her graduate work, Dr. Peattie did postdoctoral work at Stanford University.

Brandon J. Price is Vice President of Marketing and Sales at Quality Biotech Inc. He has senior management experience in the biotechnology industry and has authored many publications in technical and trade literature. His experience includes general management and marketing positions at Damon Biotech, Inc., and the Endotronics Partnership, and research and development in flow cytometry at Ortho Diagnostic Systems (a Johnson & Johnson Company). Dr. Price was also on the Oncology and Biomedical Engineering faculty at the University of Miami, Florida. Dr. Price received his B.S. and Ph.D. degrees in biophysics from the University of Michigan in Ann Arbor and completed a postdoctoral fellowship at Los Alamos National Laboratory.

Norbert Rau, a marine biologist and oceanographer, is head and part-owner of RAUCON Bioinformatics and Consulting GmbH in West Germany. RAUCON is a management information and consulting company focusing exclusively on the commercialization of biotechnology. RAUCON has provided management consulting services in countries on five continents. Dr. Rau's company has completed biotechnology-related consulting work in genetic engineering, pharmaceuticals, health care, biochemicals, mechanical engineering, environmental technology, agricultural biotechnology, fisheries and aquaculture, food industry, financing, and space technology.

Randall G. Rupp is Vice President of Research and Development at Biohybrid Technologies, Inc. Dr. Rupp has held senior management positions at Somatogen and Invitron. Dr. Rupp is a noted authority on animal cell production systems and has been responsible for the scale-up and production of many therapeutic proteins from cultured animal cells. He has authored papers on the subject and has lectured extensively. He holds a Ph.D. in biomedical sciences from the University of Texas Graduate School at the M.D Anderson Hospital and the Tumor Institute in Houston.

Roger E. Shamel is President and founder of Consulting Resources Corporation. Mr. Shamel has been involved with biotechnology as applied to human health, agriculture, chemicals, and environmental diagnostics and has led international executive briefing sessions dealing with the global status of biotechnology. He has an international reputation as a consultant, analyst, strategist, and author. He has appeared in such leading biotechnology and business publications as *Genetic Engineering News, Bio/Technology, Business Week,* and the *Wall Street Journal.* Prior to Consulting Resources Corporation, Mr. Shamel worked at Arthur D. Little, Inc., where he was a Senior Project Manager. He received his M.S. in chemistry from Ohio State University and his M.B.A. as a Humphrey Fellow at Harvard Business School.

James P. Sherblom is Chairman, President, and Chief Executive Officer of TSI Corporation, a publicly traded biotechnology company. Before coming to TSI, Mr. Sherblom served as Senior Vice President, Finance and Administration, Chief Financial Officer, and Treasurer of Genzyme Corporation, a biopharmaceutical company. He has also been a consultant with Bain and Company, a strategy consulting firm, for which he worked in Boston, London, and Munich. Mr. Sherblom is a director of the Massachusetts Biotechnology Council, Inc., which he helped found. He received his M.B.A. with high distinction from Harvard University and his B.A. from Yale University.

John M. Stewart is an entrepreneur who has a consulting practice in strategic marketing and corporate finance in biotechnology, pharmaceuticals, and hospital management. Mr. Stewart's most recent venture was as founder and President of Eygene, Inc. (now Telor, Inc.) an ophthalmic pharmaceutical company. Prior to Eygene, he headed the Biotechnology Services Division of Regis McKenna, Inc., an affiliate of KPMG Peat Marwick. Mr. Stewart has also served as Chief Financial Officer of Chiron Corporation and Director of International Sales and Marketing for Allergan Pharmaceuticals, Inc. Mr. Stewart has worked in new product planning and marketing for pharmaceuticals, diagnostics, instruments, medical devices. He recently addressed biotechnology issues at a conference sponsored by the Ministry of Science and Technology of the People's Republic of China.

Harriet M. Strimpel has authored numerous reports on a variety of technical and business issues concerning the biotechnology and investor community. Dr. Strimpel has been a consultant to the bioscience field and went on to found HMS Biosciences in Boston. She later became affiliated with Bioventure Consultants in San Mateo, California. Furthermore, she has acted as consultant to venture groups, biotechnology companies, and nonprofit institutions. Dr. Strimpel received her Ph.D. in molecular virology from Oxford University and carried out her postdoctoral studies on the baculoviruses at Hebrew University in Rehovot, Israel. Dr. Strimpel continued her research in gene structure and function at the National Institute for Medical Research in London and at Biogen Research Corporation.

Robert Teitelman is the author of *Gene Dreams: Wall Street, Academia and the Rise of Biotechnology*, published by Basic Books, New York. A veteran financial writer, he has served on the staffs of *Forbes* and *Financial World* magazines. He is currently a senior editor at *Institutional Investor Magazine* in New York, and is writing a book on technology and business in post–World War II America.

FOREWORD

As biotechnology enters the 1990s, a wide range of issues and trends will have an impact on the ability of this emerging industry to bring its products to a fiercely competitive global market. Some challenges are new ones brought on by the introduction of a new technology, others are challenges that every high technology industry must address. As an industry, biotechnology must anticipate, identify, and take on the legislative, regulatory, and public relations hurdles ahead. The long-term commercial success of our young industry depends on it.

But many successes are already taking place. In 1991, biotechnology raised over $1.2 billion in capital in the first quarter and strategic alliances continue to be formed in medical and biotechnology companies at a great rate. As a result, combined sales of U.S. biotechnology products are expected to top $3 billion in 1991. Our industry continues to work feverishly on the development of products and processes that contribute to improved diagnosis and treatment of diseases, more vigorous food crops, alternative sources of energy, and a cleaner environment. We expect the next 10 years to bring $40–50 billion in annual sales of U.S. biotechnology products and a dramatic increase in marketing and manufacturing personnel.

But it won't be easy. This book thoroughly examines the many existing and future business challenges for biotechnology, providing a unique insight into the intricate web of critical factors with which biotechnology entrepreneurs must come to terms if they wish to be successful. It is a story told by key players in the field—CEOs, marketing executives, regulatory experts, investment specialists—who live with these challenges every day.

Those of us who work in biotechnology are staking our futures on the promise of this burgeoning industry. But we know that the onslaught of bioagricultural and biomedical advances may be delayed or seriously impeded if biotechnology, as an industry, fails to come to terms with a number of critical issues potentially affecting its commercial viability. These include: patent backlogs, state and federal legislation and regulation, government funding of basic science literacy and training, health care costs, the development of biotechnology in Europe and Japan, product safety concerns, liability insurance, regulatory resources, and public perceptions.

Though we cannot directly control all trends and conditions affecting our development, such as swings in the economy, Wall Street's perspective, and biotechnology's acceptance in other countries, we do expect our successes on matters we can influence to turn the others in our favor. Already, federal and state legislators are recognizing biotechnology's enormous potential benefit to society—and their actions are beginning to reflect that realization.

The biotechnology industry must continue to lay the foundation for its products' acceptance by working to generate an enhanced understanding of our societal contributions by regulators, legislators, the media, and the general public.

Once that battle is won, the others will be easier for all biotechnology entrepreneurs.

Richard D. Godown
President of the Industrial
Biotechnology Association

PREFACE

The final decade of the twentieth century offers us a good opportunity to pause and reflect on where we have been, where we are, and where we are going. When Butterworth–Heinemann asked me to edit a book on the business of biotechnology, I felt that the timing was appropriate for those of us who participated in the formative years of biotechnology to stop, reflect, and then jot down some of our impressions of this fledgling industry.

No one will argue that biotechnology is here to stay. Bioengineered products for human health care represent just the tip of the iceberg. Environmental, agricultural, and even marine applications will be realized in the not-too-distant future. We have no real inkling of biotechnology's vast potential as we approach the year 2000 and beyond. As time marches on and technological advances continue at an increasing pace, our children and their children will enjoy the benefits that accrue from the efforts of those who came before: the entrepreneurs whose clear vision set the stage for the technology innovations, the business leaders who had the ability to translate the technology into business opportunities, and the investors who fueled the visions and dreams of these pioneers.

In planning the conceptual framework for this book, I wanted to capture the vibrancy of this industry by having the authors address key issues affecting biotechnology today. I wanted to ensure, however, that the issues were not so specific that they would be out of date by the time the book was published. Such an approach, unfortunately, prevented the contributors from going into great detail. Such detail could constitute several treatises, perhaps several books in some instances.

The authors who have so graciously given their time and effort are all pioneers in the world of biotechnology. Not one has a "typical" biotechnology background. The biotechnology industry comprises individuals with a vast array of backgrounds, all contributing their unique experiences and skills to achieving the tremendous advancements we have seen in a short period of time. Like the varied backgrounds of their authors, the contributed chapters also proved to be eclectic. I have purposely chosen not to homogenize them, but to allow the individual contributors to express their thoughts in their unique styles. Any inaccuracies, fallacies, or other incongruities are the result of my oversight and I take full responsibility for them.

This book has been in the planning and execution stage for over two years. I would like to thank the authors collectively for putting pen to paper or mouse to pad and sharing their valuable knowledge with the readers. I would also like to thank the talented staff at Butterworth–Heinemann, among them Brad Gray, Kris Smead, and Greg Franklin, for their support, encouragement, and enthusiasm for this book. It made the difference. In addition, kudos go to Andrea Seward who contacted me on a regular basis and kept the project moving as the book entered the final days of production. Dolores Wolfe at York Production Services deserves special thanks for working closely with me on a day-to-day basis to produce and review the text in a critical and timely fashion.

Finally, I want to thank my wife Tita and our children Celia and Max. No longer will the drafts of manuscripts clutter the dining room table night after night . . . at least not for this project!

R. Dana Ono
Arlington, Massachusetts
May 1991

CHAPTER 1

MIRACLES AND MONEY: THE EVOLUTION OF A NEW INDUSTRY

Robert Teitelman

A TANGLE OF ANTECEDENTS

The evolution of biotechnology is conventionally told in terms of its scientific antecedents. The industry explodes onto the scene as a byproduct of two fundamental advances, Herbert Boyer and Stanley Cohen's 1973 recombinant DNA experiments and the discovery of a technique for harvesting monoclonal antibodies by Georges Kohler and Cesar Milstein in 1975. But to trace its development in such simple terms suggests that biotechnology, as a commercial phenomenon, followed in lockstep from those technical achievements—that these particular scientific breakthroughs had a deterministic effect on the growth of the industry. And to do so is to believe that there is a fairly simple causal link between science and technology.

This is not the case. As one examines the tangled roots of biotechnology, one is struck by what a diversity of experience had to come into play for the industry that we know to appear. Some fairly prominent signposts mark the way—the war on cancer, interferon and the other immunotherapies, the debate over science and commerce—while others are hidden in the underbrush—the lessons of Syntex, the inflation of the 1970s, the development of venture capital and of the public equity markets. Each of these events provided a piece of the puzzle; each of them altered, however subtly, prevailing ways of thinking.

In the 1970s, these experiences successfully renovated two traditional mindsets—one in academia, the other on Wall Street. In academia, many biologists became convinced that the traditional wall between science and commerce served no particular purpose and had to come down. On Wall Street, the financiers came to see biotechnology as a sort of biological proxy to microelectronics and computers. The final step in the road to a biotechnology industry resulted when these two outlooks merged: business as science, science as business.

American biomedicine was undergoing a quiet, but profound, upheaval in the mid-1970s. Pressed by a proliferation of jostling biological movements, by a mounting population of researchers, by militant public interest groups, and by macroeconomic pressures and ideological politics, the traditional pillars of the biomedical community began to shake. Peer review, the sharp separation between science and commerce, the metropolitan dominance of the National Institutes of Health (NIH), and the central regulatory role of the Food and Drug Administration (FDA), all came under attack. Biomedicine was decentralizing and, in a sense, deregulating. The tripartite institutional structure—academia, drug industry, and funding bureaucracy—that had been in place since World War II struggled to adjust to the incursions of major new forces. In hindsight one can see the connections between the rise of biotechnology and the future problems of the FDA, between the ascendancy of public financial markets and the erosion of peer review, and between the democracy of finance and the meritocracy of science. Of course, they were not so easy to see at the time.

All this is not to underestimate the technical breakthroughs of the 1970s. In particular, Boyer and Cohen's first recombinant DNA experiments created a powerful tool that offered up the possibility—one that seemed, at the time, to be so very close—that molecular biology could be manipulated like any other engineering discipline. It suggested that the gap between science and technology had slammed shut, that revolutionary developments would quickly follow. That realization, in turn, unleashed two seemingly contradictory movements: first, a flurry of doubt, soul searching, and fractious controversy over moral and social issues; then, a pell-mell rush into business. But beneath all this seeming confusion runs a broader, more profound theme: academic molecular biology's attempt in the 1970s to define and then alter its view of itself and its place in the world. Was molecular biology a sort of Platonic republic of science, self-contained, self-regulating, a model of free inquiry and open communication? Could it continue to strive for that Platonic ideal as the science matured, as it became more powerful, more strategically and commercially valuable? Also, if that were the case, what were the connections, and responsibilities, to a greater nonscientific, democratic public?

ACADEMIC CONTROVERSIES

By now the controversies that erupted in 1970s academic biology have taken on an antique air, a glow one associates with adolescence. There are memories of the idealism of Asilomar, a shaking of heads at the moratorium on recombinant DNA work and the monster-in-the-sewer debates in Cambridge, and the pain of bitter academic warfare over commercial ties. It is not that the fears that motivated these events have been eliminated; rather, some of these issues have been resolved and others today seem merely irrelevant. The underlying philosophical conflicts have been reduced to guidelines, language, and bureaucracy: adjustment. Biology, even in academia, no longer belongs to the biologists alone—if it ever really did. Like many of the physicists at Los Alamos during World War II, molecular biologists naively believed that they could control the powerful forces they had unleashed and that molecular biology could operate as a large, powerful community as it had when it was a small, obscure sect.

Molecular biology had been an expansionary field—competitors for funding called it hegemonic—since Watson and Crick elucidated the structure of DNA in 1953. The long string of Nobel-caliber discoveries that followed attracted increasing funding, which, in turn, fed the growth of laboratories and manpower. Not surprisingly, the number of foot soldiers—mostly graduate students, postdoctoral students (postdocs) and technicians—at the base of this pyramid exploded. In 1971 came the war on cancer, which by 1981 was pumping nearly a billion dollars into advanced biological research through the National Cancer Institute. The age when the discipline could gather each summer around the cove at Cold Spring Harbor—described so well by Horace Freeland Judson in *The Eighth Day of Creation*—was over. Laboratory overheads were soaring; as the 1970s moved on, the tendency toward big science accelerated. No longer was it just a community of teachers and students. Instead, there were now laboratory chiefs and major investigators, technicians, and postdocs, the tenured and the untenured. The community itself split into specialties (virology, immunotherapy, the DNA and RNA crowd, the biochemists, the cell development folks, and in time, the cloners and sequencers) and factions, defined by age and class (those established and the far larger numbers seeking their funded niche, and those at the center and those on the periphery). The social and professional tension between physician-researchers and Ph.D.s would be a fascinating study on its own. At the top of this scheme sat the laboratory chief, a powerful academic entrepreneur—a veritable scientific superstar if he had a Nobel—who organized a middle-management level of section chiefs and investigators, who in turn rested on a working class of postdocs and technicians.

In the early 1970s, tensions between these "classes" were only just beginning to be felt. Then in 1973 came the successful Boyer and Cohen genetic recombinations. In June of that year, the experiment was all the talk at the first organized meeting after that event, the Gordon Conference on Nucleic Acids in New Hampshire. Fearful of the cancer-causing effects of an altered virus that escaped the laboratory, those attending the conference felt some sort of response was necessary. Thus, the two conference chairpersons, Dr. Maxine Singer and Dr. Dieter Soll, jointly wrote a letter to the president of the National Academy of Sciences and to the National Institute of Medicine expressing their "grave concern. . . . These experiments offer exciting and interesting potential both for advancing knowledge of fundamental biological processes and for alleviation of human health problems. Certain such hybrid molecules can present a health hazard for the laboratory workers and the public. . . ." The letter then appeared in *Science*. [Cherfas 1982].

The controversy, so far confined to scientific circles, was already racing ahead. No one wanted to unleash some Andromeda strain (the popular novel of that name was published in 1969, the movie appeared in 1971) on a helpless population. On the other hand, any attempt to freeze research ran against even more basic drives: curiosity, free inquiry, and the progress of science, not to mention continuing funding and fame. A committee on biohazards was formed by the National Academy of Sciences with Paul Berg from Stanford at its head, and including James Watson, Herbert Boyer, and Stanley Cohen.

The Berg report came out in mid-1974, urging certain safety restrictions when experimenting with recombinant DNA products, and calling for a moratorium on certain experiments until more was known. This was—and still is—an unparalleled exercise: A group of scientists unilaterally deciding that a broad and clearly fertile area of experimentation should not be pursued (Berg himself anticipated this moratorium when he voluntarily halted his own recombinant experiments using the SV40 virus in the early 1970s).

It could not last. In March of 1975, almost two years after the original Boyer and Cohen experiment, 140 of the most prominent molecular biologists in the world adjourned to a lush California state park called Asilomar on the Monterey Peninsula three hours south of San Francisco to decide what to do next.

Asilomar marked the height of the early, and fairly unified, idealism; it was the very embodiment of a republic of science, a sense reinforced by the Edenic setting. The meeting participants hardly agreed on everything, but they did agree that genetic recombination was a scientific issue that should be dealt with by academic scientists; the urge toward self-regulation was very strong. On that consensus, the meeting glided over serious contradictions. This was, after all, a self-selected (and international) elite, even if it did operate in a highly democratic fashion; and it was an elite, particularly in its dominant American wing, that was supported by public money. The proceedings themselves were run in the sort of labored openness that characterized student activism just a few years earlier. This was hardly surprising, since throughout the 1960s, molecular biology loosely shared a generally liberal to left-wing political stance—another key to its sense of community. The only jarring note came when a lawyer invited to speak warned that this group, for all its expertise, was not qualified to make decisions affecting the public at large. Only the legislature—Congress—can do that, even if the choices it makes turn out to be foolish. And, a second attorney added, self-regulation also implied accountability, raising the specter of liability suits. Suddenly, the outside world of lawyers and politicians seemed to press uncomfortably close [Rogers 1975]. James Watson, the spiritual center of American molecular biology, summed up the feeling a few years later in a book on the biohazards controversy he wrote with John Tooze:

> *Although some fringe groups (such as Science for the People) thought this was a matter to be debated and decided by all and sundry; it was never the intention of those who might be called the Molecular Biology Establishment to take this issue to the general public to decide. The matter was not only too technical but in a way too fuzzy for responsibility to be easily shared with outsiders. We did not want our work to be blocked by overconfident lawyers, much less by self-appointed bioethicists with no inherent knowledge of, or interest in, our work. Their decisions could only be arbitrary. Given that there were no definite facts on which to base "danger" signals, we might find ourselves at the mercy of Luddites who did not want to take a chance on any form of change.* [*Watson and Tooze 1981*]

Instead, the conference came up with a complex, rigid set of guidelines, which, in due course, was approved by the National Institutes of Health and monitored by a committee called the Recombinant DNA Advisory Committee, dubbed, with a notable lack of felicity, RAC. With these controls in place, the moratorium ended and experimentation began again, though constrained by both RAC rules and by the inevitable fact that safer laboratories meant higher costs. The moratorium did not extend beyond federally funded research: while there was little recombinant work going on in the drug industry at the time, experimentation did proceed at a handful of pioneering biotechnology companies such as Cetus and Collaborative Research (Genentech would not be formed until 1976). As a result, Asilomar, for all its idealism, underscored the reality that molecular biology was increasingly a game to be played by the well funded, and in academia that meant those well connected to the traditional peer-review system. Even as the establishment demonstrated a remarkable coherence, at least at the top, its power was eroding, particularly at the bottom. Small companies—insignificant at the time—were forming on its margins; and for all its idealism, Asilomar triggered a wave of intense academic competition for money and brains. Asilomar was the crest of the wave. The academic establishment of molecular biology could never assume so much again.

For all the attempts by the academic establishment, the public could not be kept from these deliberations. It was not only that the press found good copy in these issues, or that agitators such as Jeremy Rifkin (whose first broadside *Who Shall Play God?* with Ted Howard appeared in 1977)—who looked suspiciously like Watson's Luddites—had already appeared. It was also that academia itself was divided, increasingly so as time went on, despite the efforts of Watson's Molecular Biology Establishment to paper over the differences. For those who did not belong to that establishment, or those who did but found themselves in fundamental disagreement, the public, through the media, beckoned as another avenue of recourse, first in social and academic controversies, then, inevitably, as a source of funding.

The cockpit of the social and political debate was Cambridge, Massachusetts, home of Harvard University and the Massachusetts Institute of Technology and, today, a score of biotechnology companies. The first controversy arose over the construction of a new P3[1] biological laboratory at Harvard in 1976. Harvard scientists, including Watson, Walter Gilbert, and Mark Ptashne, needed the laboratory in order to keep up with developments from the West Coast, particularly in the race to clone and sequence the insulin gene that was heating up between Gilbert's laboratory, a Genentech team, and a group at the University of California at San Francisco. Harvard administrators had agreed it should be built. But not everyone in Cambridge, or on campus, was satisfied. The debate within the Harvard community quickly turned bitter. The dissenters argued that such a laboratory should be isolated from large numbers of students and faculty, not be located, as it was planned, on the third floor of an already crowded facility. Leading the group was Dr. Ruth Hubbard, a tenured Harvard biologist. As support to stop the project grew, she later argued in Stephen Hall's *Invisible Frontiers*, "untenured professors grew silent":

> *"Untenured people very quickly got the message,"* recalls Hubbard. *"Given the degree of vehemence that even got into the debate be-*

[1]One of four levels of physical containment for conducting genetic engineering experiments. The higher the level, the riskier the experiment.

> *tween colleagues at the same level—that is, tenured faculty—it was obvious that it was not a healthy thing for an untenured person to do." The message, as Hubbard saw it, was "that the people with the power in molecular biology were all on one side, and that if the untenured people wanted to go somewhere with their careers, this was a good subject to stay out of."* [*Hall 1987*]

As support dwindled, the critics finally decided to go outside the university community. "The sin," Hubbard told Hall, "for which there is no forgiveness."

The result was a long Boston *Phoenix* article in June laying out the dangers of recombinant DNA work in Cambridge. "This is a biohazard," Hubbard was quoted as saying. "It's not enough to stop doing the experiments. Once it's out it's impossible to shut down. It's worse than radiation." The *Phoenix* article caught the attention of Mayor Alfred Vellucci, who quickly scheduled public hearings. The debate, before television cameras and the assembled press, spun out of control. Although Vellucci's confused statements were lurid, and attention getting, some major scientific names, most associated with the Science for the People movement, also testified against the facility: Hubbard, Nobel laureate Dr. George Wald (Hubbard's husband), Dr. Jonathan King of MIT, and Dr. George Beckwirth of Harvard Medical School. In short, there was no unanimity in the scientific ranks. And on an issue of such importance, the critics were willing to seek support from a nonscientific public.

After the sensational aspects of the case receded—after a three-month moratorium at Harvard and the establishment of a joint Harvard–Cambridge safety committee—the university proceeded to build its laboratory. But a sort of innocence was lost. The theoretical democracy of the scientific community was perceived to have broken down and, as a result, the consensus on which self-regulation was built was also threatened. Moreover, the behavior of Mayor Vellucci confirmed the Establishment's worst fears, just as they were growing more confident on biohazards and more aggressive scientifically.

In 1977 and 1978, RAC progressively loosened the guidelines, and six months later, the suggestion was made to drop them entirely. In September 1979, the Asilomar guidelines were all but abandoned for considerably less stringent rules (in some cases they were replaced by state and local regulations). Many, however, recognized a new reality: Once welcomed at the party, the public could not be easily removed; instead of ignoring the nonscientific public, it was necessary to educate and manage them.

Indeed, while the biohazards debate raged on, a second set of disputes flared up. At issue: How far should universities go in soliciting commercial support, and What constituted a conflict of interest? In California,

Herbert Boyer was attacked for his ties to newly formed Genentech, though he never left his academic post at the University of California, San Francisco; there was even talk that he had lost a Nobel because of the controversy.

Similar disputes arose at other major research campuses. But the most acrimonious, and most revealing, debate took place again in Cambridge. In the early 1970s Edwin "Jack" Whitehead, the cofounder and president of Technicon, a pioneering medical diagnostics company, began to seek a way to make a major philanthropic contribution. More specifically, he wanted to set up a research institute in alliance with a university. He approached Stanford University, the California Institute of Technology, and Harvard, and actually began negotiating with Duke University in 1974; three years later, however, negotiations collapsed. Accounts vary as to why, from questions about control to uncertainty created by Technicon stock fluctuations to whether Duke could attract Nobel laureate-caliber researchers [see Kenney 1986, and Angier 1988].

In the mid-1970s, Whitehead met Nobel laureate David Baltimore of MIT and convinced him to head up the center. Baltimore, the codiscoverer of reverse transcriptase, in turn convinced Whitehead of the merits of MIT, a university that had long collaborated with corporate backers. In 1981, a formal proposal was made and the battle joined. Whitehead was willing to put up an impressive sum of money—$120 million (making his institute the greatest single research gift in fifty years)—in return for retaining control over faculty appointments; that faculty would then be offered tenured positions at MIT. Such control was necessary, the pro-Whitehead forces argued, to attract and retain a high level of talent. Those opposed to the institute argued that Whitehead appointees would add nearly a third to the MIT faculty, shifting the balance of power on the MIT biology faculty and opening it up to commercial pressures.

The appearance of these questions—and, in nearly every case, the resolution in favor of outside funding or outside commercial ties—further eroded the traditional biological ideal and the sanctity of academic research. If the biohazards question pivoted on social and moral debating points, the conflict-of-interest issue hinged on financial ones. The universities needed the money not only to support current facilities but to build new ones—that is, to compete. Increasingly in the 1980s they couched the academic argument for biological research in economic terms: They needed the funds so America could compete. Indeed, even during the Harvard laboratory debate, the university complained that it was losing talent to better-equipped institutions. Federal support, which had expanded academic research enormously after World War II, showed signs of leveling off. The universities were feeling the pressure from small start-ups—the first biotechnology companies—and individual scientists began to consult and eventually leave academia to found commercial enterprises. Baltimore even argued at MIT that the Whitehead bequest was necessary to *avoid* commercial ties.

Whitehead and Baltimore triumphed. The Whitehead Institute for Biomedical Research was founded and housed in a handsome structure on the MIT campus; Whitehead appointees became part of the MIT faculty. And, in fact, the Whitehead Institute churned out first-rate basic science. Baltimore himself went on to head Rockefeller University in 1990, succeeding Joshua Lederberg, the Nobel laureate who, coincidentally, had first introduced him to Whitehead. (The network at the top of the pyramid remains quite small.) But it is a sign of how much the academic biological community has changed that little is heard anymore about the conflict-of-issue question. Structures and guidelines have been built by universities; but more importantly, the visceral fears that outside commercial interests would distort the search for truth, would taint academic freedom, have all but disappeared. It is not that those fears and conflicts do not exist; but in a world where there is constant interchange of manpower and money between academia and commerce, no one much protests publicly anymore. The money, in most cases, has become too important. Where protests or talk of revising guidelines erupt, they usually stem from university administrators who feel they are not getting enough money for research.

While those debates were agitating campuses, the interferon bandwagon was picking up steam. Interferon took these conflicts and drove them into a new region: peer review. Interferon was a mysterious antiviral protein first discovered by Alick Issacs and Jean Lindemann in 1957. After several decades of sporadic research, little was yet known about it except that it was very rare, very expensive to isolate, and played some role in the immune system. Nonetheless, in the early 1970s it was taken up by Dr. Mathilde Krim, a politically well connected former Memorial Sloan-Kettering researcher who had helped write a Congressional committee report for the War on Cancer legislation, and who began to promote it as an anticancer agent. As the 1970s wore on, the demand for a cancer cure grew in step with research budgets. Krim, arguing that interferon had been ignored by the funding establishment, began to organize and promote the still-obscure protein and market it to a new constituency: the press and the public.

The interferon story is too complex to tell in all its convolutions here. Suffice it to say, interferon was the catalyst that triggered the first wave of biomania on Wall Street in the early 1980s; it was biotechnology's first great "concept," and attracted the first flood of public investors. Interferon was the hinge between academic and commercial biology. What is most fascinating about it, however, and what makes interferon so important historically, is how traditional means were abandoned for new ends. Krim sought to change the underlying rules by which biomedical research was conducted. She bypassed the conservative judiciary of peer review; like scientific activists in other spheres, she actively nurtured a nonscientific constituency that, in turn, could apply pressure to the funding establishment. In effect, she opened the door to public investors and invited them to draw up their own research agenda. After interferon, the role of the public—particularly the Wall Street public—became a permanent part of the agenda-setting system [Teitelman 1990: Panem 1984].

Krim did not work alone. Funding pressures within academia fed the interferon boom. The advent of immunotherapy, a research initiative peculiarly susceptible to public acclaim and distortion, provided a theoretical justification for promotion, just as it would for anticancer antibody therapy. The country in general was changing; the Reagan attack on centralized authority, from regulatory agencies to social welfare agencies, had found a broad constituency. The laetrile fad that erupted in the late 1970s revealed a deep frustration with and a profound distrust of established institutions such as the NIH and the FDA. Of course, between Asilomar in 1975 and interferon in 1980, the community of molecular biology had changed. Indeed, if there was a biotechnology revolution, it took place here. The debates, the battles, and the continuing scientific developments eroded the strict separation of science and business. The ironies multiplied. Many in biology who had once fervently believed, with Watson, that scientists should retain control, now rushed to raise money on Wall Street, effectively ceding control. And many of those quickest in the biohazards debate to seek alliances outside the community—to share the science—now warned of the dangers of commerce. Nonetheless, the result was the same: Academic science broke open like an egg; and from it emerged biotechnology.

NEW EXPECTATIONS ON WALL STREET

If a revolution in expectations took place in molecular biology in the 1970s, so too did one on Wall Street, though it unfolded more slowly and more quietly. All the molecular biologists eager to rush into commerce would hardly have mattered if the money to start companies was not available. Wall Street, like molecular biology, had to alter its modus operandi to accommodate new forces. In particular, Wall Street had to grow comfortable with the notion—incorrect, at least over the short term—that biotechnology, for all its complexity, was just like the more familiar microelectronics and computers. And to do so meant funding companies far earlier in their life cycles than public markets had ever before done on such a scale. In short, the evolution of Wall Street in the 1970s involved two movements: the strengthening belief that biotechnology constituted a technological revolution and the development of what might be called *public venture capital.*

At the time, Wall Street had considerable experience with small biomedical companies. But these tended to be small operations that required, relatively speaking, small infusions of capital to expand already existing product lines. In many ways they were no different from, say, a small machine tool firm. Not very exciting, but occasionally quite profitable, particularly if they were eventually acquired, as Whitehead's Technicon was, by a larger corporation. There was, however, one exception that proved important not only in the way Wall Street came to view biotechnology, but in the way early biotechnology entrepreneurs assembled their firms. It was a company called Syntex, and it had the distinction of being the only fully integrated drug company to muscle onto the scene since World War II.

The Syntex story begins with a now obscure figure named Dr. Russell Marker. In the early 1940s, Marker was a professor of chemistry at Pennsylvania State University. He specialized in steroids, a field just beginning to attract commercial and public attention. One variety of steroid in particular, the cortisones, were viewed as a potential cure for everything from rheumatoid arthritis to cancer to common, everyday fatigue. In language that would echo in the 1970s and 1980s, *Fortune* wrote in 1951:

> *As therapeutic agents, the hormones differ profoundly from nearly all previous drugs. They are not germ killers . . . but directly alter the complex glandular balances of the human body. But few doubt that, in the continuing story of hormones, medicine will soon be striking at some of the most deep-seated organic complaints of man. [Fortune, 1951, "The Cortisone Shortage"]*

The article also predicted that the hormone industry—still nascent—would stimulate "an economic revolution of far-reaching significance."

Raw steroids were expensive to produce. Marker had spent months tramping through fields trying to find a natural steroidal source. In 1939, he had discovered that he could extract quantities of a steroidal precursor called diosgenin from the root of the sarsaparilla plant; he was then able to turn the diosgenin into progesterone, a hormone that prepares the uterus for the implantation of the egg, helps to maintain pregnancy, and prevents a second fertilization from occurring at the same time. For all of that, even the diosgenin produced by the sarsaparilla proved inadequate to the needs of researchers and commercial manufacturers. Then Marker heard about a lumpy black root of a hillside yam that grew in Mexico called, picturesquely, *la cabeza de negro*, that contained far greater quantities of the material [see Syntex 1966].

In 1943, Marker rented a small laboratory in Mexico City. A year later, he appeared at the offices of Laboratorios Hormona, a marketer of raw, natural steroids, hugging two jars wrapped in Mexican newspapers containing 4.5 pounds of progesterone worth roughly $160,000. Hormona had been buying small amounts of very expensive natural hormones, then reselling it, mostly to U.S. drug companies, at a profit. Marker, off the street, without an appointment, was hauling around enough to equal the world's production. The firm's co-owner, Dr. Emeric Somlo, immediately offered to fund Marker's work through a new company called Syntex S.A., and share the profits. Marker agreed, and went back to his yams. In 1944, he produced several more pounds of progesterone. Then in early 1945, he argued with Somlo and walked out, confident of his own indispensability.

Marker miscalculated, and history left him behind. In Cuba, the frantic Somlo tracked down George Rosenkranz, a 29-year-old fellow Hungarian. Rosenkranz had read Marker's papers and thought he could begin producing diosgenin from *la cabeza de negro*. He had a few personal complications (he wanted to get married and Syntex looked risky, to say the least) but Somlo convinced him to give it a try. So he went to Mexico City. Within a few months, he had succeeded in emulating Marker's technique. Within five years, he and his researchers developed the means to produce synthetically all four major steroid groups—androgens, estrogens, corticoids, and progestogens—from diosgenin [*Fortune*, 1951, "Mexican Hormones"]. As for Marker, he tried to set up a diosgenin company and failed. At 50 years old he dropped steroids and started selling Mexican ceramics.

By then, Syntex was peddling bulk steroids to American and European pharmaceutical companies. Rosenkranz, however, had larger plans. Researchers had demonstrated that if you injected progesterone in rabbits, you could prevent them from ovulating. In 1950, Gregory Pincus of the Foundation for Biological Research in Worcester, Massachusetts, with Dr. John Rock of Harvard, launched a successful test on women. Pincus turned to G.D. Searle & Co., an Illinois pharmaceutical company that was a supporter of the foundation, for help in devising a more powerful steroid. At the same time, chemist George Djerassi was trying to develop a steroid at Syntex that would prevent miscarriages. Djerassi's team came up with a hormone called norethindrone, which proved to be a far more potent version of ordinary progesterone and which could be taken orally: A pill, or as it would become know, *the* Pill.

Syntex, however, faced a dilemma. The company had begun to sell a few of its own drugs in bulk, particularly for skin ailments, in the United States. But that was a losing proposition. The company sold raw steroid to Merck for, say, ten dollars a pound; Merck would package it, and sell it by prescription for as much as ten or twenty times that. High prices kept markets small, which did not hurt Merck much, but stifled Syntex, which required high volume. Moreover, as the years went by, other companies came up with similarly economical ways of making bulk steroids, forcing the price down and wiping out whatever meager profits Syntex could make. There was not much future in the commodity pharmaceutical business. After making almost two million dollars in 1953, Syntex barely broke even in 1956.

Syntex needed a partner. In the mid-1950s, Lehman Brothers, the New York investment banking house, poked around, but was distressed by the company's Mexico City location and its confused books. In Lehman's wake came Charles Allen, the founder of a private New York investment house, Allen & Co. He had read about Syntex in *Fortune* and he was far more of a gambler than the folks at Lehman. A brilliantly intuitive investor, Allen already controlled a company called Ogden Corp., where he stashed corporate odds and ends. In 1956, Allen bought Syntex from Somlo and folded it into Ogden. Rosenkranz and Dr. Alejandro Zaffaroni—a Uruguayan hormone researcher who would later found drug delivery pioneer, ALZA Corporation—were put in charge. Syntex promptly began to hemorrhage. By 1959 it was losing half a million dollars a year. Nonetheless, Allen waited patiently while Syntex's researchers worked away. Finally, in 1961, Allen announced that he would sell some 1.2 million shares of Syntex.

Allen's timing was impeccable. The Pill was taking off, the market loved it, and the stock went crazy. From a low of $11 in 1962, the stock peaked at $260, triggering a three-for-one split, before sliding to $213 on November 22, 1963—when trading was halted by the Kennedy assassination. When trading resumed the next Tuesday, the stock started out at $63 and roared to $135 ($495 before the split) before falling to $125. Rumors soon spread: Syntex is cutting prices, steroids cause cancer, insiders are dumping their stock. The stock sank. Then, the equal and opposite reaction: Syntex is announcing new products, steroids cure cancer. Up the price flew.

By early 1964, just as Syntex was getting FDA approval for a Pill under its own name, the American Exchange moved to dampen the gyrations. The Exchange, dogged by recent trading scandals, banned stop orders on Syntex, that is, an order telling a broker to sell a stock when it fell below a specified price. Widespread stop orders tended to increase volatility. The Exchange also began to poke into who was buying and selling—searching for manipulation. But although there was considerable short-term speculation in Syntex, officials never discovered overt manipulation. The real problem was that Wall Street was just wild over birth control, and that "the float"—the number of shares available to be traded at any one time—was very thin. Charlie Allen controlled about one quarter of the shares, with another 15% held by institutions. Despite the splits, reportedly urged on Syntex by and Amex to spread the stock, the mania over birth control, and the willingness to speculate on the mania, made Syntex the epitome of the go-go stock.

FORESHADOWING BIOTECHNOLOGY

A strategic planner at a biotechnology company once said, "Syntex today is a nice, respectable pharmaceutical company.

> *But if you were going to found a company to purify something out of Mexican yams, nobody would think that was a legitimate thing to do. It shows the absurdity of what can happen. Never underestimate what can come out of anybody's work."*

But there was more to the Syntex story than just a rationalization for serendipity. Syntex was a paradigm for the biotechnology companies to come. It was formed to exploit natural, not synthetic, substances. It had been a rank outsider that challenged the established, bureaucratic giants. Originally, it had no other weapon but its scientific skills. It captured the wild, promiscuous imagination of the stock market. Even the role of Russell Marker, scientific entrepreneur and former academic, echoes in the later history of biotechnology.

Syntex taught a few basic strategic lessons that were central to later biotechnology strategic planners. Rosenkranz expanded the company in discrete stages, each strategically perilous. First, Syntex sold bulk birth control materials to the major drug companies. Second, with that money and with cash from the sale of stock, it began to make its own name-brand birth control pills, eliminating a middleman. This cost more than selling bulk materials, but the returns were better. Third, at even greater expense, it formed a marketing group to sell its products, pulling another slice of profits in-house. Finally, with the birth control patents expiring, Syntex diversified, adroitly making up in new hormone-based products what it lost in Pill profits. It steadily moved toward full integration. Each transition could have ended in disaster; each called for larger infusions of capital. It required intelligence, steady nerves, and continual innovation. It also required an alliance with sources of ready capital, first Charlie Allen and, eventually, public investors. Allen provided entree to the stock market, where the real money comes from, and the allure of the birth control pill did the rest.

No one pretended that the perspective of those public markets was scientifically sophisticated, particularly as one dropped down-market toward retail investors. The markets tended to discount the fact that Syntex had nearly twenty years to develop before it appeared with such vigor on the American Exchange. To investors, it seemed to have sprung, fully formed, from the fertile brow of Charlie Allen in the early 1960s. It was an early "miracle" company—to use a label *Business Week* later applied to Merck. Moreover, the stock market overestimated the "revolutionary" impact of Syntex's technology. Although Syntex developed a business around the Pill, it hardly had the market to itself, as in most blockbuster pharmaceutical products. Indeed, in the years ahead, it made the bulk of its profits from Naprosyn, a nonsteroidal, antiarthritic drug. There was no technologic revolution here, except in the loosest definition of that term. Finally, the market may have mistaken its own enthusiasm for operational fundamentals. Syntex did join the major pharmaceutical ranks; but by the late 1980s, it was still the smallest of the major U.S. drug companies and thus exposed when its research productivity lagged in the 1980s and takeover fever within the pharmaceutical industry mounted. But despite a drought of new products, Syntex still carried the faint aura of its early research vitality and of the era during which it first offered its shares to the public.

Syntex embodied Wall Street's desire to believe in miracles. While the go-go years passed, the desire for go-go profits remained. The 1960s had been the longest sustained boom on Wall Street since the 1920s. More and more of the country's wealth had worked its way into stocks. Indeed, Wall Street brokerage firms grew so fast and so haphazardly that when the inevitable bust occurred in 1969, many firms simply went out of business, unable to handle the flood of orders. Many of the survivors, in turn, were forced to seek larger partners wealthy enough to build the necessary computerized back-office systems. Wall Street adjusted and survived, of course, but in the process it became a harder, faster, more grasping arena, perhaps with more of a tendency to overreach. The old school atmosphere was passing, and a sort of meritocracy of ambition and aggression took its place. One of the institutions to emerge from the debris was an automated, reasonably well regulated, over-the-counter market that fit somewhere between the more established New York and American Exchanges above and the shadowy world of very cheap stocks, the so-called penny stocks, below. The new market was driven by the growing interest of investors (from individuals to traditionally more conservative institutional investors) in companies—often high technology companies—which, if they took off, provided blockbuster returns, venture capital–type returns.

These changes on Wall Street mirrored the evolution of the venture capital game itself. Before World War II, financing for new enterprises came from a very limited number of sources. Bank loans or equity offerings might be available in bullish economic times, but only if a company had already achieved a degree of maturity—and at a fairly high price. Bank loans loaded heavy interest payments on a developing company; equity offerings by small companies often took place in a shadowy, capricious and unregulated marketplace. Perhaps the best source of risk capital was the wealthy benefactor who would provide steady backing in exchange for a stake in the company—but even that left the entrepreneur dependent on the whims, and the resources, of a single, often eccentric or rapacious, individual.

This haphazard system of venture, or risk, financing was only regularized, and institutionalized, after World War II. The impetus came from above. Wealthy investors such as John Hay Whitney and Laurence Rockefeller pioneered venture funds that concentrated on the riskiest class of investments, often those with a strong technological component. Whitney put $5 million of his considerable fortune into J.H. Whitney & Co. in February 1946 (he added $5 million more a few months later), opened an office at Rockefeller Center with four other partners, and backed companies as disparate as Minute Maid, which was using vacuum technology developed during the war to make orange-juice concentrate, and Spencer Chemical, a fertilizer company that grew out of a wartime munitions factory [Kahn 1981]. In Boston, a small group of New England financial leaders, including Boston Federal Reserve president and engineer Ralph Flanders and Karl Compton, president of MIT and a powerful figure in the federal research community, set up American Research and Development. ARD actually went out and raised money in the investment community, a key development that tapped larger pools of institutional capital. ARD was run not by a wealthy benefactor, but by arguably the first venture professional: Harvard Business School professor and fervid promoter of entrepreneurism, General Georges Doriot, and made its biggest killing in 1957 when it backed a small computer company called Digital Equipment [Bylinsky 1967].

For several decades, venture capital was a relatively minor game. A company like Syntex would struggle to maturity, and a public offering, the old-fashioned way: through bank loans and private investments. But in the 1960s, money began to pour into venture investing. As the high returns some venture capitalists made became apparent—as big successes such as DEC, Polaroid and Xerox were heavily publicized—corporations and European "fund of funds" (a 1960s-era mutual fund that invested primarily in other funds) began funneling their money through U.S. venture companies organized as limited partnerships. The big shift, however, came in the early 1980s when, for the first time, Congress allowed the pension, insurance and endowment funds to invest up to 10 percent of their considerable caches of capital into venture deals. Now venture capital could tap into the largest pools of money in the economy. Other fac-

Other factors also channelled capital toward new, technologically oriented firms. The inflation of the 1970s depressed real returns among established corporations, and manufacturing, beset by foreign competition, took a nose dive [Florida 1990], particularly in the recession of the early 1980s. Investors began to look toward technology as a way of breaking out of hard economic times.

As a result, enormous capital flows were redirected to hoped-for high returns. "The $100 to $300 million per year flowing into venture funds during the 1960s and 1970s rose overnight to billions in the early 1980s and has remained at $2.5 to $5 billion ever since. [Kunze 1991]." From 1980 to 1989, the volume of venture capital increased ten times, from $3.5 to $33 billion. The result: Some 300 new funds entered the field, producing a vast new army of newly minted venture capitalists combing the landscape for deals. This sudden spike in the demand for new companies created some of the distinctive characteristics of biotechnology. In the scramble for higher returns, venture investors reached for riskier, less mature opportunities. The line between venture investments and equity offerings began to blur; the emphasis on "blockbuster" technologies increased. And, in the long run, so much money created an enormous glut of companies, and of competition, which eventually depressed the fabled high return of venture money.

In short, by the 1980s, both the equity markets and the venture funds were prepared to assume more risk for greater potential returns—a mentality that would also fuel another financial product, junk bonds. Patience fell from favor and the venture capital cycle accelerated. Traditional venture capital featured the careful incubation of new companies. By feeding the young company increasingly larger amounts of capital, the venture capitalist hoped to build a viable operation with enough management depth and product flow to survive life outside the private nest. The key transitional moment—the coming of age—of any venture capital start-up occurs with the first public offering; at that point, the backers can cash out and move on, transferring the financial responsibilities to a larger base of public investors. The timing of that offering is, in turn, determined by two factors: The perception of backers and management that the firm is mature enough to survive publicly, and the perception of the market that the firm would provide a suitable return over time. Strategically, the public offering works best when the company requires the capital the least—that is, when it has products ready to generate a self-sustaining take-off. Like Mother Nature, the public markets can be arbitrarily, almost whimsically, cruel; and even the market realizes that the less dependent a firm is on its generosity, the better.

In the late 1970s, the over-the-counter markets were becoming less discriminating, as a number of brokerage houses sent large numbers of relatively immature firms boasting exciting "concepts" into the world. Investors eager to get in earlier in the investment cycle, prone to conceptual siren songs, snapped them up. Thus arose a phenomenon known as *public* venture capital, a trend that would arguably climax with the computer and biotechnology offerings that went public from 1981 to 1983. Whereas some biotechnology companies would go public earlier in their life cycles than others—a few firms went public just months after incorporation papers were filed—no biotechnology company sold public stock after it was actually selling products. The exception here might be Hybritech (based in La Jolla, California), which was founded in 1978 and did not go public until October 1981, the year it received approval for its first antibody-based diagnostics. These products, however, did not reduce its dependence on the market all that much; they only made future public financings easier.

With its speculative fever rising, Wall Street still required a charismatic figure to push it over the edge. In computers that figure would be embodied in Steve Jobs, the former house visionary of Apple Computer. Later, in junk financing, Michael Milkin served as the omnipotent prophet; the man in the vanguard. Their spheres of action were different, but both served as objects of the deepest faith—an entrepreneurial faith.

Biotechnology really had no single dominating personality; instead it had a more generic concept that represented the point at which academia and business crossed: The scientific entrepreneur, the Ph.D. CEO. The financial markets, and the public at large, seemed to demand such a figure. American industry was widely viewed as slow moving, dull, and bureaucratic. The antidote seemed to be the kind of entrepreneur that, like some corporate Luke Skywalker, wielded a technology that could slay the dark forces of bureaucracy and capital. These entrepreneurs carried the aggressive idealism of the 1960s into the new calling of business. They were said to possess the "entrepreneurial spirit," which could transform the individual while reviving the economy. The entrepreneur radiated youth, both literally and figuratively; and the entrepreneurial start-up, with its creativity and flexibility, was often contrasted to the arthritic stiffness of larger, older organizations. The entrepreneur was cast by the media as a sort of revolutionary—able to turn the stereotypical tide, leap the abyss of high risk, break old rules, release pent-up energies, produce scientific miracles, and restore the economy. As one typical view went:

> *The future success of the U.S. economy hinges on its ability to participate successfully in this Fourth Industrial Revolution. . . . Entrepreneurs, like other revolutionaries, are committed to a new way of thinking and acting that is alien to members of the business and financial establishment.* [Doerflinger and Rifkin 1987]

That message proved seductive both for Wall Street and for certain segments of academic biology. Both parties, in turn, made miscalculations of the other based on ignorance. Wall Street was happy to view these entrepreneurs as technological magicians who could harness unlimited powers. In molecular biology, this new kind of company—academic science (so many mistakenly thought) with stock options—provided an outlet for mounting frustrations and thwarted ambitions. In the 1970s, the biological community had argued its way toward an accommodation with the public. Turning to public investors for financing simply took that a step further. Besides, there were two aspects to the undeniable public scientific ignorance: one that believed the horror stories of Jeremy Rifkin, antibiotechnology crusader, and the other that accepted the tales of miracles and money. In the early 1980s, Wall Street proved to be believers (it would not last, of course, but no one wanted to know that then) and biological entrepreneurs eagerly accepted the resulting largesse as if it had fallen from heaven. Few considered the old Wall Street adage that Wall Street itself so often forgets: There is no free lunch. And if there was a resulting erosion of the central structure of academic science, it seemed worth it. After all, biotechnology was only a technology, not a basic science.

THE ACTORS EMERGE

What of the actors in this drama? A few small companies had appeared prior to 1974, most notably Cetus (based in Emeryville, California) and Collaborative Research (based in Lexington, Massachusetts)—both near academic centers. The continued financing of these early companies may well have hinged on a legal decision: The Chakrabarty decision in 1980 that allowed the patentability of living things. Patents would continue to be a source of uncertainty, bitter dispute, and huge legal bills in the years ahead, but Chakrabarty cleared the ground of the largest, and most fundamental barrier—and it gave a boost of confidence to early venture investors.

The most important start-up in biotechnology came when Genentech was formed in a San Francisco bar in 1976. Genentech contained all the various ingredients we have been discussing: a noted academic scientist (Boyer); an entrepreneur (Robert Swanson); a source of professional venture capital (the firm of Kleiner Perkins, of which Swanson had been a junior partner); and a powerful technology (recombinant DNA). Genentech, of course, would trigger biomania in October 1980, when its public offering price shot up from $35 a share to a record-setting $89 on the first day. (The firm was not, however, technically the first biotechnology public offering. New York City–based Enzo Biochem had sold shares to the public a few months earlier.) In the years ahead, Genentech would pioneer many of the major commercial and technological trends in biotechnology, from limited partnerships to aggressive patent litigation to the first recombinant products—human insulin (licensed to Eli Lilly), alpha interferon (licensed to Hoffmann–LaRoche), and human growth hormone—to the first true pharmaceutical product, tissue plasminogen activator. Genentech also proved the most dramatic victim of Wall Street's moods. Arguably overvalued throughout the 1980s, Genentech found itself shunned when those overheated expectations were not achieved. In 1990, Genentech announced a deal that could result in a merger with Hoffmann–LaRoche.

Of course, Genentech alone did not create biotechnology. The term *biotechnology* became common currency because of the hundreds of companies that followed in its wake, a process that continues today. Biotechnology as a commercial concept exists because of those large numbers. No one declared a new industry when Syntex broke through in the 1960s, despite chatter about steroidal revolutions. The fact is that biotechnology, as a commercial concept, may eventually lose its utility. After all, the original thread—that these companies alone shared a set of technological tools—no longer applies. Genetic engineering has now diffused far and wide, and is just one method of research and development attack. Even the far looser link to academic molecular biology means very little in a world where every drug company now hires molecular biologists by the carload. Moreover, as a business phenomenon—a small world of entrepreneurial research-based biomedical companies floating somewhere between academia and the drug giants—biotechnology threatens to fragment into a variety of other "sub" industries, from drug delivery companies to neuroscience firms and diagnostic houses, just as molecular biology did in the 1970s. In terms of control, many biotechnology companies have been coopted by larger firms that possess the financial and market power to commercialize their products. The Genentech acquisition is simply a more visible symbol of a process that has been accelerating since the 1987 market crash: The gradual reabsorption of many biotechnology companies by larger corporations possessing the advantages of capital and size. In the years ahead, the term *biotechnology* will either fade from the scene—or, fueled by a new wave of breakthroughs—create a new cohort of companies.

One could go further. The most authentic revolution in biotechnology took place not technologically, but intellectually, in the hearts and minds of academic biologists during the great debates of the 1970s. We do not yet know the full consequences of that revolution, but we do know that a significant transformation swept through molecular biology in the 1970s, toppling traditional verities, and spawning a new set of relationships with the outside world. For it was that transformation, combined with the enthusiasm on Wall Street—an enthusiasm as old as greed itself—that created the conditions for the rise of biotechnology in the 1980s.

REFERENCES

Angier, N. 1988. *Natural obsessions: The search for the oncogene.* New York: Houghton Mifflin.

Bylinsky G. 1967. "General Doriot's Dream Factory," *Fortune*, August 1967.

Cherfas, J. 1982. *Man-made life: An overview of the science, technology and commerce of genetic engineering.* New York: Pantheon.

Cohen, S. and Boyer, H. 1973. Construction of biological functional bacterial plasmids *in vitro*. Proc. Natl. Acad. Sci. U.S.A. 70:3240.

Doerflinger, T. and Rifkin, J. 1987. *Risk and reward: Venture capital and the making of America's great industries.* New York: Random House.

Florida, R. and Kenney, M. 1990. *The Breakthrough Illusion: Corporate America's Failure to Move from Innovation to Mass Production.* New York: Basic Books.

Fortune, 1951. "The Cortisone Shortage," May: 83.

Fortune, 1951. "Mexican Hormones," May: 86.

Hall, S. 1987. *Invisible frontiers: The race to synthesize a human gene.* New York: Atlantic Monthly Press.

Judson, H.F. 1979. *The eighth day of creation.* New York: Simon and Schuster.

Kahn, E.J. 1981. Jock: *The Life and Times of John Hay Whitney*, New York: Doubleday & Co.

Kenney, M. 1986. *Biotechnology: The university–industrial complex.* New Haven, Conn.: Yale University Press.

Kohler, G. and Milstein, C. 1975. Continuous cultures of fused cells secreting antibody of predefined specificity. *Nature* 256, 495.

Kunze, R.J. 1991. *Nothing Ventured: The Perils and Payoffs of the Great American Venture Capital Game.* New York: Harper Business.

Panem, S. 1984. *The interferon crusade.* Washington, D.C.: Brookings Institution.

Rifkin, J. and Howard, E. 1977. *Who shall play God?* New York: Dell.

Rogers, M. 1975. "The Pandora's Box Conference," *Rolling Stone*. June 19:28–40.

Syntex 1966. *A corporation and a molecule: The story of research at Syntex*. Palo Alto, Calif.: Syntex.

Teitelman, R. 1990. *Gene dreams: Wall Street, academia and the rise of biotechnology*. New York: Basic Books.

Watson, J. and Tooze, J. 1981. *The DNA story*. San Francisco: W.H. Freeman.

CHAPTER 2

CAPITALIZING ON NEW OPPORTUNITIES: ENTREPRENEURSHIP IN BIOTECHNOLOGY

John M. Stewart

The biotechnology revolution provides a unique microcosm within which to study the entrepreneurial/venture capital interface and from which to begin to define paradigms for entrepreneurial development. Several characteristics of the biotechnology revolution distinguish its industry life cycle from that of other technology-based industries. These must be factored into its transferability to other fields of broad observations regarding entrepreneuring:

1. The biotechnology revolution's beginnings were comparatively clearly defined, with the implantation of a selective genetic marker in *Escherichia coli* by Cohen and Boyer in 1973 and with the demonstration of hybridoma technology by Kohler and Milstein in 1975. Thus, biotechnology entrepreneurs may have benefited from a sharper distinction of biotechnology from its predecessor technologies than would be the case in a more dispersed or pervasive type of technological evolution. For the biotechnology entrepreneur, product opportunities may have stood in sharper relief than for electronics.

2. Biotechnology's growth was accelerated by the policies of the Reagan administration in regard to the transfer of technology from the academic sector to the commercial sector. In 1982, for the first time, universities could take title to and could transfer exclusive commercial rights to technologies developed under federal funding, for example, National Institutes of Health research grants. For the venture capitalists, this opened a storehouse of intellectual property. For the industry, it overcame a severe limitation—in the late 1970s only some 12,000 Ph.D.s were skilled at biotechnology, and the vast majority of these were entrenched in academia. The liberalization of technology transfer policies, together with the standard practice of allowing academics to dedicate 20% of their work week for consulting activities, had a multiplier effect on the available talent pool.

3. The prodigious capital appetites of the biotechnology industry were sated largely by managed pools of venture capital and by offerings of stock to the public. The availability in the early 1980s of $20 billion in funding arose primarily out of changes in pension fund regulations allowing these funds to allocate a portion of their investment portfolio to more speculative issues. Reductions in capital gains taxes, disposable incomes expanded by a decade of inflationary salary increases, and venture capitalists' tales of Eldorado found in the microcomputer industry contributed to the willingness of the lay public and venture capitalists, for the first time in history, to invest in basic research. For the most part these investors were unable to assess the costs, timing, or probability of technical success of the technologies they were fostering. One early observer writing for the *London Financial Times* likened the feeding frenzy in biotechnology to that surrounding single-celled protein in the 1960s. He recalled that "popular science" and financial analysts projected that we would all be eating cubes of single-cell protein and swearing it was filet mignon; the author pointed out that only one such plant remained, producing cattle feed at an economic loss.

For the start-up biotechnology entrepreneur, the early 1980s (and 1983 in particular) were an unprecedented era of capital availability. As might be expected, the availability of so much capital led to excesses—some product concepts stretched the credible to angstrom thinness—but it also led to entrepreneurial expectations of balancing the cash-flow equation over a decade through repetitive equity infusions. In any other era, with its surrounding circumstances, that expectation may not be workable.

4. Because biotechnology investments represented investments in research occurring at the molecular level, technical due diligence was difficult to verify, and there were no applicable valuation models at the outset of this industry. Thus, venture capitalists, investment bankers, and entrepreneurs dealt with almost totally unverifiable values. A number of objective models of valuation were proposed, including ones based on the greater-fool theories of "x times sunk research cost," "x times the number of Ph.D.s in research," and the Drake model,* which sought to discount

* After Peter Drake, a former leading biotechnology analyst with Kidder, Peabody & Company, Inc., who later cofounded Vector Securities International, Inc., an investment banking firm specializing in the life sciences.

appropriately future sales projections to a present value comparable to public companies. As a corollary to this issue of ethereal valuations, initial public offerings of biotechnology stocks have largely been a transfer of wealth from the investing public to the founders and venture capitalists. The definition of a successful biotechnology company varies greatly from the venture capital community to the investing public to the chartists, to the fundamentalists. The latter are unlikely to find many biotechnology investments that have developed over the past decade to the point of generating adequate cash flow from operations to replace its depreciating assets and to provide returns to investors. Thus, the characterization of one entrepreneur as "successful" versus another as "unsuccessful" may vary greatly depending on the perspective of the critic; also, these labels may still be premature for many companies in the biotechnology industry.

5. Before searching for any parameters of a paradigm for entrepreneuring in biotechnology, it would be worthwhile to narrow the scope of definition of "entrepreneur" in applying it to biotechnology. In terms of entrepreneurs coming to the biotechnology field, there have tended to be observable waves of participants having various characteristics:

Pre-1981	Visionaries	Ron Cape at Cetus George Rathman at Amgen Robert Swanson at Genentech Howard Birndorf at Hybritech
1981–1983	Tinkerers	Franklin Pass at Molecular Genetics William Rutter at Chiron
1983–1988	Start-up venture capitalists	Robert Johnston at Genex Gary Takata at Oncogene Science Isaac and David Blech at Genetic Systems
1984–1987	Professional managers	David Hale at Hybritech James Vincent at Biogen
1985–present	Double dippers	David Hale at Gensia Pharmaceuticals Howard Birndorf at Progenix George Rathman at ICOS

The visionaries held some view of a future reality that they were trying to create or move toward, based on a technology that was still being defined (e.g., whether yeast cells would prove superior to mammalian cells). The tinkerers were primarily academics who understood the technology and were anxious to see it applied in any or all of its potential markets. The start-up venture capitalists represent a group that did not intend to maintain a daily operating role in the companies they created, nevertheless they were the principal organizers and early financers of the companies. Like the visionary, they hold a view of the future, but it tends to be more defined by what their retinue of follow-up investors will finance than by the desire to create an enterprise or to see a technology commercialized. These venture capital entrepreneurs also dare to cross technology boundaries more in the style of arbitrageurs than as studious scientists; they will surface in industrial enzymes at one time and in oncogenes two years later, often with an uncanny sense of the economic significance of obscure discoveries. The professional managers were those whom the venture capital community brought out of Fortune 500 companies, largely Baxter Travenol and Abbott Laboratories and later SmithKline Beckman (now SmithKline Beecham), to start or to fill key positions in established biotechnology companies. They were more accurately described as employees of the venture capitalists and not entrepreneurs in the sense of being the principal organizer, although many used the experience and personal net worth gained in this initial position to go on to found a second-generation company. These became the "double dippers," who used one success, usually in a secondary capacity (i.e., other than as the principal organizer) in one venture-backed situation to become the principal organizer in a subsequent start-up.

For the purposes of this discussion, the focus will be on the visionary, the tinkerer, and the start-up venture capitalist as being the principal organizers of their enterprises. The lore that surrounds these contemporary heroes is replete with characterizations as risk takers, as having a "fire in the belly," and as being undauntingly self-confident. The latter may hold up under the present scrutiny, but venture capitalists tend to back those who express a clear and rational vision that creates economic value at minimal risk and over the shortest time; rather than a fire in the belly, they seek a stable team player who has made money for someone else in the past—somewhat of an organization man. Interestingly, however, the principal organizer is often not what the venture capitalist would have chosen. The organization is too well formed by the time the managed venture capital funds come to the deal for the venture capitalists to install their own management.

THE ROLE OF THE PRINCIPAL ORGANIZER: CONCEPT, COLLEAGUES, ZERO-STAGE CAPITAL

If there is any area in which a great contribution could be made to the understanding and fostering of entrepreneurship, it is in the range of activities that a principal organizer must undertake in creating a new enterprise. Based on observations of the process in the biotechnology industry, a generalized process for approaching such activities can be developed. Professionals will advise on legal, accounting, and tax issues facing a principal organizer, but the present focus is on developing a concept, selecting an initial cadre of colleagues, and approaching zero-stage capital.

The principal organizer who has made or is making the decision to become a "zero-stage CEO" must have a vision or a concept of the business that he is trying to build. He uses that vision to recruit moral and economic support and start-up colleagues from among friends and relatives. This is before any thought of venture capital has formed.

How is this vision delineated? Market researchers cite statistics to the effect that most new products are incremental improvements on existing products. The pharmaceutical industry states that the largest source of new products come from compounds of known activity. Thus it might appear that entrepreneurs simply seek to eliminate some irritant in their daily life or to overcome some deleterious aspect of existing products. In dealing with new technologies, however, and particularly those with such long product development times as in the health care arena, which absorbed 80% of the funding in biotechnology, a different and more forward-looking model must be at work in the mind of a principal organizer developing a previously unknown technology.

The proposed model builds on the empirical observations of Richard N. Foster [1986] in his book, *Innovation, The Attacker's Advantage,* and on the general principle arising out of the Harvard Business School Studies on the Profitability Impact of Marketing Strategies: "It's easier to succeed in fast-growing markets." Like the moth to the flame, entrepreneurs may be inevitably drawn to areas of rapid growth (i.e., to frontiers, be they social, geographical, or technological). In the vocabulary of a Jungian psychologist, they may fit the rarest of the four prime character types—intuitors—exercising their natural proclivity for spending much of their mental effort involved with the future time frame and with projecting the convergence of seemingly diverse forces on it. The following expansion of the Foster model of innovation attempts to synthesize a framework for considering how a biotechnology entrepreneur might go about constructing a business enterprise out of what for others may seem like idle musing. Once the entrepreneur has established his business concept, then he is ready to deal with the considerations of recruiting colleagues for this venture and of raising capital.

S CURVE OF INNOVATION

In his book, Richard Foster proposes an S-shaped curve (formally a Gompertz curve) that relates the incremental advancement of technology, usually in terms of increased profitability from its application, to incremental amounts spent on research to further that technology (Fig. 2-1). In the

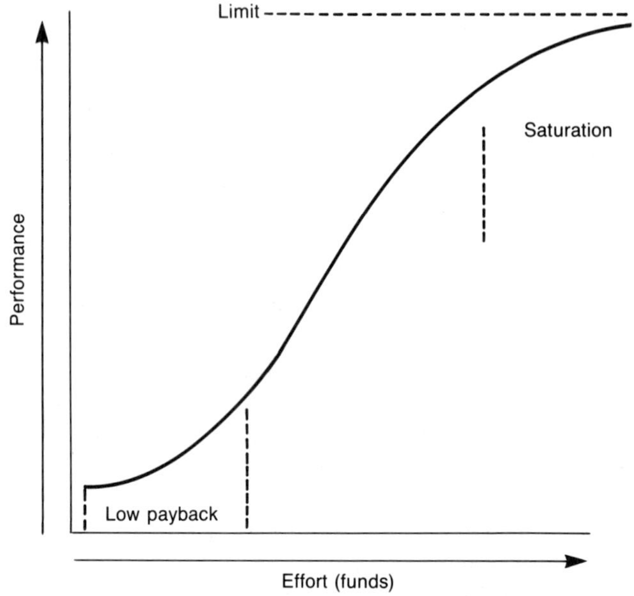

FIGURE 2-1 Foster's innovation's S curve of research. Simply stated, technology advances (as measured in terms of increased profitability) are related to the amount of research funding to further that technology. (From Foster, R. N. 1986. *Innovation: The Attacker's Advantage.*)

early development of a new technology incremental spending on research produces little measureable advancement as the preferred embodiment of the technology is still being resolved. False starts, such as the effort to develop *Bacillus subtilis* as an expression host, may even result in expenditures that are counterproductive insofar as their contribution to the evolution of the preferred embodiment.

Foster further postulates that, in its early stages, spending on the new technology as a whole may be less productive than incremental spending on the old technology. There would be a net downward shift in productivity in converting from research on the old technology to more tentative research in the new, unfamiliar technology (Fig. 2-2). As a result, Foster explains, corporate research managers are reluctant to make the necessary budgetary allocations to begin shifting the disciplinary basis of the corporation's technocracy. These corporate managers envision the adoption of the new technology as leading to the obsolescence of its installed capital base in research and manufacturing facilities, as well as threatening their own technical ability to manage. The promoters of the shift to the new technology are more likely to come from outside the executive ranks

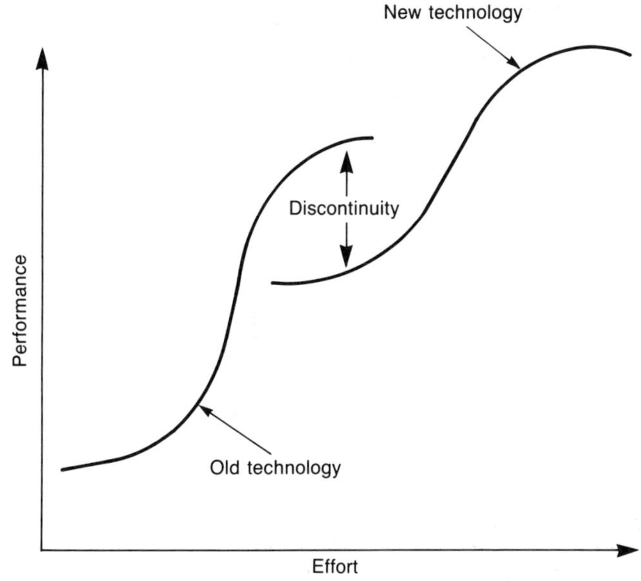

FIGURE 2-2 In early stages, spending on new technology overall may be less productive than incremental spending on old technology. A net downward shift (discontinuity) in productivity would result in converting from research in old technology to more tentative research in new technology. (From Foster, R. N. 1986. *Innovation: The Attacker's Advantage.*)

of industries most directly affected by the technology (i.e., entrepreneurs). Foster goes on to prescribe ways for corporate research and development managers to keep pace with emerging technologies.

The Foster S-curve model seems particularly applicable to the pharmaceutical industry in its early response to biotechnology. The former was driven by medicinal chemistry and pharmacology; the latter represented a shift to molecular biology and biochemistry. Small wonder that it took a decade for the entrenched management of pharmaceutical companies to begin to participate in biotechnology in any major way. The S-curve model also implies a role for federal funding of basic investigations prior to the first inflection point of the curve, after which the preferred embodiment becomes known, and a role for venture capital between this point and the point at which the productivity of research spending on the new technology begins to equal that of the old technology (Fig. 2-3).

A third plane of consideration in the Foster model of innovation is the methodology that principal organizers of biotechnology companies appear to have applied in formulating their vision of the future. That vision becomes the core of their business concept. The third plane is a social

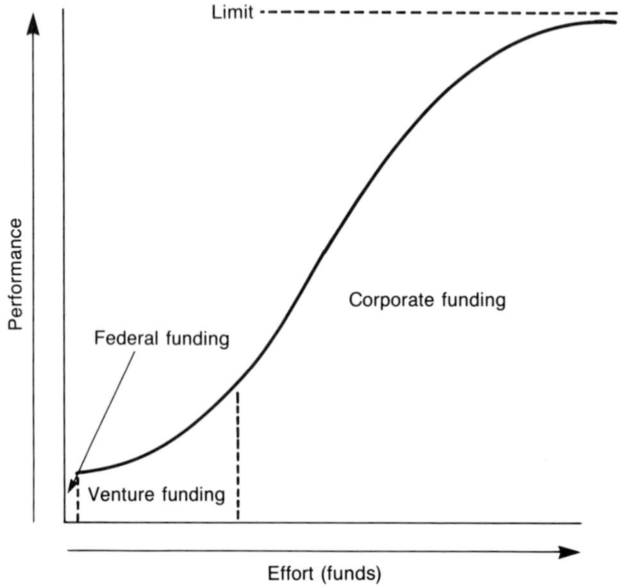

FIGURE 2-3 S-curve model as fitted to the pharmaceutical industry's relative response to biotechnology (initially as "corporate funding"). Also, the role for federal funding of basic research in biotechnology is shown as an early source of capital to the biotechnology companies. (From Foster, R. N. 1986. *Innovation: The Attacker's Advantage*.)

dimension that projects parallel to the z and x axes of Foster's curves (as though viewing Foster's curves from above) and is tangent to the S-curve of the new technology at its point of zero slope (Fig. 2-4). This point may be conceptualized as a "point breakthrough" in a basic technology, such as the biotechnology achievements cited above, or it may be signified by the emergence of a unifying theory of a natural phenomenon or disease process. This point breakthrough sends a wave radiating through society as the technology comes to bear on various aspects of social needs that result in market opportunities (Fig. 2-5). This radiating wavefront is similar to the basic physics model of the wavefront or envelope of waves emanating from a point of light in a darkened room. However, not only technology breakthroughs send waves through the fabric of society; other events—changes in government regulation, the emergence of new diseases, the graying of the population, and the increasing incidence of two-income families—can also initiate wavefronts. Thus, in any one plane of society, wavefronts generated by technological breakthroughs are colliding with wavefronts from other origins in the social plane (Fig. 2-6). These waves are a microeconomic version of Alvin Toffler's concept of agricul-

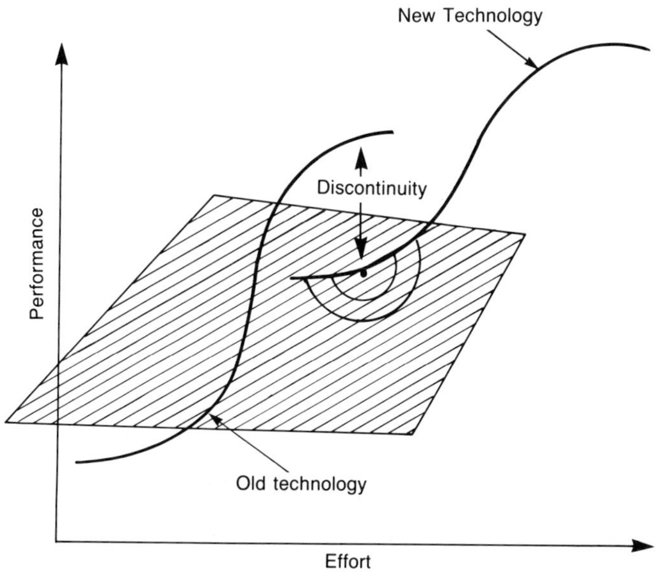

FIGURE 2-4 The "point breakthrough" in a basic technology (at the discontinuity) represents the "vision" of the principal organizers of biotechnology companies. Companies are created around such breakthroughs. (From Foster, R. N. 1986. *Innovation: The Attacker's Advantage.*)

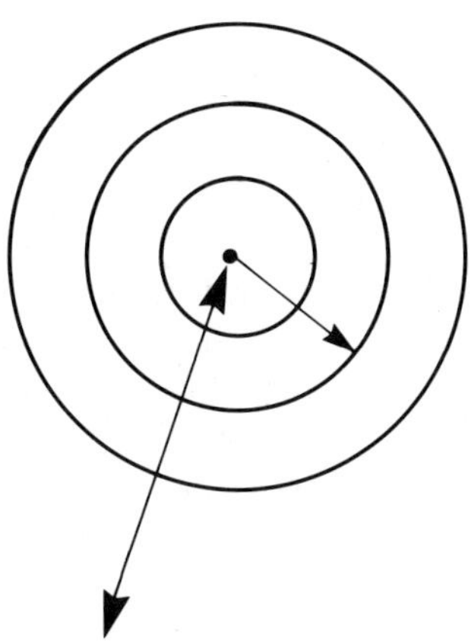

FIGURE 2-5 The technical "breakthrough" creates a wavefront that radiates through the industry, resulting in new market opportunities. (From Foster, R. N. 1986. *Innovation: The Attacker's Advantage.*)

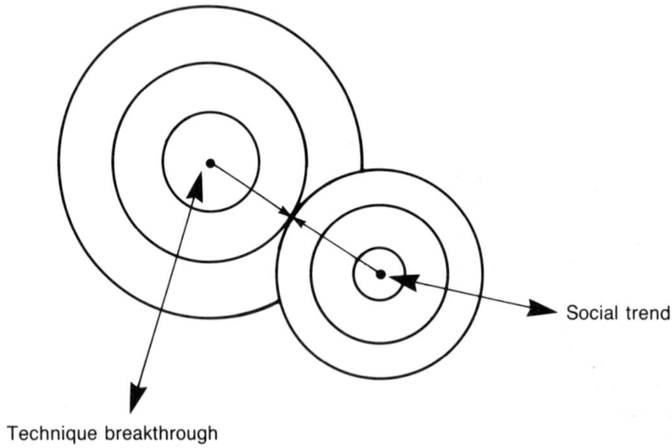

FIGURE 2-6 The technical breakthrough wavefront can collide with wavefronts from other sources (government regulations, socioeconomic issues, and emergence of new diseases within the plane. (From Foster, R. N. 1986. *Innovation: The Attacker's Advantage.*)

tural, industrial, and informational waves that shape society, as presented in his book *The Third Wave*. By comparison, the magnitude of the biotechnology wave is of the ripple category, but it nonetheless will have a substantial impact on the pharmaceutical, diagnostic, industrial enzyme, and agricultural industries, as well as on the lives of the patients who will benefit from its discoveries.

Although we do not yet know where to draw the origins of these point breakthroughs relative to one another, we can conceive of the point at which two such generated wavefronts collide and reinforce one another. Where the collision is of the "constructive interference" type, a "window of opportunity" is formed. Looking from the origin represented by our technology breakthrough along a vector to this window of opportunity, the length of the vector represents the development time to produce and bring to market a product to meet the needs of that opportunity window (Fig. 2-7). Looking down the vector from the window end, the length of the vector to its origin represents the market development time (i.e., the time for the market to reach a critical size for its infrastructure to develop for the delivery of goods or services and for collections). In our mathematical model, the length of the vector must be the same measured from either end. In the real world of a principal organizer, the two perspectives may not give equal answers; the entrepreneur learns that one can hit the wall just as hard by aiming ahead of a window of opportunity as by aiming too far behind it.

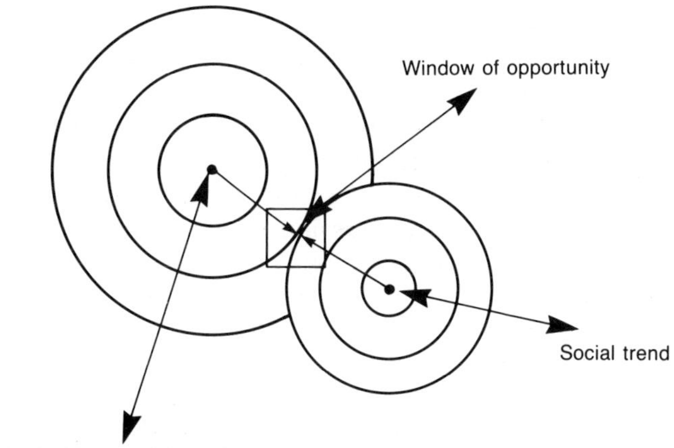

FIGURE 2-7 When wavefronts collide, "windows of opportunity" can form. (From Foster, R. N. 1986. *Innovation: The Attacker's Advantage.*)

Thus, principal organizers seem to be able to identify significant technology breakthroughs and to anticipate their impact on a developing social need. In this, they operate in an additional plane to those of technocrats operating on Foster's S-curve. That plane is characterized as correlating what may to others may be unrelated social phenomena and as dominated by a future time orientation. Principal organizers use this perspective to identify windows of opportunity, and these become their central concept for rallying others to share the vision. It may well be that the clarity with which these entrepreneurs see their vision accounts for their reduced perception of risks associated with undertaking their venture.

PARADIGM AND REALITY

Although it is true that Genentech adhered to a vision of creating a pharmaceutical company to serve the needs of hospital-based physicians, and that Amgen's initial public offering prospectus several years ago mentioned erythropoietin, the principal organizers of these companies also demonstrated a remarkable willingness to defer the fulfillment of their vision. They became practitioners of "the art of the incremental": to the extent that they were unable to sell, communicate, or implement the dream, they were willing to accept the direction and suggestions of their venture capital investors. Ed Goodman, Managing Partner of Hambro International Ventures, ranks this as one of the key attributes their firm looks for in assessing entrepreneurial candidates. Thus the principal organizer's vision serves more as a touchstone than as a Holy Grail to be preserved in its original form.

In the case of the tinkerers, the wavefront model may still apply, but it applies *frequently*. In Chiron Corporation's case, the company's vision underwent substantial metamorphoses: from a growth factor company to a vaccine company, from a yeast expression company to a diagnostics company. This reflected the principal organizer's aversion to committing to a single strategic direction. Rather, he moved toward the opportunity that loomed largest on the horizon until another eclipsed it, then he moved toward that one. The result was a portfolio of opportunities vying for resources and providing various avenues for success.

Where, then did principal organizers acquire the early outlines of the vision that they projected toward a window of opportunity? For the technically competent organizers, opportunities usually arose out of their own research or from that of colleagues. For the start-up venture capitalists and others, it came principally from reading summary articles in *Science, New Scientist (UK)*, or *Science Week*. The lore behind the founding of Genentech traces the genealogy of the vision to an article in *Scientific American* that caught Bob Swanson's attention. Of course, the literature search of the principal organizer usually extends much deeper into the technical publications, usually via a reference source such as the *Index Medicus*.

COLLEAGUES

Thus armed with his vision of a future reality, tempered by an appreciation of the art of the incremental, the principal organizer must enlist a cadre of key participants embodying specific talents, expertise, and prestige in their fields. For the technically competent entrepreneur, the challenge is to find business skills: finance, administration, marketing, product development, manufacturing, and regulatory affairs. In selecting candidates for such positions, the critical importance of task-relevant experience is often underemphasized in favor of frugality. This leaves the company in the position of financing learning curves and losing momentum as decisions have to be more completely researched than would be necessary with more experienced managers.

A second common error has been to overemphasize the transferability of experience between industry segments. Thus, marketers of hospital disposables became quick-studies in pharmaceutical promotion and distribution, and regulatory affairs staffs who specialized in devices learned new procedures in the Bureau of Drugs.

For the business-competent principal organizer, the challenge is to find a chief scientific officer who has adequate skills in the techniques of molecular biology to be able to critique the architecture of experimentation carried out by subordinates, and who also has enough knowledge of the disease mechanisms addressed by the company's portfolio to select viable intervention strategies. As applied here, the term "architecture of experimentation" refers to designing a set of experimental procedures that uniquely confirm the hypothesis while ruling out the alternative possible hypotheses. It is an art among researchers, and one that may be difficult to find in combination with disease-specific knowledge. The response to this limitation generally has been to establish a matrix organization having cells of specialized skills—DNA synthesis, cDNA cloning, DNA sequencing—reporting to a technology officer, and a disease/project organization reporting to a product development officer. The project leaders vie for technical resources, and the CEO sets priorities under the advice of the scientific advisory board.

In the circumstance in which the skills in experimental architecture and project selection cannot be combined in a single position, the staffing of the scientific advisory board becomes critical to the principal organizer. In addition to lending a technically prestigious imprimatur to the company's masthead, the board's members must also be prepared to offer insights about approaches to the disease process. These individuals are more likely to be Ph.D. researchers than M.D. clinicians, although exceptions in both directions abound. Fierce competition also abounds in the realm of academic research; the principal organizer must aggressively seek to eliminate incompatibilities within the scientific advisory board while maintaining the broadest manageable range of opinion as to the direction of the company's research. Usually this is resolved by bringing together people who have collaborated or otherwise worked together in the past.

For the start-up venture capitalist, however, the task of identifying scientific talent is less trivial. He most likely starts the search with a review of the literature in the target field. A lead article identified through the *Index Medicus* cites other references. These referenced publications designate their authors and institutions. The mastheads and editorial boards of professional publications and associations provide a cursory overview of "opinion influencers" in the field. A more diligent searcher may write to the Freedom of Information Office at the relevant institute within the National Institutes of Health, and presumably at other federal funding agencies, and request a listing of research and teaching grants awarded. One researcher may have been awarded grants in a number of areas, but the ones with the largest aggregate funding in the research areas of interest to the organizer are prime candidates for some role in the company.

Further insight into the "pecking order" of the research community may be gained by referring to the *Science Citation Index* published by the Institute for Scientific Information in Philadelphia. This reference identifies how frequently a researcher's published work is cited by other authors in their own publications; that is, it signals whose work has been seminal in spawning the progress of other researchers. The *Index* is, however, a historical perspective and may not reflect a researcher's recent contributions or his ability to continue to contribute in a major way. The *Index* also gives equal weight to all citations, rather than weighting on the basis of the stringency of peer review applied by that journal in selecting articles submitted for publication. In any field covered by two or more journals, there is generally a recognized hierarchy of publications, ranging from the most technically stringent to nonreviewed journals.

Having culled the names of his potential scientific officers, scientific advisory board members, and research staff from these sources, the principal organizer is now prepared to begin making contacts in the field of his targeted technology. Some contacts are by telephone, some in person. In either case, the objectives are to obtain confirmation of the vision, to solicit preliminary interest in participation, to identify intellectual property available for license, and to build a file of the curriculae vitae of key researchers. The best setting for making such contacts is generally at a professional meeting at which the focus is the presentation of research results, such as scientific meetings and poster presentations. However, attendance at such functions can become costly in terms of both time and money.

ZERO-STAGE CAPITAL

The principal organizer of a zero-stage company gets his first lesson in practicing the art of the incremental when initially going after capital. Although the uninitiated will immediately begin drafting a two-hundred page plan to impress the most cynical venture capitalist, the effort would be better expended in raising the first $30,000 that will be needed to publish and promote the plan to seed-stage venture capitalists.

If intellectual properties need to be licensed, the licenses should be fully negotiated and executed prior to approaching investors beyond the seed round. The implication of this is that the zero-stage capital sought should include expenses for the principal organizer to operate long enough to execute the initial options (six months) and to pay the expected option fees. The seed round, following the zero-stage funding, then must be large enough to complete the staffing of the company and to absorb any fees that may be due on signing of a definitive license.

Thus, the principal organizer's initial priorities should be directed toward raising the first $100,000 to $200,000. This is where the "double-dippers," or executives whose net worth benefited from one successful biotechnology venture, enjoy an advantage, not to mention their improved visibility and access to venture capitalists. However, it is generally not the managed funds of these venture capitalists that go into zero-stage ventures. Rather many of these partners are individual investors in a pool of personal funds that operates like an investment club. These funds frequently go into zero-stage deals. The challenge to the principal organizer is to obtain an introduction to a key partner in the club. The organizer's professional acquaintances—bankers, attorneys, and accountants—may be able to arrange introductions, or he may choose to go to a finder, who arranges introductions to sources of capital, but often at very high cost, taking a percentage of the funding raised, plus founder's equity, plus an ongoing consulting contract, and a board of directors seat.

The alternatives are to approach a Silver Angel—a wealthy local individual who makes small investments in start-ups, or one of the start-up venture capitalists identified above, or one of the handful of firms like TA Associates in Boston who sincerely like to take a deal from its inception forward. In making such an approach for zero-stage capital, it is only necessary to have prepared a terse and cogent description of the vision, good scientific literature support, and market data of the quality available from not-for-profit fundraising organizations ("The Society for the Prevention of . . ."). This is in addition to the skeletal operating budget and intellectual property costs discussed above. These documents, usually comprising twenty-five pages or less, are substantially less burdensome than the full business plan that will be needed at the seed stage, but they are the valuable initial embodiment of the Vision of a principal organizer. With luck that Vision will become the core of a new enterprise, whether in biotechnology or in other fields.

REFERENCES

Cohen, S., and Boyer H. 1973. Construction of biological functional bacterial plasmids *in vitro*. Proc. Natl. Acad. Sci. U.S.A., 70:324.

Foster, R. N. 1986. *Innovation: The attacker's advantage*. New York: Summit Books.

Kohler, G., and Milstein, C. 1975. Continous cultures of fused cells secreting antibody of predefined specificity. Nature 256, 495.

Toffler, A. 1980. *The third wave*. New York: Morrow.

CHAPTER 3

THE LIFEBLOOD OF BIOTECHNOLOGY: UNIVERSITY–INDUSTRY TECHNOLOGY TRANSFER

Lita L. Nelsen

The university was the birthplace of the biotechnology industry, and it continues to be the source of most of the basic new technology that fuels the industry. Moving this technology from the university laboratory into the company is the major challenge of the university technology manager. It is a cross-cultural job, in which the technology manager must represent the attitudes, goals, and measures of accomplishment of the "pure" academic scientist in terms that the industrialist can understand and accept. The job also requires a great deal of judgment and diplomatic skill to balance the often-competing needs of the university and industry. In addition, for the technology transfer process to be successful, the university technology manager relies on the presence of an equally sophisticated partner on industry's side of the table—one who understands these competing objectives and is willing to compromise in order to ensure a successful long-term relationship.

In this chapter, we discuss the university-industry technology transfer process in the context of a variety of types of collaborations. These range from ostensibly the simplest—the straight, single patent licensing agreement—through the most intertwined research and development collaborations. All of these collaborations share the problems of the difference in cultures between academia and industry and of balancing the competing objectives of academic freedom and the public mission of the university with the need of industry for defined outcomes and protection of property.

THE NATURE OF THE UNIVERSITY'S "PRODUCT"

The first major difficulty confronting the university technology manager is the nature of the product. University technology is by its nature basic and early, and nowhere earlier than in the biological sciences. Frequently, the "invention" is little more than a discovery of a basic mechanism, supplemented by deductions of its utility. The invention may not yet have been tested on rodents, much less humans. Neither the academic researcher/inventor nor the technology manager may have a clear understanding of the potential commercial utility of the invention, or even of the different fields to which it may be applied. And, though some university technology managers have the training and background to conduct the market assessment studies that would enhance such understanding, few if any have the time to devote the three to six months per invention that would be required to conduct an adequate study.

As a result, university technology is usually presented as is:

> *Here's what we have found, and here's what we are trying to protect (either as patents, copyrights, or perhaps as "tangible material"). We are not certain it will work, and we are not terribly sure what it is good for. We aren't even sure that any patents will issue. We do know, however, that this invention is based on very good science in a promising field, and it appears to be a potential solution (or at least a clue toward a solution) to an important problem. Our researcher is a leader in her field, and would like to see someone develop the practical side of the invention. She is continuing research in the pure science aspect of the technology, but would be happy to help on a consulting basis with practical development.*

This embryonic nature of university technology is critical for the industrial technology scout to understand. He should not approach the university for a list of "products available for licensing," expecting inventions that require little more than bottling and labeling. Such expectations are bound to lead to disappointment and are a waste of time for both sides. Instead, the company approaching a university for technology should expect to be at least a partner in the applied research required to bring the invention even to the prototype stage. The company should further realize that it will be carrying essentially all of the burden (and risk) of development of the invention into a marketable product.

Fortunately, the early stage of the research or the invention is a positive factor in the risk equation. Because it is basic and early, the invention represents a head start in the field; it is also unlikely to be a "me-too" product. The intellectual property (usually patents) that results from this early invention has the potential for broad, strong coverage. If the company can get exclusive rights to such an invention, it may be able to dominate the field of any products resulting from it. The company's affiliation with the inventor allows the company to remain up to date (if not somewhat ahead) in a rapidly changing area.

It is this potential to "get in on the ground floor" of new technology that is leading so many biotechnology, pharmaceutical, and chemical companies—both large and small—to strengthen their relationships with universities. These companies are adding new technology acquisition departments manned by scientific professionals with strategic knowledge of the market and their companies' objectives, and capable of interacting both with academic scientists and with the university technology managers. They can recognize the potential of new technology (which often comes to them without warning and without invitation). Working with the new breed of technically trained, entrepreneurial university technology managers, the technology acquisition professionals from industry can also help devise and negotiate the flexible-by-necessity agreements with the university needed to develop the embryonic inventions.

WHAT IS THE COMPANY LOOKING FOR— AND WHAT CAN IT HOPE TO GET?

The recent upsurge in industry's interest in university technology appears to be driven primarily by the realization that technology is moving too fast for any one company to keep ahead on its own. The famous "Not Invented Here" syndrome heretofore prevalent in many companies (and particularly notable in the past in much of the pharmaceutical industry) has faded in the face of rapid technological change. No single company can afford to begin the hundreds, perhaps thousands of separate scientific initiatives that may someday lead to the products needed for the next decade. The success of many Japanese companies in acquiring technology through patient, persistent monitoring of external developments has further led American companies to realize the necessity of extramural awareness.

For the biotechnology industry, and particularly for that segment concerned with therapeutics, the university is the primary source of new product ideas. The typical new product concept arises directly from a scientific finding: the scientist observes a new effect, isolates a new gene and/or protein, and then hypothesizes its function; from there the scientist (or someone else) hypothesizes a use in the human body. In this process, the road from "pure science" to a vision of a product is unusually short: (Of course, as is well known, the road from the vision of a product to a product on the market in the biotechnology fields is also frequently unusually long as compared with conventional technology.)

There are a large number of world-class scientists now working in the biotechnology field in academia. Because of intense worldwide competition among academic researchers in this field, and because of the unusually rapid development of new tools in the field, the science is progressing at an astonishing rate. Significant new findings are occurring daily, and a company simply cannot afford to be out of touch with the academic laboratory for long.

As a result, biotechnology companies are expanding their liaison activities with universities. Even the smallest biotechnology company will have a "technology acquisition specialist" spending much of his time looking for new technology in universities and seeking to forge alliances. The challenge for these people and their companies is to define what kind of alliance is best at a given time and for a given technology.

There are a number of reasons why a particular company may seek specific alliances. These include

- A desire to get access to the technology early—before it is preempted by others. Early access may allow the company to get broad, exclusive rights. And, because the true potential of the technology is still unknown, such rights can sometimes be acquired relatively inexpensively.
- A desire to collaborate with a leading researcher in a particular field. In this way, and particularly if the collaboration is long-range, the company is kept continuously up to date on developments in the specific laboratory. If the academic researcher is truly at the forefront of the field and reasonably diligent in keeping abreast of the literature, the company will also become informed of technology throughout academia that may affect its technology. If each party in the university–industry collaboration is doing research and sharing results with the other, the results of the two parties working together can be greater than the sum of the individual contributions.
- Unique resources available at the university but not in the company. This may include special equipment, expertise on the part of the investigators, and unique biological materials. The company needs it, and the university has it. Frequently, such expertise becomes available to the company on a more permanent basis when a graduating student from the university project joins the company full time. Consultation between the faculty member and the company may provide additional access to university know-how.
- Particular intellectual property rights (notably patents) are owned by the university, and the company wants a license.

- Particular development work (for example, the next stage of development of patented university technology) is needed, and the university is best prepared to carry out that work.
- The company perceives that it is cheaper to do the work in the university than in the company. (*Warning:* This is usually a poor reason for a collaboration. The savings in manpower rates and overhead are usually more than offset by the company's lack of control over the specific tasks of the project. The differences in the types of projects best carried out by a university and those best carried out in industry are such that if the company can do the work well itself, it probably should.)
- The company is not sure why it is funding the work, but it thinks a relationship with the university is a good idea. (*Warning:* Experience shows that this is the worst reason for an alliance, and one that is bound to lead to disappointment on one or both sides. Expectations should be clear—and articulated to both sides—before the alliance is begun.)

WHY DOES THE UNIVERSITY WANT ALLIANCES WITH INDUSTRY?

It is not always primarily for money, though financial considerations certainly play a big part in the university's interest in industrial collaborations. American education is in a tough state financially. The costs of education and research are increasing much faster than inflation, while support from the major source of funding—the federal government—is increasingly hard to come by. Even some of the leading universities have run in the red during the past few years. Alliances with industry, including sponsored research, licensing of intellectual property, and/or investments in start-ups are seen as substitute sources of funds.

This is a potentially attractive approach, but success may lead to trouble. In the sponsored research arena, if industrial funding becomes more than about 20 to 25% of the total university research budget, the university may find itself compromising its basic research identity. More and more projects may be accepted that are little more than fee-for-service contracts, in which specific well-defined development or testing tasks are contracted for and little actual research goes on. As the proportion of the total university budget coming from industry increases, the university may find itself increasingly unable to turn the money down and say "That's not the sort of work we do."

In the long run, this is good neither for the university nor for industry. The university may find itself simply a cheap source of labor, losing the independence and freedom of action that allowed it to attract the best talent, turn out good students, and serve knowledge as a whole. Industry may then find itself having killed the goose that could lay the golden eggs: as the university turns itself more and more to short-term focused projects to please its industrial sponsors, it ceases turning out the truly innovative research results that stem from a long-term focus and freedom to follow research results wherever they lead.

There is also the danger that licensing activities may lead to an overly great focus on industry's needs, but mostly this is kept under control. A great danger to the success of such licensing operations seems to be unrealistic expectations. University administrators are looking to licensing revenue to substitute for government support of research. The facts simply do not support this expectation.

With one or two exceptions, even the most successful university licensing offices receive licensing revenue equal to only 1 to 2% of their universities' total research budgets. Most of these universities, and especially the exceptions with slightly higher percentages, rely on single "blockbuster" patents for the majority of their revenue. Such an exceptional patent may come along, if the university is very lucky, once per decade, and when that patent expires total licensing income will drop precipitously. The challenge is therefore to develop a middle tier of licensable technologies, which will bring in a moderate income over time. The university should not expect to get rich quick.

If the expected income from licensing is so small, why are universities putting so much emphasis on this aspect of technology transfer activities? Many of the reasons apply to all phases of the university technology transfer program. They include

1. *The university's mission to transfer technology for the public good.* For many universities, this is the primary motivation behind their technology transfer programs. These universities realize that their basic research results will benefit the public only if these findings are effectively communicated to industry in a manner that will induce industry to invest in product and market development.

The U.S. Congress explicitly recognized the need for increased technology transfer from basic research and increased incentives for investment in Public Law 96-517, which went into effect in 1980, granting title to the university for inventions made in university laboratories under government funding. Over the past decade, this initiative was strengthened through additional legislation and executive orders that removed almost all restrictions on the university's licensing of such inventions, now allowing the university to grant fully exclusive licenses, if the license will commit to development.

2. *The desire of researchers to see something real come out of their labors.* The good researcher is a practical animal. In many fields, science is seen as "proven" only if it works, and the best indication that it works is successful translation into practical products. The best researcher is also an idealist: she wants to do something useful for society, both in expanding human knowledge and in bettering the lot of humankind. This latter motivation is particularly keen in the biological researcher.

Successful transfer of university technology to industry provides the opportunity for the researcher to see products of her labors without having to leave the university. And, a suitably structured licensing agreement also allows the researcher to share in the profits of such labors while remaining in academia.

3. *The uses of licensing money.* Universities allocate a certain fraction (ranging from 15 to 50%) of their licensing revenue to inventors. This money is given as personal income, and can provide a significant incentive to faculty, graduate students, and other researchers to report inventions and to cooperate in the patent filing and technology transfer processes.

The remainder of licensing income is usually split (sometimes through very arcane formulas) among the individual departments and/or schools of the university and its general fund. The real value of these usually small sums comes from the fact that they are denominated as "discretionary funds," money not preallocated to particular programs but that can be used on an as-needed basis for such purposes as student scholarships, seed money for blue sky projects, much-needed equipment, and facilities repair. This discretionary money is the scarcest resource in any university.

4. *Bringing practical industrial problems into the university.* Scientists in universities, and particularly engineering scientists, have a continuing need to identify practical problems, or else they risk coming up with irrelevant solutions to "nonproblems." Collaborations with industry through sponsored research and licensing agreements help bring reality into the academic laboratory. The research benefits, and the students benefit through working on the most up-to-date problems.

5. *Opportunities for consulting.* A good number of sponsored research agreements and licenses also result in consulting opportunities for the faculty inventors. In most, but not all universities, consulting agreements are regarded as private, not connected in any formal way with the university–company agreements. These consulting agreements allow additional know-how to be transferred in a very personal manner, and allow the faculty member to become aware of the real applications of his inventions.

6. *Opportunities for graduates.* This is self-evident, but important. The company often gets to know the graduate students and postdoctoral fellows through a sponsored research program or a license negotiation. The student gains an opportunity to work for a company directly in her field, and the technology is transferred to the company more completely. "Technology transfers better in bodies (and heads) than on paper."

7. *Local economic development.* University spin-offs have built Route 128 outside Boston, Massachusetts, Silicon Valley around Palo Alto, California, and Biotech Alley in La Jolla, California. And state governments have begun to notice. Many are now explicitly encouraging collaborations with industry in order to build high value-added jobs in their depressed regions.

THE CULTURAL CLASH: THE DIFFERENT WORLDS OF THE UNIVERSITY AND INDUSTRY

Universities and companies are two very different worlds. The cultural differences between these two worlds are so great that sometimes the only thing allowing the two worlds to communicate at all is the fact that most company managers have spent some time at universities—if only as students—and the fact that, at least in the United States, most university faculty and administrative managers have spent some time in industry—if only at summer jobs during their student years or as consultants to industry.

The differences between the two start with their reasons for being. The university sees its mission as "fostering the public good," with two major objectives: to educate students and to discover and disseminate knowledge. All else, at least in theory, must be subordinate to these objectives.

The mission of the company is proprietary rather than societal: to make a profit and return to its investors a profit on their capital. The economic theory of our capitalistic society proclaims that the sum total of companies seeking these proprietary objectives will best foster the economic public good.

The differences in these somewhat theoretical mission statements lead to very concrete differences in the two entities' ways of operation and underlie many of the clashes between the university and company technology managers seeking alliances between their institutions. In the following sections we will highlight some of these differences and the disagreements to which they may lead. For clarity's sake, the differences are taken to be those between the typical large, research-oriented university, and the product-oriented company. Hybrid institutions, such as the "research institutes" in some multinational companies have been ignored.

Goals: The Work Statement

Industrial research and development is inherently goal oriented. The industrial researcher is expected to know why she is pursuing a particular project, and what products are expected to come out of it. She is expected to define the project in advance, lay out the tasks to be done, and know very clearly how success or failure will be measured. She is expected to be willing to abandon the project altogether if the results show the project not to be cost effective or if the market is shown to be too small. She is also expected to change the direction of the research if a more profitable avenue opens up.

In contrast, the academic researcher sees his work as "discovery" rather than "development," and therefore as open ended and largely unpredictable. He is reluctant to define his goals other than fairly broadly and generally, and is particularly reluctant to commit to specific tasks in advance since "who knows where the research will lead." He requires the freedom to change the direction of his research if a more scientifically interesting avenue opens up. Measures of accomplishment are vague, because, for example, a "failed" experiment (in terms of the original goals of the project) may lead to more interesting discoveries than a "success" might have done. In general, the academic researcher makes no promises, because "Mother Nature makes no promises," and "if we know how the project will turn out, we shouldn't be doing it."

Now imagine how these two attitudes clash when a company looks to sponsor a research program at a university and wants to know "What are we going to get for our money?" The company is expecting a well laid-out research plan, perhaps with a classic PERT (Program Evaluation and Review Technique) diagram showing goals, milestones, and even a defined "critical path." Meanwhile, the professor believes that a single paragraph defining the area of the work to be done and a reasonable first approach to it ought to be sufficient. Compromise, diplomacy, and cross-cultural understanding are needed if the project is ever going to get started.

The Individual

The industrial researcher is expected to be an "arm of the company." As an employee, she is expected to subordinate her own wishes to those of the company (assuming that the company's behavior is ethical), and to be a willing member of a team, rather than a prima donna. It is expected that the company can, within limits, define how she will spend her time and can make promises to others about her performance "on behalf of the company." If the individual assigned to the program cannot deliver as promised, the company is expected to provide another employee who can.

In marked contrast, the academic researcher fits few of the expected characteristics of the "employee." He is judged and rewarded as an individual, and his loyalty may be as much (or more) to his profession as to his university. He expects his accomplishments to be judged by his scientific peers, rather than his administrative "superiors" at the university, and teamwork with other senior researchers is the exception rather than the rule. If the principal investigator leaves the university, it is expected that his project at the university will fold.

Because of the preeminence of the individual researcher at the university, the university technology manager is quite limited in what she can promise on behalf of the university about performance in an individual agreement. She can agree to use "best reasonable efforts," for example, to have "complete" quarterly reports delivered to the sponsoring company; but if the principal investigator does not write them, there is little that the technology manager can do besides nag and plead. It is therefore critical that the company sponsoring work at the university have a good relationship not only with the university administration, but with the individual researcher. Far more than in industry, individuals are the key to success.

Publication Rights

Nowhere is the difference between university and industry cultures wider than in the two institutions' views of information. Most companies have a portfolio of jealously guarded trade secrets that they hope to hold from their competitors for as long as possible. In addition, almost all companies require prior review of their employees' planned publications or other public disclosures in order to protect their trade secrets or, at a minimum, their ability to file for patents.

In contrast, freedom to publish without restriction is the sacred cow of most major universities. Many universities will proclaim that "they have no trade secrets"—except for the period (typically on the order of a few weeks or months) required to file patents on a particular invention. Any general requirement to review manuscripts or presentations prior to publication would put an intolerable (and probably impossible) burden on administrators, and would probably foment rebellion among the faculty. Again, the problem is most severe in the biological professions, in which publication follows so rapidly after discovery.

So what can the sponsoring company do to protect the information arising from the work it sponsors? Some limited compromises are available. Most universities will agree to submit manuscripts to the sponsor prior to publication or other public disclosure. (Here, again, it is critical to get the assent of the faculty member in advance, since his cooperation will be critical.) The period prior to publication usually ranges from thirty to sixty days. Shorter periods may be demanded by the faculty member in "hot" areas of research, and longer periods (usually not over ninety days) can sometimes be negotiated by the sponsor in less critical fields. The sponsor is given the right to require confidential information from the company be removed from the manuscript and the right to ask that a patent application be filed on the work prior to publication.

This compromise will usually work well, with little resistance from either side, if

- The university has been consistent in protecting its rights to publish, so that the company understands that prohibition of publication is "simply impossible" and not a subject for negotiation; and
- The company does not try to use its economic power to override the university's publication policy. Such bullying techniques over publication rights are the single greatest occasion for conflict between university administrators and companies, with the university administrators complaining that the companies "just don't seem to understand who and what the university is all about"; and
- The university has gotten prior consent from the researcher to abide by the agreement for presubmission and has educated the researcher to understand that any nonconfidential disclosure outside the university (preprints, poster sessions, even an oral presentation) can endanger patent rights; and
- The company has in place a mechanism for rapidly reviewing publications and returning its comments to the university with minimum bureaucratic delay; and
- The university has in place a mechanism for rapidly filing patent applications so that publication is minimally delayed.

The process is helped further if there is continuing close communication between the researcher and the project coordinator at the company, so that the sponsor is aware, *before* the manuscript is completed, that an invention has been made and should be protected. This presupposes a trusting and trustworthy relationship between the researcher and the company, so that the researcher need have no fear that he may be "scooped" if he pre-discloses his information to the company. Again, individuals are key.

Property Rights and Freedom of Action

In contrast to their limited control over individual performance, many universities (although by no means all) have well-defined intellectual property policies that specify that the university owns and, at least in theory, controls the intellectual property (patents, biological materials, software, etc.) arising out of its research. In federally sponsored research, these internal policies are supported by the government's regulations that give title to such property to the research institution.

Several primary principles guide what promises universities will make with respect to their intellectual property:

1. The university considers itself to be a unique resource, not simply a "place for hire." Thus, the ownership of any intellectual property that arises at the university must recognize the unique contribution of the institution, not just the funds used for the individual project from which the invention arose. For this reason, most universities will not assign patents to the sponsor, nor will they grant royalty-free exclusive licenses just because the sponsor paid for the research. In general, the "product" that the sponsor can expect from the research is the knowledge and the advance in the field. Patent and other intellectual property rights are seen as "extras," for which the university should be additionally compensated.
2. The relationship between the university and the company will be for only a limited term, and the university must protect its freedom of action for the long-term future. Thus, for example, the university will be very careful about what promises it can make about future "improvement" patents after the company's sponsorship is over. A compromise that some universities make is to grant to exclusive licensees options to improvement patents dominated by the claims of the original patents licensed "since they can't be licensed to any other party anyway."

This principle of preserving freedom of action is often in conflict with granting of exclusive licenses, since it is unlikely that any second company will sponsor research in an area in which another company holds an exclusive license to an earlier, dominating patent. There is no easy answer here. The university should seek to limit the field of use in early exclusive licenses, but this is not always possible when it is seeking exploitation of embryonic, early, very risky technology. At a minimum, both the university technology manager and the researcher/inventor should discuss among themselves the implications of such an exclusive license, and the fact that it may very well foreclose future commercial sponsorship from anyone except the license.

3. Individual researcher rights should not be mixed. Senior researchers at most universities operate quite independently, seeking support for individual projects without detailed coordination from the administration. The university must therefore be careful not to inadvertently mix the intellectual property of Professor A with that of Professor B, unless they are personally collaborating. Thus, any "related technology" or "improvement patents" to which a university grants a license or option in conjunction with another license, should be restricted to the work of the individuals from which the original license arose.

4. The individual researcher should have a say as to how his inventions are licensed—within the limits of ethics and common sense, and reasonable cooperation on the part of the researcher. This is clearly different from how a company handles its intellectual property, seeing it as a property of the firm in which the individual inventors have little or no rights.

The actual administration of this "faculty preference" is a gray area, and one that periodically makes university technology managers squirm. Although most universities give the individual faculty member a significant fraction of the royalties received from their inventions, no university will grant the individual the final say in how his technology is licensed, since the potential for arbitrariness and/or conflict of interest is great. At the same time, for both practical and political reasons, and from a sense of intellectual fairness, the university technology manager strongly prefers that the inventor be aware of and happy with arrangements made for his technology. And the company licensing the technology usually wants, and sometimes critically needs, the active cooperation of the inventor in order to transfer the technology effectively out of the university laboratory.

In practice, the university technology manager usually consults with the inventor before beginning to license his technology, and periodically throughout the licensing process. Usually, the inventor will not be consulted on the detailed terms of the license, but his advice will be sought in choosing among companies and on any promises about future technology. At the same time, the university manager has the responsibility to ensure that respecting the inventor's wishes does not result in "pipelining" of inventions to companies for which the inventor consults or in which he holds ownership, nor that any other "sweetheart deal" is made. A delicate balance must be maintained on all sides.

From a company's point of view, the preference of university technology managers to accommodate the faculty inventor's preferences in licensing puts a premium on getting to know university researchers early, making them aware of the company's interests, and ensuring competence in the researchers' fields.

TYPES OF ALLIANCES

The types of alliances between academia and industry are myriad and range in an ill-defined continuum from distant "one-time sales" to long-term close collaborations. Generically, they can be divided into the categories discussed in detail below, each with its particular strengths and dangers. In each case, well-defined "prenuptial" discussions and agreements are advised as a safeguard against future misunderstandings. However, the two sides must also keep such prealliance discussions from degenerating into bureaucratic nit-picking and preventing the alliance from ever getting started. Flexibility and goodwill are a must.

The types of alliances frequently found between universities and companies are discussed below.

Research Collaborations

In a true collaboration, both sides are cooperating in the research, each agrees to share results with the other, and it is expected that the results will be jointly published. Such collaborations are usually begun because each side has particular knowledge and expertise that the other needs; for example, one party may have a particular receptor and screening technology, while the other has drugs to be screened. Money may or may not change hands.

The major disagreements that arise from such research collaborations seem to stem from unfulfilled expectations: One party feels that the other has not put in sufficient effort. This may be because no such effort was intended, because the results did not come as quickly as intended and more effort was needed than was "reasonable," or (and usually the industry side is most at fault here) priorities changed in the company and resources were taken off the project. Other, rarer misunderstandings arise because the property rights were not defined in advance. Who will own the patents if any arise? What about software? And biological materials? What are the company's prior commitments? Is any third party sponsoring the work at the university, and are there any commitments to that third party? Are both parties clear about their rights to publish? And who will be first author?

Not all misunderstandings can be prevented, but the majority can be headed off by discussing these uncomfortable issues before beginning the collaboration. The technology managers in the two organizations can make critical contributions to these discussions by defining the major issues, avoiding getting trapped in the minor issues, settling the issues quickly, and using their negotiating and diplomacy skills to help all parties reach agreement and get on with the work.

A simple clue to preserving harmony is to define the end point of the collaboration—preferably by date—as clearly as possible, in advance.

Industrial Liaison Programs

Industrial liaison programs go by a number of names, but are basically subscription and coordination services. For an annual fee, the company is introduced to the new happenings at the university that are relevant to the particular company. Meetings are set up with faculty members in the company's areas of interest, reports and articles are distributed, and special seminars may be held. Usually, nothing proprietary is given to the company; instead the emphasis is on information, which is well-screened, well-selected, and delivered in a timely manner; and on continuing informal access to faculty.

These services are simply described, but can be very complex to administer well. The interactions between a large, multidisciplinary company and a large, multidisciplinary university occur on many levels. It usually takes a technically knowledgeable liaison officer at the university to manage these interactions, neither flooding the company with irrelevant material nor ignoring its needs. Ideally, the liaison officer is made privy to the strategic plans of the company so that she can better identify information of relevance.

Equally important is the liaison person at the company. This person should be

- interested in the liaison program (not just appointed),
- a believer in its relevance to the company,
- knowledgeable about the university and its mission,
- knowledgeable about the needs of the company for information—both long and short range,
- knowledgeable about the company's technology,
- knowledgeable about the company's structure and influence chain, and

- sufficiently powerful to influence the company to consider new relevant technology

The liaison person at the company should be a natural "gatekeeper" or "technical gossip"; someone who naturally makes connections between disparate bits of information and brings them to the attention of others in an effective manner. He also should not be afraid to make demands on the university—to ask more from it than it might tend to give if not asked.

It cannot be overemphasized how important this internal liaison person can be. If he is just a paper-shuffling coordinator, the company will soon wonder why it is paying such high subscription fees to the university "when nothing ever happens." In contrast, companies with effective internal champions of the liaison program find these programs to be exceptionally effective, sometimes keeping uninterrupted memberships lasting two to three decades or more.

The wise university seeking new liaison members will demand that the company place influential coordinators in place before beginning their programs. People make all the difference.

Sponsored Research

This is perhaps the most common of the alliances between industry and the university—and one likely to increase in importance as government funding of university research becomes scarcer and as companies seek to expand their research base. Its potential for lasting, close relationships is great, but, similarly, its potential for misunderstanding is perhaps the greatest, as the two cultures interact for prolonged periods of time with fairly high expectations of each other.

Typically, the company will contract with the university for the latter to do research in a defined area for a certain period. In return for its funding, the company will expect to get data, perhaps tangible property in the form of biological materials or drawings, and perhaps software. It will also expect to get options to license patents that may come out of the research.

A successful sponsored research program depends on the following:

1. Defined reasons for why the company is entering into the program and what it hopes to get out of it. This is seldom put in writing between the university and the company, but it should be documented at least at the company to make sure that there is a basis for later evaluation of the "success" of the project. Making this document available to the university can be a good way to check the realism of the company's expectations.
2. A defined responsibility at the company for coordination with the university project.

3. A defined work statement for the university: What are the boundaries of the research area? (This is particularly important in a large laboratory, where there may be a number of sponsors and where it is important that rights to intellectual property do not overlap.) Can tasks be specified in advance? What are the "deliverables," if any? And how can the direction of the research be changed if new results demand a change in direction?

4. Defined responsibilities of the company (if any besides simply paying the bills!).

5. Defined intellectual property rights: Who owns the patents? What are the sponsor's rights (options, licenses, etc.)? How is confidential information to be handled? Who decides what to publish? How will joint inventions be handled? (A secret here: joint invention ownership should be defined in such a way that neither side loses a great deal if the other contributes to the invention. Openness between the university and company scientists about possible solutions to the problem at hand is the best guarantee of maximum creativity.)

Licensing of Intellectual Property

A surprising range of intellectual property is licensed by universities: patents, copyrights, software code (both patented and copyrighted), cell lines, other biological property, trademarks, maskworks, drawings, know-how, and (sometimes) trade secrets.

Ostensibly, this is one of the least complicated forms of industry–university alliance in that it can be a simple, single transaction: defined intellectual property is licensed under agreed-on terms, such as exclusivity, royalties, terms of license, and diligence by licensee. In theory, once the agreement is signed, the university licensor simply sits back and waits for the money to roll in.

In fact, of course, things are much more complicated, primarily because of the very early stage of the technology licensed. This is especially true in biotechnology, in which the parties negotiating the contract must agree on precisely defined terms and conditions when neither side knows with any precision, for example,

- what the products will actually be or look like
- whether the technology will actually work for the purpose intended
- how the science at the university from which the patents arose will evolve—to improve or perhaps obsolete the technology being licensed
- whether the patents will issue
- how important the patents will be to other distantly related products and technology

- how long it will take to develop the first product
- how much money it will take to develop the product
- what other products will emerge from the technology
- how much the products will cost to make (and whether they can ever be made in practical quantities)
- what the price of the products will be
- what the size of the market will be
- what the primary mode of marketing will be (direct, distributors, sub-licensing, etc.)
- what other new technology (including others' patents) will be needed to make the products successful
- whether the regulatory authorities will approve marketing of the products

In the midst of such massive uncertainty, negotiating precise terms and precise wording of license agreement clauses can have a surreal quality. Successful consummation of such agreements requires people on both sides of the negotiating table who can deal flexibly with uncertainty, defining conditions as precisely as possible while leaving room—and goodwill—for future renegotiation if conditions change.

Some Particular Issues in University Licensing of Biotechnology

Know-how and trade secrets. Licensing of know-how and trade secrets is a controversial issue in university licensing. Such know-how can be particularly important in manufacturing processes, in which slight changes in procedures can make major differences in reproducibility, quality, and yield, and in which the "little secrets" that make the difference are seldom documented in scientific journals and may also be omitted (because not yet discovered) in patent applications filed early.

Universities' policies differ about whether know-how can be licensed exclusively and whether there should be such a thing as a nontangible trade secret in an open university. The debate centers around the issue of free dissemination of information from the university. By definition, an exclusive license to a trade secret implies the obligation not to reveal the information to any other parties. If university policy states that "all results must be publishable," then such exclusivity cannot exist. Many universities therefore refuse to grant exclusive licenses to any nontangible property.

Biological materials. The convention of academic publication requires that the manuscript describe procedures and results in sufficient detail that the author's peers can reproduce the experiment—in order to confirm the results or to build on and improve the knowledge reported. Biotechnology presents a special problem in that often the results involve particular materials (for example, antibodies, large genes, or cell lines) that, given the current state of the art, cannot be adequately described in such a way that they can be precisely replicated. Or, if the materials can be precisely replicated, the effort to do so is so enormous that it is not considered practical.

As a result, it has become customary (and is now required by the National Institutes of Health and many of the more prestigious scholarly journals) that after publication, academic researchers provide the biological materials reported to other researchers on request. Frequently (and particularly when the results and materials have commercial value), such materials are provided under biological material transfer agreements, which stipulate that the recipient will use the material for "research purposes only" and not for any "commercial" purpose.

Therein lies the rub. What is "commercial use"? Is a scientist doing basic investigations in a company laboratory engaged in research? Probably. Is it "commercial"? Probably. Does it make a difference if the results will be published? And what about if the scientist does not yet know what, if any, commercial product will result? And what are the implications of the material transfer agreement if the biological material is used only to test or screen new products, so that the final commercial product does not contain any of the biological material or use it in its manufacture?

These problems are exacerbated when the major value to be gained from the material is through an exclusive license. If the material (or its use) is patented, should the university grant an exclusive license to a company such that the company can prevent other "researchers" from using the material? (Contrary to popular opinion, there is no clear cut "research exemption" to U.S. patent law.) And, if the license is granted prior to issuance of the patent, is the university acting properly if it distributes the biological material to other researchers in commercial companies, thus damaging the head start that its exclusive licensee hopes to gain by use of the material prior to patent issuance?

At the moment there is no consensus among universities (nor, in general, even within individual universities) as to how to resolve these contradictory objectives of inducing development of inventions through licensing and fostering the progress of research through widespread distribution of biological materials. Custom and policy are evolving on a case-by-case basis, usually, though not always, in consultation with the researcher who developed the materials. Generally (though by no means universally), it is agreed that it is acceptable for a university to charge a "reasonable sum" for use of biological materials in research by companies, while transferring such materials "at cost" to other universities. There is, however, no consensus on the acceptability of

- Withholding such materials from all commercial entities (except the exclusive licensee) on the grounds that all work in such entities is "commercial."
- Requiring the recipient (including other universities) to get permission from the giver before filing patents on the results of work with the material.
- Requiring the recipient to grant the giver (and its licensees) licenses to any improvement inventions (material and/or patents) it makes using the material.
- Placing any restrictions (or asserting any ownership rights) on products identified, isolated, or screened using the material, but which do not contain the material.

The blockbuster patent: is exclusivity justified? This is one of the great theoretical terrors in the hearts of university technology transfer managers. It is often expressed as "What if the Cohen–Boyer patent had arisen out of an industrially sponsored research agreement?"

The dimensions of the terror are as follows:

- Most industrially sponsored agreements for university research are, much like government grants, fairly broadly defined and covering basic research projects for which the results are unknown, and "anything" that might develop.
- Most such research agreements grant the sponsor an option to an exclusive license to any patents that may come out of the research project.

What would happen then, if, quite serendipitously, the research project resulted in a new finding (along with a patent) so unexpected and so broad that it had the potential to create (and dominate) an entirely new field of endeavor, comparable in magnitude and import to humanity of the genetic engineering revolution now dominated (at least in the United States) by the Cohen–Boyer gene-splicing discovery? Would the sponsoring company then have the opportunity to hold the whole new field "at ransom" through its option to an exclusive license? Or, worse still, would it slow the development of the field down to a crawl as it restricted the use of the patent to its own limited resources to develop the technology?

No university has fully come to grips with this terror. Most have ignored it as hypothetical and unlikely. A few have put in override provisions in their sponsorship agreements that allow the university to refuse exclusivity (on an exceptional basis) for a particular invention if granting exclusivity to that invention would be counter to the public interest. And a few have finessed the question by granting sponsors an option to "a license" to research inventions, without specifying the degree of exclusivity.

Start-ups and Equity Participation in New Ventures

Start-ups are the hottest topic in university licensing today. As the universities have become more active in protecting and licensing their intellectual property, they have simultaneously found it increasingly difficult to find licensees among established companies. This difficulty appears to stem from the embryonic nature of university technology coupled with the realities of industry. Acceptance of a university invention, unless perchance that invention is directly related to an ongoing project in the company, usually means redirecting already-committed and scarce research and development resources inside the company to work on the new invention. Few companies are willing to do so. In addition, the well-known tendency, if not necessity, for many large American companies nowadays to favor short-term profits over long-term investments makes it less likely that they will invest in early, high-risk technology that will take many years to come to market—regardless of the long-term potential of the technology.

Universities have therefore begun turning to venture capital sources to form new companies specifically devoted to particular university technologies. Fortunately, despite the apparent disillusionment of some investors with biotechnology stock, there are still enough seed-stage investors interested in the very high returns possible from successful biotechnology ventures to fund a number of new companies each year.

Technology and start-ups. It should be remembered in universities' enthusiasm for new ventures over established companies for licensees, that only a minor fraction of the inventions arising from a university are suitable for a venture-capitalized start-up. At MIT, my colleagues and I find that only approximately 3% of the inventions disclosed (at most a dozen of the over 300 disclosures received in a year) have the requisite characteristics for a start-up:

1. The technology must be basic and broadly based, and the patent (or copyright) position must be strong.

2. The technology must have a high potential market (usually at least $100 million per year or higher in biotechnology) to compensate for the high risk of development.

3. The technology should lead to a portfolio of products, not just a single one. Preferably, there should be some potential products that can be developed within a relatively short time, some intermediate time-frame products, and some long-range products with very high potential.

4. Preferably, there should be at least one inventor with excellent scientific credentials who is enthusiastic about participating in the company—either as an employee, a consultant, or a scientific advisory board member. If that inventor is famous (perhaps a Nobel laureate) or charismatic, all the better, but enthusiasm and dedication to the development of the technology is more important.

Advantages of a start-up. The formation of a company specifically around a university's invention offers several advantages to the university. These include the following.

1. Development of the technology is, and will continue to be for a long period, the "first priority" of the company. Changes in company strategy or personnel are unlikely to change this priority.

2. The inventor/faculty member will usually have an advisory role in the company, will remain close to the technology, and will help guide its development. (This is in contrast to some licenses to large companies, when the inventor may feel that the university has "given away" his technology—for a price.)

3. Frequently, and particularly in biotechnology, a small company will have an incentive to develop several products in parallel from its core (university) technology, rather than one at a time, as a large company might do. These parallel-development projects are usually through strategic alliances with larger companies, who pay development funds to the venture. (Given this probable scenario, it behooves the university to grant licensing terms to the venture, which encourage sublicensing.)

4. If the university receives equity in the new venture, it will share in any major appreciation in the value of the company. Faculty member/inventors will have similar opportunity and will have an incentive to support the technology enthusiastically.

5. A combination of license fees, cash royalties, and equity (a frequent mix has the usual cash payments halved in return for 5 to 10% of the company) allows the university to profit under a number of different scenarios in the development of the company. Even if the direction of the company diverges from the initial technology, the university will benefit from stock appreciation.

6. Start-ups frequently result in the creation of new jobs in the region surrounding the university. The benefit is virtually immediate in providing opportunities for new graduates, but rapidly spreads to the community at large if the company is successful. Eventually, a critical mass of technical and managerial talent may be built up in the community, leading to more new companies and faster economic growth.

7. Start-ups are visible. They result in good public relations for the university and increased support by alumni and by funding agencies.

Risks. Of course, there are also risks in licensing of valuable university technology to new ventures. The financial health of any start-up company is fragile—affected by many factors inside and outside the company's control. By far the greatest risk to the university is not that the company will go broke, but that it will stay barely alive in a chronic, underfunded state. If the company dies completely, the licensed technology will usually revert to the university. However, if the company just limps along, exclusively licensed technology may be woefully underexploited while the university and the company argue for years over whether the company is meeting the performance clauses of the license. In theory, this problem of underexploitation could be solved by carefully crafted performance clauses in the license. In practice, however, such performance clauses "with real teeth" are extremely difficult to draft when the technology is embryonic.

Conflict of interest. The other major risk, currently leading to much soul searching in universities, is conflict of interest—or even the appearance of such conflict. Universities feel, almost instinctively, that an equity participation by faculty and/or the university in a licensee company is inherently different from a promised royalty stream and inherently more fraught with a possibility of conflict of interest—even though the two positions might be thought very similar from a theoretical economist's point of view. The reason for the difference is that, in contrast to a royalty stream, which delivers its promise only after the product is successful in the market, an equity participation offers the possibility of substantial gain in wealth (through a public offering or acquisition by another company) even before it is known whether the technology will be successful. The potential impact of these "visions of sugarplums" on the behavior of scientists and administrators is a major worry.

What kinds of behavior changes are the universities worried about?

They range from the subtly misjudged to the criminally fraudulent. (The latter, though rare, have been highlighted by a few recent, highly publicized scandals.) Among the activities that conflict-of-interest policies and guidelines are designed to prevent are

1. Redirecting of research away from academically interesting areas into projects of specific benefit to the company.
2. Inhibiting collegial sharing and publication of research results in order to keep such results confidential for the company.
3. Exploiting students on company projects in the university laboratories and/or favoring students working on company projects.
4. Making tenure or promotion decisions based on the faculty member's contribution to the company (and indirectly to the university's wealth) rather than her contribution to teaching and academic research.
5. Funneling new inventions arising from government (or other nonrestricted) funds directly and at below market price to the company, when other firms might better exploit the technology.
6. Investing university funds in projects helpful to the company or investing them directly in the company, with decisions made by university managers whose judgement is biased by being too close to the company. When university funds are directly invested in the company, there is the additional danger of "throwing good money after bad" if the company suddenly needs additional funding to stay alive. In the worst of such scenarios, the university keeps making additional investments as the fortunes of the company decline, in order to try to salvage its initial investment.
7. Concealing or deliberately changing research data in order to avoid a drop in the company's stock. This is widely considered to have the most potential for abuse in clinical trials run at the university.

Conflict-of-interest guidelines. With the clear intention and understanding that conflict-of-interest policies are designed to help honest people decide how to act, and that such policies and guidelines neither are designed to nor are capable of preventing deliberate unethical or criminal behavior, MIT has set forth the following guidelines for technology licenses involving equity to either the university or the faculty member/inventor. Some of these are generally followed by many other universities, while others are unique to MIT.

1. MIT will not invest its own funds in the early rounds of funding of the company. Thus, we allow market forces to decide which of our technologies are worthy of support.

2. Any equity obtained from a license agreement is transferred immediately from the Technology Licensing Office to the Treasurer's Office. Thus, decisions on whether to invest university funds in future rounds of funding or whether to hold or sell the stock when it becomes liquid are made by professional portfolio managers who were not involved in the technology transfer.

3. We will not accept sponsored research from the company into the inventor's laboratory if the inventor holds stock in the company. This is to prevent confusion as to the inventor's role at the university.

4. The inventor who remains an MIT employee while taking stock in the company and serving as a consultant or advisory board member of the company is asked to distinguish in writing between the work he is doing for the company and his research at MIT. This distinction and the inventor's relationship with the company is reviewed on an annual basis by the department head and/or a departmental committee.

5. The inventor must sign a "Conflict Avoidance Statement," in which he agrees not to use students for research and development projects for the company, not to restrict or delay access to information from his MIT research, and not to take direct or indirect support from the company for his research activities at MIT.

6. The fact that MIT owns shares in a company should give that company no special access to future technology from MIT, regardless of whether such technology originates in the inventor's laboratory. Technology licensing officers are expected to prevent any such "pipelining" of technology to the company. One small exception is our willingness to grant the company options (at a fair price) to new technology dominated by the patents of the original exclusive license, since it would not be licensable to others.

How ventures are formed. Individual universities' approaches to funding new ventures vary widely. Some have set up separate foundations staffed by professionals who identify promising technologies, conduct market research, write business plans, seek venture capital, and may even act as an interim management team for the new start-up. This approach is expensive, is a relatively high risk, and is limited in the number of ventures it can handle, but offers the advantage that the university may be able to command a higher percentage of the company's equity because the presenting package is more complete.

A few universities have set up separate "captive" venture funds, some with a portion of the capital coming from the university, some with at least a portion of state funds, and some with all outside capital. Such funds are often willing to invest in "pre-seed" technology, perhaps funding feasibility research at the university before deciding whether to build a company. One disadvantage of such funds is that they and the university become somewhat isolated from market forces in deciding which technologies to invest in and therefore may make poor investment decisions. An additional disadvantage to the university is that, even if the captive fund does not have exclusive rights to the university's technologies, other potential investors assume that "all the good deals are taken" and no longer come shopping for new technology; the university may then find itself with less access to capital than it might otherwise have had.

Either of these two modes of funding may also lead to an overemphasis on new ventures and the neglect of conventional licensing. University inventions that are not suitable for start-ups become "second-class citizens," and the majority of the university's technology remains unexploited.

At MIT, we take the position that we do not start companies ourselves, but catalyze the start-up by others. Over the past few years, we have established relationships with several dozen venture firms interested in seed-stage technologies. When a technology is identified as a potential start-up, we will contact several firms whose portfolio needs the technology appears to fit. We have found that venture firms vary greatly in what they find attractive. Some are interested in only very high potential ventures, and are willing to consider technologies that may take six to eight years to come to market; others are interested only in those that will "go into the black" within eighteen to twenty-four months and that require less than $2 million to get there. Some will consider basic molecular biology–based ventures, while others want nothing more complex than medical devices. Some find diagnostics attractive because they can reach the market relatively quickly, while others consider diagnostics to have margins that are too low. And some put a greater premium on the presence of an in-place management team than others.

Any venture firm interested in one of our technologies, however, will have to be prepared to do a lot of work; often considerably more work than is required by the typical "zero-stage" venture. In contrast to that typical zero-stage venture, ours frequently lack any management team (or even any founder willing to commit to full-time employment) and usually lack a business plan. Because of the lack of business plan and management team, our ventures have been dubbed by one venture capitalist as not "zero stage," but "minus-two stage."

What we do offer is some world-class technology, usually a strong early patent position, and almost always, an enthusiastic faculty member. We almost never have done the market research on the venture, but our experience shows us that the "due diligence" capability of the experienced venture capitalist will result in a much better picture of the market than we would be able to produce. The lack of a management team results in more work for the venture capitalist but offers her the opportunity to assemble her own team. Frequently, one of the faculty member's graduate students, post-docs, or a junior inventor may be willing to join the company full time. Finally, if the initial investment needed is large enough to require several investors, the lead venture capitalist may work with the faculty member to write a business plan to attract a syndicate of investors.

This is a labor-intensive effort for the lead venture capitalist, but does offer her the opportunity for significant influence on the shape of the company that is formed. The labor involved usually means that no single venture firm can take the lead on more than one minus-two–stage venture in a given year. As a result, we find ourselves working with a number of different venture firms in any given year. We are pleased to report, however, that after 3½ years of these start-up ventures, a good fraction of the lead venture capital firms have come back to us for "the next one," after their first MIT venture is standing on its own. It will take more time, however, to determine how profitable these ventures will be for all concerned.

A final note on new ventures. As for licensing in general, one of the major dangers to the success of university endeavors in start-ups is unrealistic expectations. In some cases, the "visions of sugarplums" have visited not only individual inventors, but university managers and regional planners, and university venture offices have been expected to become self-supporting and to provide major additions to university laboratory endowments within three to five years. This is just unrealistic. Start-ups are risky businesses. Some go broke, some grow slowly, and only a few make a great deal of money. The average university will start no more than one to three ventures a year (MIT's current average is about six, if we count only those with at least $1 million invested in them.) Except for the few ventures that raise their initial rounds of funding from public markets, most will not result in liquid capital returns for up to five years or more.

Given time to grow, these ventures may very likely result in a number of benefits (not all in cash) to the university, the investors, and the region. And, given time to grow, the university technology transfer offices that gave birth to these ventures will produce a continuing stream of such ventures and other profitable licenses. However, unrealistic expectations and lack of support for the long run will kill off both the ventures and the technology transfer offices before either can produce the fruits they have promised.

TWO EXAMPLES OF UNIVERSITY TECHNOLOGY TRANSFER

Example 1. Licensing of the Cohen–Boyer Patents on DNA Cloning

The DNA cloning invention by Stanley Cohen of Stanford and Herbert Boyer of the University of California is the key technology underlying the genetic engineering revolution. Stanford University's original strategy for broad, nonexclusive licensing of the patents from this invention has resulted in very wide exploitation of the technology, only a few disputes to date, most of which were relatively easily resolved, and substantial income to the two universities.

In the following section, Katharine Ku, formerly Associate Director of the Stanford Office of Technology Licensing, and now Vice President, Business Development at Protein Design Labs, Inc., Mountain View, California, describes this strategy [Ku 1983].[1]

Stanford's most well-known licensed "product" is the Cohen–Boyer DNA cloning invention. This invention has launched more than a product in the marketplace; it has, in fact, launched an industry and is transforming many existing industries planning to use this technology.

After attending a United States–Japan joint meeting on bacterial plasmids held in November 1972, Herbert Boyer of the University of California and Stanley Cohen of Stanford discussed their research at a delicatessen in Waikiki, Hawaii. Boyer had been doing research in the area of restriction enzymes and Cohen has been working with plasmids. Together they conceived the DNA cloning technique.

By March of 1973, Cohen and Boyer had shown the basic feasibility of DNA cloning, thus reducing the invention to practice. They presented their results at a scientific conference (called the Gordon conference) in June 1973 and described it in a publication that appeared in November.

In May 1974 the work of Cohen and Boyer came to the attention of science writer Vic McElheny at *The New York Times* [May 20, 1974] who wrote an article describing this recombinant DNA research. The article

[1] Printed with permission from the Licensing Executives Society (Revised excerpt from Les Nouvelles, June 1983).

included quotes from a number of scientists describing the practical applicability of DNA cloning. Stanford News Director Bob Byers called the article to the attention of Niels Reimers, Stanford's Director of Technology Licensing. Reimers contacted Cohen and proposed that the university would apply for a patent on the DNA cloning procedures. After much discussion because of Cohen's initial reluctance, Cohen finally agreed to cooperate in seeking patent protection.

Stanford approached Boyer, who was willing to cooperate. Stanford eventually was able to obtain approval from the three research sponsors—American Cancer Society, National Science Foundation (NSF), and National Institutes of Health (NIH)—to administer the invention under Stanford's Institutional Patent Agreement with NIH just in time to file a patent application on November 4, 1974, one week before the one year U.S. publication bar was to occur. The two universities agreed that Stanford would manage the licensing, that 15% of any gross income would be deducted by Stanford for administrative costs, and that after out-of-pocket patent and licensing expenses were deducted by Stanford, the remaining net revenues would be split 50/50.

In December 1974, scientists were invited to an international conference to be held in February of 1975, sponsored by the National Academy of Sciences with funding provided by NIH and NSF. The purpose of the meeting, held at the Asilomar Conference Center on California's Monterey Peninsula, was to review the progress, opportunities, dangers, and possible remedies associated with the construction and introduction of engineered recombinant DNA (rDNA) molecules into living cells. After the meeting rDNA research was resumed but under strict, self-imposed laboratory safety guidelines.

In May 1976, a meeting was arranged with Stanford scientists and administrators to discuss the policy and practice of patenting biotechnology discoveries, particularly the rDNA patent. There were concerns that patents would interfere with scientific communication, and that there might be a perception that Stanford would have a conflict of interest with respect to rDNA safety if it were to hold a proprietary interest in rDNA work. It was decided that Stanford would discuss the issues at a high, public-policy level in Washington, and with those at NIH and NSF, specifically.

NIH reaffirmed its approval that universities in general should patent and license rDNA inventions provided that industry licensees comply with standards set forth in its "Guidelines for Research Involving rDNA Molecules [June, 1976]."

Overlying all this national and Stanford activity was the now famous General Electric patent application claiming microorganisms *per se*. In 1972, Ananda Chakrabarty, a microbiologist working for the General Electric Company, made a bacterium capable of breaking down multiple components of crude oil. Chakrabarty's organisms were not a result of gene-splicing and cloning but of conventional genetic manipulations wherein a genetic mutation is "forced." Because of the oil-eating property possessed by bacteria that do not occur naturally, Chakrabarty believed his invention to have significant value for the treatment of oil spills.

GE filed a patent application with claims of three types: first, the method of producing the bacteria; second, the bacteria combined with a carrier material; and third, the bacteria themselves. The patent examiner allowed the process and combination claims but he rejected claims for the bacteria *per se*, saying that the microorganisms are "products of nature," and that as living things, they are not patentable subject matter. GE appealed and kept appealing all the way to the Supreme Court. After eight years, on June 16, 1980, the Supreme Court held 5–4 that a living, humanmade microorganism is patentable subject matter. The Supreme Court based its decision on the fact that Congress used expansive terms in writing the patent laws, and therefore they should be given wide scope. The court cited the evidence that Congress intended statutory subject matter to "include anything under the sun that is made by man."

Supreme Court Chief Justice Burger, writing for the majority, stated that "the patentee has produced a new bacterium with markedly different characteristics from any found in nature and one having potential for significant utility. His discovery is not nature's handiwork, but his own; accordingly it is patentable subject matter under Section 101."

The Cohen–Boyer patent application originally filed November 4, 1974, covered both the process of making, and composition for, biologically functional chimeras. During the course of prosecution, the examiner indicated that he was willing to allow process claims that described the basic method for producing transformants, but that he was not willing to allow claims on the biological material *per se*, i.e., "products." It was decided that the original patent application would be divided in two. The "process" patent was issued December 2, 1980; the "product" application, which had been held up awaiting the Supreme Court decision, is still pending in the Patent Office.

Niels Reimers, Director of Stanford's Office of Technology Licensing, was really the person behind the marketing of the invention, the connective tissue who tied it all together. In a memo dated July 11, 1976, Reimers wrote: "At the onset, we must acknowledge that we are going to have to act on imperfect information. Ten months or 10 years from now we'll know what we should have done."

Stanford was trying to license an invention for which products had never been sold, and which would apply to many diverse, established industries, in addition to the then newly emerging biotechnology industry.

Objective/philosophy behind market strategy. Stanford and the University of California wanted the terms of the license agreement: (a) to be consistent with the public service ideals of the universities; (b) to provide the appropriate incentives to industry to bring genetic engineering technology forward to public use and benefit, in an adequate and timely manner; (c) to minimize the potential for biologically hazardous development; and (d) to provide income for educational and research purposes. No genetic engineering research, academic or industrial, was to be inhibited by Stanford's licensing program. It was recognized, however, that commercial organizations might not conduct research in the area unless there was some assurance that they could receive a license to market if a successful product were developed. Although exclusive licenses in the limited fields of use were initially considered, it was decided early on that nonexclusive licensing of this broad, far-reaching invention would best serve the public and the universities' interests.

As noted, the license needed to be suitable for many industries, including the new companies specializing in biotechnology, and existing companies in the chemical, agricultural, pharmaceutical, mining, and oil industries. It was necessary to make the terms of the license reasonable for companies big and small, rich and not-so-rich. It was recognized that if the patent(s) was important, which it was expected to be, Stanford would have to accept and anticipate the likelihood that the validity of the patent(s) would be "challenged," particularly if it became more economical for companies to litigate rather than pay royalties. In addition, since a patent grant is limited to 17 years, and development, testing, and regulatory approvals could take up to 10 years or more, there was also the possibility that the patent could expire before useful, royalty-bearing products would reach the marketplace.

Strategy. Beginning from the first public notice that a patent application for gene-splicing had been filed, Stanford presented the discovery as "underlying the entire field of genetic engineering." Stanford consistently stated that the patents, if granted, would be licensed broadly, nonexclusively, and at modest royalties. As mentioned before, the market targeted was and is very large; therefore, the fundamental strategy was directed to encouraging companies to take nonexclusive licenses early on.

Credits. The biggest incentive for encouraging companies to take a license early on was to offer companies who signed up by December 15,

1981, a five times credit on the $10,000 minimum annual advance royalties, i.e, for each $10,000 payment, the licensee received a $50,000 credit against future earned royalties. This credit was accruable for five years or until the first calendar year in which $1 million of end product was sold. It was believed that the offer of five times credit for a limited time would encourage companies to take a license sooner rather than later (which could be much later) when products were finally on the market.

Certainty. The availability of the license was announced in early August 1981. Companies were advised that the terms of the license were guaranteed only for companies taking the license by December 15. Any possible future changes were not divulged, so if a company wanted certainty, the company should sign up. The license was printed in order to emphasize to potential licensees that the terms were standard.

Opened patent prosecution. Given Stanford's intent to license as broadly as possible, challenge to the patent(s) in the courts seemed certain. As a strategic move to enhance the validity of the patent(s) Stanford decided to make the patent process open to the public. Anyone who was aware of factors that would affect the patent's validity was asked to make them known to the Patent Office. Any company seeking to challenge the validity of the patent after its issue would then have the burden of justifying why they had not raised those issues before the patent issued.

Personal contacts/media blitz. In the period from the August announcement of the license until the December 15 closing, companies throughout the United States, Europe and Japan were visited to explain the license. It was decided that earned royalties should be modest so as to encourage companies to take a license rather than contest the patent(s). It was also decided that a step-down earned-royalty schedule would reward companies whose sales were high without unduly penalizing the smaller company for smaller sales volumes. Discussion with several companies of differing sizes and markets were held while license terms, particularly earned-royalty terms, were being formulated. By doing this, the license was "presold" and unrealistic license terms were avoided.

A logical question would be: Did the initial licensing strategy work? It did! Seventy-three licensees (including many foreign companies) signed up by December 15, 1981. After December 15, 1981, the DNA story continued to unfold. Stanford is determined to monitor the license thoroughly. At present, there is a full-time market research person and a full-time licensing person basically dedicated to the case. If other staff are needed, they will be added.

A second patent application, pending in the Patent Office, received a notice of allowance in early 1982. Al Halluin, Exxon attorney at the time, had written a treatise entitled "Patenting the Results of Genetic Engineering Research: An Overview" [Halluin 1982], based on a talk that he had given at Cold Spring Harbor Laboratory during a conference also attended by Niels Riemers, director of OTL. A footnote to the treatise, published in the *Banbury Report*, raised some questions about the availability of the plasmid pSC101 and caused the Patent Office, without explanation, to suddenly rescind their Notice of Allowance for this second patent application.

Although the patent prosecution had been opened to the public previously, Stanford decided to close the file in order to continue the prosecution without media hype. Stanford was willing to share the prosecution communications from the Patent Office with licensees on an individual personal basis only. This closing of the file eventually prompted the law firm of Wegner and Bretschneider to petition the Patent Office to reopen it. This petition was denied.

The composition of matter claims specifically referring to prokaryotic host organisms issued on August 28, 1984, with another continuation filed to continue the prosecution of claims specifically referring to practice, continued to prepare for potential litigation by hiring Donald Dunner of Finnegan, Henderson et al. to assist in enforcement of the patents. In April 1988, the third and final patent specifically referring "cell" hosts was issued.

Stanford has been actively licensing the patents throughout the post-1981 period and has 105 active licensees as of June 15, 1990. The royalty terms have been adjusted on a regular basis to reflect the increasing value of the patents over time. The patents have generated over $16.5 million in cumulative revenue and the university continues to sign on new licensees.

Example 2. The Start-up of Matritech

Matritech was one of the first companies founded out of the MIT Technology Licensing Office after MIT revised its technology licensing policy to allow faculty to receive equity in such companies and to allow MIT to receive equity partially in lieu of royalties. In the following section, Steve Chubb, founder and president of Matritech, describes the on-again/off-again start-up of the company [Chubb 1990].

Frequently the technical achievement that provides the foundation for a dominant commercial company occurs not by evolution but by revolution. This was certainly the case for Matritech, Inc., a company founded in 1987 to produce products for the detection of cancer. Although Professor Sheldon Penman was an MIT faculty member in the Biology Department, he had actually obtained his doctorate in physics. During the 1960s he had become fascinated with certain controversial issues that had long puzzled cell biologists. What accounts for cell type and identity in view of the observations that cell constituents such as carbohydrates, lipids, and DNA appear to be quite similar from cell type to cell type? What accounts for and controls cell differentiation? What accounts for cell malignancy? Perhaps most importantly, what is the importance of spatial relationships of internal cell components and the consequence of altering these relationships?

Most other investigators in cell biology had employed conventional methods such as extraction and fractionation to attempt to solve these complex questions. The revolutionary vision of Professor Penman and his postdoctoral fellow, Ted Fey, included a characterization of cells more as relatively rigid three-dimensional structures than as sacs of biological liquids. Proceeding with these theories, they routinely employed unconventional techniques such as electron microscopy to examine cells and to confirm their hypotheses. By introducing several innovative enhancements to these techniques, they produced images of the internal structures of cells that resemble the fiber density of Brillo pads rather than the lightly packed structure that had comprised previous models. In addition, when Dr. Fey isolated certain proteins of the cell nucleus he immediately observed two striking results. The first finding was that each cell type routinely displayed a unique protein composition; a second and more commercially useful observation was that when cells become malignant, the composition of these insoluble proteins changes reproducibility.

Penman and Fey, realizing the potential commercial significance of their discovery, contacted MIT's Technology Licensing Office (TLO) in 1985 to determine if it was appropriate to seek patent protection. In both 1985 and 1986, MIT, through the TLO, filed broad patents on behalf of the inventors claiming the methods for extracting the proteins (by then dubbed *nuclear matrix proteins*) from cells as well as their use in a variety of conventional clinical detection immunoassays employing monoclonal antibodies. This was the relatively rare situation in which the theories and methods employed by certain scientific investigators were so far afield from those of their peers that there was virtually no prior technology, and, as a result, their original patent applications have been granted with only minor alterations.

The MIT Technology Licensing Office appreciated not only that the Penman and Fey nuclear matrix protein (NMP) technology was a scientific triumph, but that a dominant commercial business could result from it if properly managed and funded. They therefore began making contacts with existing diagnostics companies and divisions in order to identify and sign up licensees. Initially, the diagnostic division of one large company expressed interest; however, the discussions were intermittent and lacked senior technical sponsorship.

At about this time in late 1986, I became aware of the NMP technology through friends who introduced me to Penman and Fey. I was also able to renew contact with the MIT Technology Licensing Office with whom I had become acquainted while at T-Cell Sciences. I came to this work with a broad background in biotechnology management, having already spent 12 years in the human health care industry. During my seven years as an executive at Baxter International I gained experience in marketing and manufacturing functions. Later, as President of Hyland Diagnostics, I was able to turn the previously troubled division around from the severe losses it had experienced in prior years. In addition, I had the benefit of having previously been chief executive officer at two biotechnology start-ups, Cytogen Corporation and T-Cell Sciences, each of which was public and worth more than $50 million. Therefore, I was able to fit the TLO's requirement for a seasoned biotechnology general manager. The stage was now set for the licensing process and for company formation to begin.

In analyzing NMP technology, I quickly concluded that it easily conformed to my requirements for a commercial winner. My personal preference was to back technologies that elegantly involved fundamental structure and function. In my view, the identification of a first-order molecular structure present in malignant cells and absent from normal cells would remedy the deficiencies of existing cancer diagnostics. At that time cancer tests and detection methods had relied on indirect structures such as those found on cell surfaces. The fact that the Food and Drug Administration had approved virtually none of these was a clear indication not only of the fundamental flaws in these approaches, but of the enormous market opportunity as well. It was now clear what we wanted to do, the question was how to put it all together.

The TLO was quite supportive of me as well as of Penman and Fey throughout 1987. They quickly prepared and obtained MIT support for a technology license to my venture (which Lita Nelsen of the TLO routinely referred to in those days as Chubbco). The license was thoughtfully drafted and fit the needs and resources of a start-up. The up-front payment was modest, the royalties consistent with industry standards, and, to ensure long-term congruence of goals, it included equity ownership by MIT. The major features of the deal changed very little from the time the TLO and I first discussed it in late 1986 until a signing, approximately a year later. There were the typical nits to pick, but the major deal points were amazingly stable from the beginning.

On the financial front matters were much more straightforward than in most other venture capital-backed deals. I had unusually close relationships with Philadelphia Ventures and The Venture Capital Fund of New England, the principals of which had been lead investors in my first start-up, Cytogen. In addition, I was an investor (limited partner) in First Stage Capital in Cambridge, Massachusetts, so it was not difficult to get the partners there to focus on the deal. My plan included lending "walking around money" to the company until the license was obtained and initial hiring was underway and then to obtain a seed round of $1 million to keep things rolling. With this card ready to play it was time to consider the roles of the founders.

It was apparent from the first few meetings that there was a fundamental incompatibility between some of the founders and the MIT scientific team. From my experiences at Cytogen and T-Cell Sciences, I knew firsthand of the damage and inefficiencies that can be caused by strong personalities and their unwillingness to compromise. As plans for the new company began to take shape, minor irritants in the form of differences of opinion became major concerns about the compatibility of the players. By mid-May, things had reached the point that, despite my conviction that the technology was sound, the friction that was already developing among the various founders, scientists, and prospective managers caused me to withdrawn from the venture. This action was not a bluff. Had the remaining group been able to overcome its differences and incorporate, I was willing to remain uninvolved with the deal.

What developed was a classic standoff, one group with technology, one group with management, and another group with neither. The MIT Technology Licensing Office displayed true courage at this point by being very fair but at the same time scrupulously exacting about the qualifications of prospective licensees. For approximately five months, all the groups reassessed their respective positions. The TLO was the last to blink. Several of the founders, at last recognizing that with them there would be no company, gracefully withdrew from any direct involvement with the venture, with the exception of stock ownership.

Toward the end of this cooling-off period, I was introduced to a splendid scientist with over thirteen years of clinical diagnostics experience both as a laboratory supervisor and development scientist with Ciba-Corning Diagnostics. Graham Lidgard received both his undergraduate and doctoral degrees at the University of Manchester in the United Kingdom and progressed through the ranks of Ciba-Corning gaining invaluable technical, management, and business experience along the way. Having recently completed a major project there and itching for new challenges, he was at the perfect stage from the standpoint of both age, energy, and experience to undertake the opportunities and demands of a start-up. In

December 1987 it all came together; the license, the commitment for the initial funding, the agreement of the MIT scientists, and the initial management team.

The major lesson from this experience is that it is better to take time to do it right the first time than to do it over. At this writing Matritech has raised over $6 million in venture funding, has close to thirty employees, has received funding from two corporate partners for the development of serum-based cancer-detection assays, and introduced its first product in June 1990.

REFERENCES

Chubb, S. D. 1990. The founding of Matritech.

Guidelines for Research Involving Recombinant DNA Molecules. June 1976 (41 Federal Register 27902).

Halluin, A. 1982. Patenting the results of genetic engineering research: An overview. *Banbury Report* 10:67–126.

Ku, K. 1983. Licensing DNA cloning technology. *Les Nouvelles, XVIII*(2):112–115.

McElheny, V. 1974. Gene transplants seen helping farmers and doctors. *New York Times,* Vol. CXXIII (No. 42,485), p. 61.

CHAPTER 4

THE SEARCH FOR DOLLARS: LIFE CYCLE OF A BIOTECHNOLOGY COMPANY

Christina L. Lowell

Ten years ago a brief chapter such as this would have chronicled the financing aspects of the biotechnology industry—the life cycle of a "typical" biotechnology company. For a venture model, one looked at the computer industry. Over the past ten years, the realities of the time to market for a biotechnology product, the maturing of venture capital investment, the growth of corporate strategic partnering, increasing university interest in technology transfer, and the emergence of international capital investment have all contributed to the complex organization that is the biotechnology company today.

In the 1970s, the emerging biotechnology companies had, in fact, been in existence more than ten years with no marketable products close at hand, although there was enough promise in these few companies to raise significant amounts of capital, attracting venture investors and public market attention. The goal of these early leaders such as Genentech and Cetus was to become fully integrated pharmaceutical companies. Application of the new technologies and biotechnology methods to industries other than health care was limited. A few biotechnology companies such as BioTechnica hedged their bets by following a dual track—agricultural applications research was balanced with research into health care applications.

Investors and industry watchers assumed that most, if not all, of these early research and development companies would eventually be absorbed by the multinational pharmaceutical companies who understood and dominated the established channels of approval and distribution. These companies had the necessary resources to guide products from research through development and the regulatory maze. They could saturate diverse distribution channels quickly to succeed in a highly competitive environment.

What has become clear over the past ten years is that while the computer industry is not an adequate growth and financing model, many lessons from that industry can still be applied today to the growth of a biotechnology company. A new industry has come into existence. From a handful of companies, the biotechnology industry has grown to more than 1000 companies in the United States alone. A comparable number are developing in Europe, the Far East, and countries such as Australia and Israel. The focus of applications research is now on agriculture, bioprocessing, drug delivery methods, cosmetics, and biomaterials, in addition to the more advanced areas of health care.

A NEW INDUSTRY

From basic research to the development of a pipeline and market introduction of new products for a biotechnology company in the 1990s requires tens of millions of dollars. Various capital strategies have emerged to solve the financing needs of this industry. Key to understanding the financing aspects of a biotechnology company is an understanding of its unique growth path and the challenges faced. For the biopharmaceutical company, the route is the most complex. This chapter will focus on this one model, but all biotechnology companies face similar regulatory, market, and production hurdles. All of this is further complicated by the public's confused and sometimes fearful reaction to the presence and growth of this industry.

The time from basic research to market-accepted product is lengthy. Early patent positioning is crucial, although not clear—as recent court battles have shown. An average therapeutic drug product may take from seven to ten years to reach the marketplace, while numerous projects must be carried on simultaneously to increase the likelihood of achieving one blockbuster product and to avoid the trap of becoming dependent on any one product. For many companies, the development and sale of fast and accurate diagnostic products produce early cash flow. For certain diseases, the diagnostics themselves are a breakthrough, as no adequate or reliable means of diagnosis exists. New diagnostics may mean that the progress of a disease can be measured and earlier and new treatments effected.

Public awareness of biotechnology has increased over the past ten years. The desire to realize breakthrough promises is counterbalanced by fear of the unknown. This contradictory stance is reflected in our regulatory environment, where complaints about slow and cumbersome procedures and approvals are offset with calls for stricter controls on laboratory and production sites and on test releases. The United States has one of the most stringent sets of regulations covering not only diagnostic and therapeutic drugs but also food products, consumer protection, and environmental impact. With this regulatory backdrop, the biotechnology company must work against the half-life of its patents. The bottom line is that new products must capture significant portions of the market in their early years. There is no room for the slow building of market acceptance and upgrading of existing product lines, as has been the experience with computers and advanced medical equipment. The price of new products must be high enough to recover research and development costs and to fund new projects, but also be acceptable to various health care cost-reimbursement programs. The electronics industry can price new product introductions high to initial users and then reduce prices to capture broader segments of the market. Today's health care focus is on cost containment. New drugs enter the market only after demonstrating safety and efficacy. Pricing decisions must be made in light of existing products and the ability to demonstrate that new products are significantly better to justify much higher prices. Agricultural biological products will also enter established markets with their own unique pricing pressures. As to cosmetic pharmaceuticals, the sky may be the limit for new products that can hold back the effects of aging, but this is also a market already dominated by a few multinational players.

How does a new biotechnology company emerge and thrive given this background? The life cycle of a biotechnology company can be looked at in five stages: Stage 1 begins with breakthrough basic research at the university level. Stage 2 involves university–corporate technology transfer, moving initial research results into the commercial world. The company builds on the "sweat equity" of the founders (forgoing salary in order to see the new venture started) and grant funding obtained from the university and government, with the infusion of a seed round of venture investment. In stage 3, additional venture capital is raised. At this point, corporate alliances are formed to obtain the funding and resources necessary to continue product development and begin regulatory and market strategies. Stage 4 culminates with successful completion of the regulatory process, full product introduction, and scale-up production. The company must also have established international operations. Stage 5 encompasses production and sales of multiple products by the maturing com-

BioTech Company

Stages	*Activities*	*Years to Maturity*
One	Basic Research	(10 Plus)
Two	Incorporation/Start-Up	0-3
Three	FInancing/Development Approvals	2-7
Four	Production	6-10
Five	Expansion	10 Plus

Years to Production/Sales Dollars

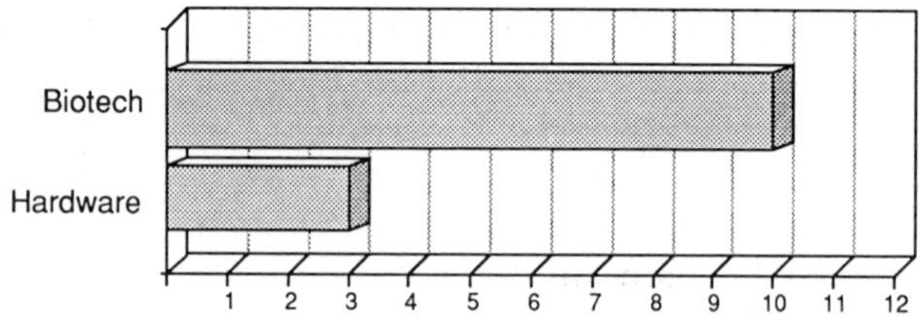

Hardware/Instrumentation Company

Stages	*Activities*	*Years to Maturity*
One	Incorporation/Prototype	0-2
Two	Financing/Production	1.5-3
Three	Rapid Growth	3-6
Four	Expansion	4 Plus

FIGURE 4-1 Comparison showing the growth stages and years to maturity between a biotechnology company and a hardware/instrumentation company.

pany or perhaps as the merged or acquired division of a larger, established entity.

Figure 4-1 compares the growth stages and years to maturity of a biotechnology company with a hardware/instrumentation company. The years to maturity, of course, are broad parameters. All investors have their skyrocket examples and also those companies they would rather not talk about.

Stage 1: Basic Research

Basic biotechnology research is accomplished in the university with funding provided primarily by government agencies. Building on many years of research breakthroughs and linkages, the researcher identifies the elements necessary to focus on possible applications. Continued research may then take place within the university. Corporations, based both in the United States and in foreign countries, interested in specific lines of study or attracted by publications and desiring access to patents, often provide funding at this point in exchange for various rights to the research results. The federal government also continues as a major funding source.

Often, the research work itself begins to move off campus. For many researchers this is the beginning of a period with a foot in both camps, university and commercial. While continuing to conduct research efforts in the academic laboratories, the scientist may also be consulting with the start-up company in its applications research and product development efforts.

Stage 2: The Start-up

The start-up company begins to take shape. Founders with both science and business backgrounds join forces. Funding is makeshift. One of the founders may bring some financial resources, either his own or a connection to a wealthy individual with an interest in seed financing. The primary source of funding is the founders' sweat equity. There are also a small number of venture funds that provide early-stage dollars and assistance with two of the critical steps at this stage—the writing of the business plan and assembling the scientific advisory board.

The business plan is the foundation of the new company. In addressing the major concerns of any commercial enterprise—organization and structure, product and production, the regulatory framework, market and sales, growth projections, and capital needs—the biotechnology company establishes its initial credibility with its business plan. It is this document, constantly revised and fine-tuned, that focuses the direction of the company and makes a statement to potential investors, management, and partners. The business plan provides the basis for due diligence on technical progress, regulatory acceptance, market assumptions, and burn rate (the rate at which cash is spent to support operations).

A second major project for the stage 2 company is the assembling of a scientific advisory board. The emergence of the biotechnology company has introduced a new strategic concern in financing. As the science has advanced in quantum leaps, it has become increasingly difficult for management and the company's outside business advisors to assess the technical progress of a project, to direct and focus research efforts toward marketable products, or more importantly, to decide the critical point at which to abandon a line of research. At the same time, the continued expansion of the definition of director's liability has made academic experts reluctant to join the Board of Directors of a risk-laden start-up enterprise. Assembling a high-caliber, internationally recognized committee of scientific advisors is the industry's answer. The advisory board often includes individuals involved in the basic research and other experts in the field from major academic and research institutions.

An emerging trend over the past few years is increasing university participation in and facilitation of technology transfer. Traditionally, university participation has been limited to channeling corporate funding into research in exchange for licensing and other technology rights. Today, several universities have established research institutes and small business parks to house start-up companies into which they have placed university technology. Certain institutions, such as Stanford University, MIT, and the University of Texas at Austin, are actively involved in matching the technology, its developers, and experienced business people to found new companies. Universities are also becoming active seed-capital funders and facilitators, often in cooperation with state governments.

Another aspect of technology transfer is the establishment of targeted research centers within the university, with funding provided by a corporate sponsor in exchange for "first-look" rights at research results available for further corporate development. The commercial development may then take place in a new company established with a mixture of corporate and venture financing.

Stage 3: Financing and Development

The emerging biotechnology company must now acquire its first significant round of capital funding. The development plan is taking shape, and a few products may be entering the regulatory maze. The founding management team needs to be expanded and strengthened. Additional researchers need to be hired. Funding at this stage comes from several sources: venture capital, corporate sponsors, government grants, and continued sweat equity.

Venture funding increases from seed funding to the first sizable infusion of capital. Typically, this and subsequent rounds of funding will be made by a consortium of venture capital partnerships. One venture firm will take the lead position, often helping to locate other venture backers and acting as the primary negotiator with the biotechnology company as to stock value and percentage ownership. The form of investment is usually preferred stock. The preferred stock agreement may encompass two volumes of paperwork, however, the major provisions include seats on the Board of Directors, conversion of the preferred stock to common stock in the event of a public offering, registration rights and rights of first refusal to subsequent rounds of financing, and liquidation rights. In addition, the preferred stock agreement may place certain restrictions or review rights on equipment acquisitions, salary levels, and other corporate expenditures. The lead venture firm will typically provide one of its partners who has industry expertise and growth-company experience to serve as a board member.

Corporate funding at this stage takes the form of research and development contracts whereby the start-up company receives funding to continue research in an area of joint corporate interest. The results of the research typically belong to the corporate sponsor, but the contract usually provides sufficient profit for the biotechnology company to fund a portion of its own research. The area of research itself usually furthers other company projects and product applications. The ultimate use of the research results is negotiated between the large corporation and the start-up. The biotechnology company may give up the right to certain geographic markets or certain product applications in exchange for a future stream of royalty revenues. Although the start-up company will usually retain the right to market certain products itself, it may find itself with a formidable competitor to whom it showed the way. The biotechnology company at this stage must be careful not to barter away all its basic research and risk becoming trapped as a small research house.

A third source of financing is government grants. Through its Small Business Innovation Research grant program, the various agencies of the U.S. government grant up to $500,000 for research in exchange for the right to purchase or arrange production of resulting products for government use. Numerous other government programs exist. Additionally, funding is available through private research foundations, also in the form of grants.

The other continuing source of funding is sweat equity. It is at this stage that the company adopts some form of incentive stock program. In order to hire the qualified management and research people it needs, the company will issue options to purchase common stock at its current low valuation for an extended number of years. The stock has registration rights in the event of a public offering. The options vest over time, for example, 20% per year for five years. An employee stock option plan provides a way to attract and retain talented people, to provide incentives that include "golden handcuffs."

Stage 4: Production

To survive and begin to build a solid base, the early years of the biotechnology company are a constant search for funding. The technical side moves from the laboratory to clinical trials or field tests while the company's focus expands from funding strategies alone to begin to address marketing and production. A quick glance at the financial statements now would show continuing contract research income, maybe some licensing income, and increasing operating expenses. Costs for regulatory procedures and organization building are growing. At the same time the company must continue to spend significant amounts on research for a pipeline of potential products and applications. To cope with this, management must still devote a major portion of its time to raising capital during this stage.

In addition to the funding previously discussed, several other funding strategies are employed. Companies who invested early development time in diagnostic products may begin to see cash flow from the sale of these products. Revenue may come from direct product sales or royalties as products licensed through earlier corporate agreements are distributed by established domestic and foreign corporations.

Another source of cash flow is the sale of related products. These early product sales do not necessarily involve biotechnology, for example, sales of generic drugs or established seed varieties, or even repackaging of high-quality fruits and vegetables. Not only do sales of these non-biotechnology products generate cash flow but they also provide market presence and a chance to build a sales force while gaining knowledge of and credibility with distribution channels.

By now the company has completed a second or third venture capital financing and if market conditions permit, the company is contemplating an initial public offering of its common stock. At the same time, the company is monitoring its earlier corporate relationships while negotiating new alliances and possible joint ventures.

Over the past ten years, stock market conditions and the public's perception of high technology and biotechnology in particular have made the timing of an initial public stock offering difficult to predict. One trend has become clear: the biotechnology company contemplating going public today must have substance, not just future promise. The candidate for a successful offering must demonstrate the ability to generate contract and royalty revenue, have products close to market introduction, and have a full research and regulatory pipeline. A complete, experienced management team must be present. The scientific advisory board should be visible and actively participating in product development and clinical trials.

The biotechnology company also derives credibility from the reputation and diversity of its corporate partners. Alliances now change from being primarily a means to generate cash flow to complex alliances with each party contributing expertise and gaining market presence. Whereas the biotechnology company may have had to trade technology for cash in the past, it now capitalizes on its achievements and retains a greater share of product ownership.

Licensing agreements with European and Japanese partners are important because the foreign partner is more knowledgeable and better equipped to serve local markets and cope with divergent regulatory authoroties. A few recent examples include Japan's Kirin Brewery forming a joint venture with Calgene, and T-Cell Sciences' marketing agreements with SmithKline Beecham plc. Inc. and Yamanouchi Pharmaceutical Co. Ltd. The biotechnology company requires a host of international partners to gain public acceptance through disparate distribution channels and to establish rapidly a credible market position for particular applications.

Several types of agreements are common. Under production agreements, the biotechnology company supplies products to be marketed under a joint label or under the partner's brand. A separate joint venture may also be established with each side contributing to the new enterprise. Resulting products are marketed under the joint venture's label. The biotechnology company may also work with a partner who supplies the product. Contract production permits the start-up company a further growth period during which resources are not diverted to designing and building its own production facilities. Whatever form the corporate alliances take, the company's reputation and ability to expand are dependent on the profitable functioning of these relationships. Disputes are time consuming, wasting executive attention and company resources.

Successful strategic partnerships follow a few deceptively simple appearing principles. The current spate of public disputes and lawsuits have shown that these principles are not so simple in practice. For most biotechnology companies, initial contacts are made at the technical level, although corporate business development and licensing departments are expanding as more people who combine science and business backgrounds and skills become available. But, as early as possible in the development of an alliance, decision makers in both organizations must become involved and remain involved over the life of the agreement.

Basic elements of an alliance agreement include a specified term. In addition, while the expertise being exchanged and the contributions of both sides are usually clear at the start, what remains undefined is the consequences of a dispute, the method for handling disputes, and a means of terminating the agreement equitably if disputes cannot be resolved. Established benchmarks and timetables for the joint project, and guidelines for negotiating next steps are a necessity.

Just as the alliances themselves can vary so, too, can the form of investment the biotechnology company receives from its corporate partner. In addition to research dollars and experience and expertise, the company may also receive a capital investment structured similarly to a venture investment that is, preferred stock issued at some multiple of the common stock price, with additional rights and covenants.

Stage 5: Growth

Today, a handful of biotechnology companies have products on the market with sales levels of any significance—well on their way to becoming fully integrated companies. Many more have products that will enter the market in the next year or two. All these companies have numerous projects underway to generate future revenues. Large-scale production issues are now a reality instead of a future contingency.

One solution to production scale-up problems has already been discussed, the forming of corporate partnerships, particularly with partners experienced in fermentation techniques. Another scale-up problem is capital requirements. Several biotechnology companies such as DNA Plant Technology and Centocor have successfully joined with European countries such as The Netherlands and Great Britain to build research and production facilities overseas. These countries provide economic incentives, including grants, low- or no-cost loans, tax holidays, and worker training, to attract biotechnology companies who are making decisions about where to locate production facilities. Local government offices work with companies to ease the way through the regulatory process and provide a welcome. The benefits of overseas production facilities are not limited to economic assistance; proximity to regulatory agencies, less-stringent clinical requirements, and a ready market are also considerations. Promises of a unified European regulatory framework and access to a potentially unified market in the 1990s require the biotechnology company to look to Europe. Similar economic and market incentives are also available in various parts of the Far East to launch products in markets of even larger proportions.

Not all companies have continued into production and distribution as independent entities. We have seen a few companies acquired by their larger, well-financed pharmaceutical partners that are now functioning as more or less autonomous corporate divisions. Many others have not been so fortunate and have ceased to exist, their technology sold off to stronger competitors. The industry has just begun consolidation: Hoffmann–La-Roche/Genentech, Chugai/GenProbe, American Cyanamid/Praxis, XOMA/INGENE, and Genzyme/Integrated Genetics. In response to many companies targeting similar markets and lines of research, biotechnology companies are merging and with combined resources and talents refocusing on better-defined opportunities (Fig. 4-2).

For companies that continue to thrive, there is a more certain understanding of the tremendous amounts of capital needed to establish a biotechnology company firmly. A cash reserve of $100 million for each major new drug product is not an unrealistic goal. Several financial vehicles have emerged to fund these "war chests." Need and the creative responses of the financial markets will trigger the development of others, as we have already seen. Public research and development limited partnerships with stock options attached are used to fund specific product development. Several of these partnerships have already been bought out for cash and stock, returning to their investors more than forty to fifty times their original investment as the funded products prove to be winners. Product preferred stock with conversion rights has also been tried for specific product development.

The latest financing vehicle for the biotechnology industry is SWORDS (stock *w*arrants for *o*ff-balance sheet *r*esearch and *d*evelopment *s*ubsidiary). Several biotechnology companies, including Alza, Centocor, and Immunex, have successfully raised money with SWORDS or unit financing. Under the SWORDS model, the existing biotechnology company forms a "Newco," which is then sold to the investors. Newco contracts with the existing biotechnology company to perform research and development. The existing biotechnology company has the option to buy out Newco. The buy out is not guaranteed, thereby ensuring the off-balance nature of the research and development contract. The investor purchases a unit, one share of redeemable common stock in Newco and one warrant to purchase a share of common stock in the existing biotechnology company.

These funding vehicles attempt to tie the return on an individual investment to a particular line of research rather than raising financing through a stock offering with the resulting dilution to the company as a whole. The risk to the present stockholders is also minimized. The key to the success of these limited offerings is the reputation of the biotechnology company and the belief that an eventual payoff may be significant while the warranty promises at least some return.

More traditional capital funding methods have also been successful—

FINANCING ALTERNATIVES

Type	*Pros*	*Cons*
Common Stock/ Preferred Stock	• Broad public appeal particularly in a "hot" market • No immediate requirements for cash to investors	• Immediate dilution
R&D Limited Partnership	• Off-balance sheet • Non-dilutive • Redeem for cash and/or stock	• Limited tax advantages to investors to offset risk
Product Preferred Stock	• Risk of failure of one product area isolated for funding purposes	• Usually has conversion rights attached
SWORDS/ Unit Financing	• Off-balance sheet • Investor can monitor a portion of his return (warrant)	• Limited in availability to public companies • Typical buyer is institutional investor— no broad public appeal
Venture Leasing	• Below market lease rates and lower payments • Greater availability of funds for high-risk companies	• Warrants to purchase common stock are part of lease package
Debt	• Non-dilutive • Interest is tax deductible	• Impact on cash flow of interest payments

FIGURE 4-2 The pros and cons of financing alternatives available to the biotechnology companies.

second and third round public offerings of common stock. Voting and nonvoting preferred stock with conversion and dividend rights are also possible at various times in the public markets. Strategic alliances continue to play a significant part in corporate strategy. Bank debt is supported by growing receivables. Facilities construction and equipment acquisition is financed through combinations of bank debt, leasing, and public debt. Another creative alternative is venture leasing, in which the biotechnology company leases facilities or equipment at below market rates in return for an equity kicker, usually in the form of warrants to purchase common stock. We have also seen public biotechnology companies complete an additional round of private financing in an attempt to bridge dry spells in the public markets or to provide funding until the company can show enough progress to reenter the public market at a reasonable price. Finally, biotechnology companies have taken a hard look at themselves, realizing that early research and development efforts were carried out across too many unrelated areas, targeting too diverse markets. Technology that does not fit the redefined tighter corporate mission has been put up for sale. With all these methods, biotechnology companies have experienced varied degrees of success.

CONCLUSION

Biotechnology has clearly demonstrated its uniqueness as an industry during the past decade. The route to a public identity and the market was a long one for the early leaders. However, the number of public companies, the interest of analysts and investors, the pervasiveness of biotechnology alliances throughout diverse segments of our largest corporations, and growing public acceptance of biotechnology methods and hopes for new products are proof that the industry has ceased to be regarded as a group of research companies clustered around new scientific techniques and is now looked at as the source of the corporations of tomorrow. As we have learned in these growth years, biotechnology is an international industry which requires large amounts of investment capital. The start-up cycle covers many years and challenges management's and the financial markets' creativity. No one series of steps will provide all the capital needed, but it is safe to say that the successful biotechnology company's financial profile is multilayered and includes venture financing, government grants, multiple corporate and university partners, and public and international financing. It is also safe to say that the biotechnology company's financial profile is a blend of new financing vehicles whose strength is yet to be tested and, increasingly, more traditional sources as markets solidify and product sales increase.

CHAPTER 5

FUELING DREAMS INTO REALITIES: A VENTURE CAPITALIST'S PERSPECTIVE

Allan R. Ferguson

From a financial perspective, the biotechnology industry has reached a critical position. We are past the days when the financial community would throw money at any cell biologist and her mouse. "Have clone, will travel," does not get the interest of the venture or investment banking community. The potential of biotechnology remains undiminished, only the investor's perspective has changed.

A change that has occurred is that the expectations of the limited partners of venture funds' returns from the venture industry will be high and quick, as they were in the early 1980s. Recent biotechnology investments have not shown either the speed or the rate of return of the Genentechs or Hybritechs of the world. As a result there has been a reduction in the amount of capital being invested in biotechnology. Investors have moved to later-stage investments in other industries, where they feel they can get better returns and show a return more quickly to satisfy their investors.

From 1987 to 1989 (Table 5-1), the biotechnology industry received roughly the same amount of capital from the venture community. However, fewer companies are being started and more capital is being applied to existing investments. The public market, since the crash of October 1987, has been virtually closed to all but a few biotechnology opportunities. As a percent of investment, start-ups and high technology investments are down. The amount of capital invested in biotechnology is flat, and the number of biotechnology financings closed is down. However, the average deal size has increased. Good deals will, and are, getting funded,

TABLE 5-1 U.S. BIOTECHNOLOGY FINANCING, 1985–1989

	1985	1986	1987	1988	1989
Millions of dollars invested	100	180	255	300	260
Number of financings	70	100	135	140	120

albeit at lower valuations than they commanded in the early 1980s. A secondary factor that should be understood by individuals or companies looking to raise capital, is that the venture capital industry itself has been affected by the expectation of high, quick returns. The venture capital industry has raised less capital in each of the past four years (Fig. 5-1). In fact, the total amount of capital raised on an actual basis is lower than it was in the early 1980s. This puts increased pressure on the venture capital funds to find investments that allow them to deploy their capital quickly and to see rapid returns so that they can demonstrate to their investors that the venture investment rewards justify the risks.

As in all industries, there is a cyclical nature to the venture capital industry. This is compounded by the fact that the industry has gatekeepers who make determinations as to where the best investment opportunities are for their clients. These are the managers of pension funds and other monies. As mentioned earlier, over the past four years, the total amount of capital raised by the U.S. venture capital industry has declined each year as investors have become dissatisfied and disillusioned with

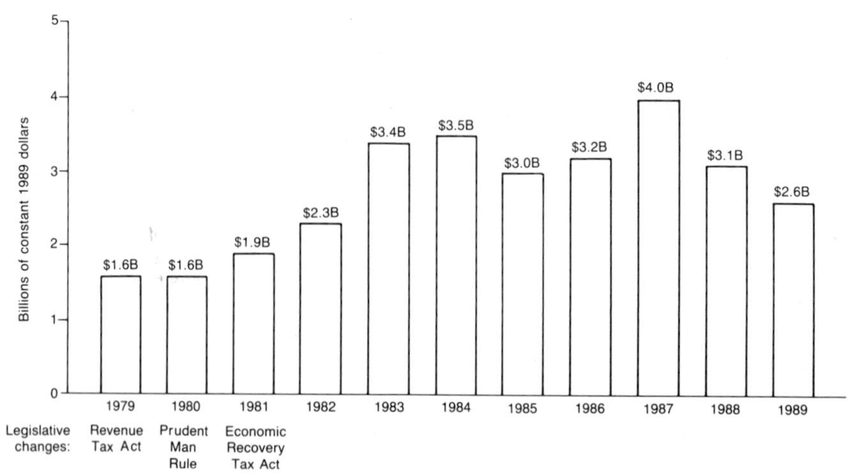

FIGURE 5-1 Annual investment by the U.S. venture capital industry. 1990 investment was $1.3B; 1991 investment is expected to be $1.0B. (From Venture Economics, Inc.)

the returns available in venture capital. As these funds leave the market, the amount of venture capital available will decline, the number of deal opportunities will remain generally constant, and we will have fewer dollars available for investment. The cost of capital will go up, thus bringing down prices and valuations of companies and improving investors' overall returns. As the returns improve, funds will flow back into the industry, thus driving returns back down and repeating the cycle. This is illustrated in Figure 5-2 and can be seen if you refer to Figure 5-1, which shows a

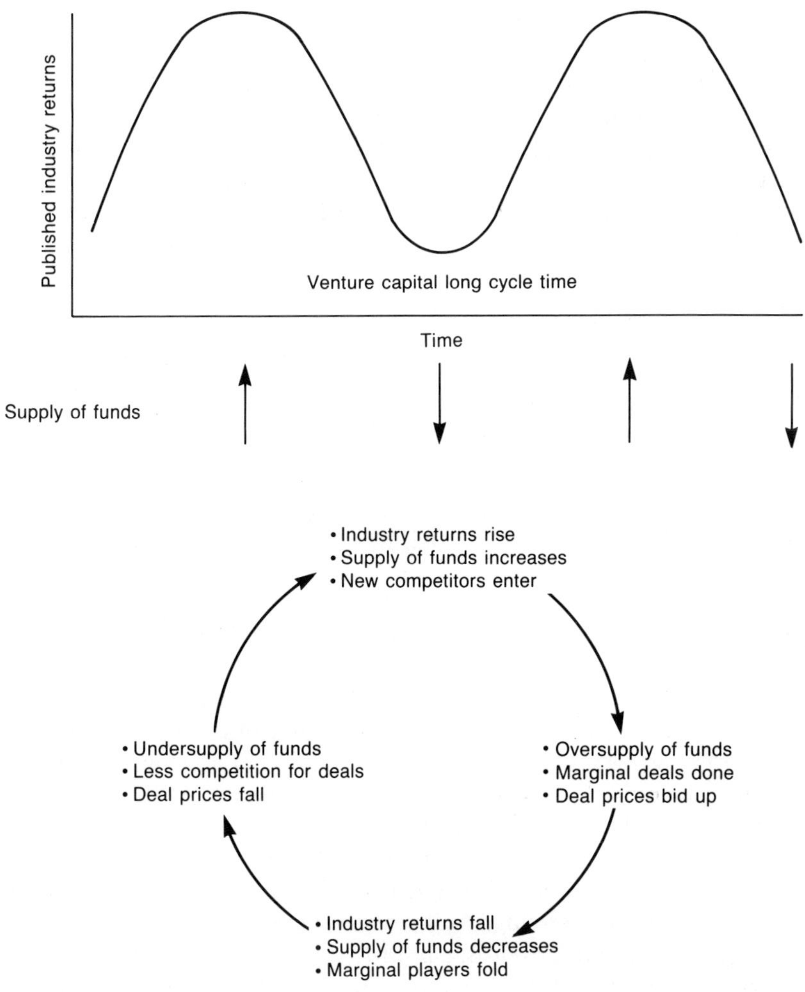

FIGURE 5-2 The cyclical nature of the financial return to the U.S. venture capital industry.

cyclical pattern in fund raising over the past five years after a period of rapid industry growth.

The European venture market is still on a growth curve, it has grown an average of 18% per annum over each of the past two years, although the total market is still smaller than the U.S. market. At present, there is approximately $30 billion being managed by U.S. venture firms, while European funds have approximately $20 billion under management.

European venture capitalists have the same pressure as their American counterparts for venture-type returns in three to four years. As a result, they have not shown an aggressive approach to biotechnology investments. Secondly, Europeans are jaded by the view that companies come to Europe when they cannot raise capital in the United States, under the guise of needing "international money."

Additionally, in the United States, we are seeing a reduction in the number of firms raising capital, and the firms that are raising capital are raising significantly high percentages of that capital. A typical cycle for a venture fund is to be raising money approximately every four years; thus, 20 to 25% of the funds should be in the fund-raising mode each year. If we look at the number of funds raising capital over the past three years, we see that they have also declined in number. The conclusions that can be drawn from this are

- A number of firms may be getting out of the business.
- Venture firms are taking down less of their capital and trying to stretch it out by reducing investments and preserving a larger percentage of their capital to protect existing investments.

A review of the funds raised over the past few years also shows that the older and larger firms are raising the bulk of the new funds. The net result of these factors is that as a customer of the venture capital fund, it is important that you understand where a fund is in its fund-raising cycle. It will affect your ability to raise capital, and it also helps determine which firms you should approach. Early in a fund's life, there is pressure to put capital out and you have more of an opportunity to be in the driver's seat in setting terms. When the total amount of capital raised by the industry is on the high end of the cycle, you also have an opportunity to influence the pricing and selection of the syndicate. At the bottom end of the industry cycle, the competition for capital is greater and your leverage is likewise less. Venture funds in the later phases of their investment cycles are generally not interested in early-stage deals and are more inclined to invest in later-stage follow-ons or mezzanine financings, as they are concerned about providing liquidity to their limited partners. This should be a factor that one considers in determining which funds to approach for potential investment.

Biotechnology, with emphasis on the technology aspect of it, has demonstrated that it has a place in both the health care and agriculture markets. Biotechnology companies today are providing approximately 60,000 jobs and generating revenues in the area of $1 billion. It is estimated that by the year 2000, we will see revenues of $5 billion.

Table 5-2, which was provided by Consulting Resources Corporation in Lexington, Massachusetts, shows the areas of high growth potential in markets served by biotechnology. It should be recognized by both investors and companies that biotechnology in itself is not an industry, but a tool to facilitate product development in a number of existing areas such as health care and agriculture. Ironically, we in the financial community and the biotechnology companies themselves have promoted the lore that we are creating a new industry. Biotechnology companies, when they grow up, are going to be diagnostic companies, pharmaceutical companies, or agriculture companies, and they will be looked at by the investment community ultimately as companies in these markets and be measured by the same performance criteria. As a result, the interest of venture investors has shifted from funding basic research to more specifically defined market areas such as diseases of the central nervous system, autoimmune diseases, and infectious disease and investing in core technology. Other areas of interest are oligonucleotides, blood products and blood substitutes, immunotherapy products, rational drug design, single-chain antibodies, transgenic animals, vaccines, lipid pharmaceuticals, and drug delivery systems. Funding in the area of human monoclonal antibodies has continued to drop over the past several years. Investors are also showing interest in environmental applications such as bioremediation and biopesticides. The investment community has moved from a technology-driven approach to a market-driven one. The initial questions now are: What is the market? What problem does this solve? and Who are the competitors?

TABLE 5-2 BIOPHARMACEUTICAL MARKETS*

	1995	2000
Cancer	$730	$1,900
AIDS (acquired immunodeficiency syndrome)	100	500
Cardiovascular disease	680	1,500
Monoclonal therapeutics	200	490
Growth factors	670	1,500

*Estimated in millions of dollars.

Source: Consulting Resources, Inc.

In summary, venture capitalists have maintained an interest in the development of biotechnology-based companies. They are, however, looking for technology that can build real business, with a product or family of products that can generate profits. This is in comparison to the early years, when there was excitement over basic research and concepts. In addition, investors have learned that experienced management and a fully integrated management team, which is committed to both excellent technology and the commercialization of that technology, is critical. In the early years, investment in scientists was considered acceptable, but it has been demonstrated over the past several years that it is integrated teams that understand the commercialization pathway from the laboratory to the marketplace that are key. Manufacturing skills and strengths are becoming more and more important in the company's attempt to scale up the products they have just brought through the regulatory process.

When looking for capital, a company should go about the search using many of the techniques outlined in Michael Porter's book, *Competitive Strategy* [Porter 1980]. You are selling a product; you are competing with other biotechnology companies for the same end customer—the providers of the capital. Understand what the venture capital funds are looking for. An excellent way of doing this is by talking to some of the other companies and CEOs in your industry, as well as to your current investors. *Pratt's Guide to Venture Capital* [Pratt and Morris 1988] is an excellent overview, which gives insights into the interests of most of the funds. Intermediaries, such as accountants, consultants, and lawyers, generally know what particular ventures funds are doing and what their interests are. By doing this early legwork, you increase your chances of success by contacting the right venture funds. In addition, you generally will find that funds are more responsive to deals that have been referred to them by intermediary sources than they are to those that are sent directly. As time is the most precious commodity to a venture capitalist and to you, the company's management, using these intermediaries and doing legwork early on, will shorten the fund-raising cycle and improve the chances of success as you attempt to close with new investors. An introduction of your company to the venture capitalist will enhance the likelihood that you will get their attention.

In looking for new investments, the following aspects are evaluated by most investors. Overall, there is a concern for the financial requirements: What is the company's financial strategy? How do they feel about corporate partners and joint ventures? Does the technology being developed have licensing opportunities or possibilities? Is there a possibility for research and development limited partnerships? And finally, how do the company and its founders view themselves in terms of an exit strategy?

In analyzing and determining an exit strategy, management should attempt to assess objectively the probability that their business will reach the objectives in their plan, and at that point, which of the options—merger, acquisition, or public offering—will provide the appropriate liquidity for both their investors and themselves. It is unlikely that many of these start-up companies will grow within the health care industry as fully integrated, pharmaceutical manufacturing companies, although Genentech has done an excellent job in this regard. The recent acquisition of Genentech by Hoffmann–LaRoche shows that the cost to maintain a sales force cannot be supported by two products, and the cost of ongoing research and clinical trials cannot be supported by these smaller companies. Thus, the exit strategy becomes key not only for liquidity reasons, but for the company's long-term success. Although an initial public offering has "sex appeal" and gets the excitement and interest of most people, it is unlikely to be the route of exit for most companies. Even in good markets, over 60% of exits for founders and investors have been through the acquisition or merger route. Most venture investors prefer to consider an exit strategy early on so that investors and management are working toward the same end.

The next criterion you should evaluate as you look for investors is the venture capital fit: Do the company and the venture investors share common objectives? Do syndicate partners share the same objectives and interests in developing the company? For hands-on investors, there is a post-investment role that needs to be evaluated to make sure there is compatibility between the company and its investors. Management should determine whether the investors they are looking for possess a hands-on style, one that would be effective and work with management, or a more arms-length style. What about international contacts? Some venture funds are highly regional in their approach and network, others have more of an international appeal. At different times, both are important to the company's growth and development.

Since only three or four out of every 100 investment opportunities reviewed by any given venture investor is actually financed it is important that you consider the above questions and develop a strategy that will improve your odds of funding your particular project or company. The more obvious reasons for rejection are questions about management, market, technology, or competition; in many cases, the turndowns occur due to an entrepreneur's lack of understanding of the tastes and requirements of particular venture capital firms. The quick screens of venture capitalists should be in terms of the amount of capital you require and your stage of development. Many of the funds state preferences in terms of upper or lower limits in the size of their investment. In addition, many funds have a preference as to the stage of development, i.e., those who prefer only seed capital areas and will not invest with valuations above $10 million, or those who are willing, and actually prefer, to be later-stage investors. Additional screens that will help you zero in on the funds that are more likely to be interested in your investment are industry specialization, geography, and leadership.

Leadership and geography may be the greatest factors today in putting together an investment syndicate. If you cannot attract an investor who is relatively close to your company geographically, and one who is willing to serve as a lead, you will have significantly greater difficulty in raising capital. There are hundreds of firms investing on a national basis, however, the closer a venture capitalist is to his investment, the easier it is for him to monitor and provide value, especially in the early stages. There are funds with active investors willing to lead the financing, whereas other firms will serve only as passive investors and will look for another firm to provide the leadership role. Clearly, you will need a lead venture capitalist, and it is inappropriate to approach passive investors until after you have identified a lead investor.

The last criterion that should be considered is that of industry specialization. If a venture capital firm has an investment preference, not only is it more likely to understand your business, but it is clearly in a better position to provide added value to your business. Often, the industry expertise of a particular fund is respected by other venture capitalists, who may provide additional funding based on the confidence in the industry specialist.

Using the above criteria, you can prepare through a venture capitalist to approach targets for your investment. Information on how different funds fit into these categories is readily available from sources such as *Pratt's Guide to Venture Capital Sources*, CEOs and CFOs of other start-up companies, as well as intermediaries such as your banker, lawyers, and accountants or industry consultants. In addition, there are often venture capital funds that are targeted at generating investment opportunities that fit the firm's investment interests. A simple telephone call to the firms often can provide you with a brochure that will help you further understand the interests of these firms. Also, these brochures often show portfolio listings that will allow you to identify investments that the fund has made in biotechnology and will thus provide you with additional opportunities to understand the firm and its expertise.

In these times, especially, the availability of capital is clearly a consideration that management should evaluate in their potential investors. As there are long-term needs, reserves should be provided as follow-on funding by your potential investors. The company and its management should also consider the policy of your investors regarding follow-on investment. Many funds have limits in terms of either dollar amount or as a percentage of the fund. Do your investors have deep pockets, or will you have to go back on the fund-raising trail to find new investors to carry the company the rest of the way when the next round of funding is required? In early-stage companies, the operating experience of the venture partners is important because they understand the roles of management; they have done it before and they have networks that may be able to assist in finding staff as you build your team.

What is the current status and outlook on the venture capital community? Biotechnology companies have over a dozen therapeutic products and numerous diagnostic products already in the marketplace. The next wave of products has begun in the 1990s. Biotechnology started with successful companies such as Applied BioSystems, a molecular biology instrumentation company which provided the tool kit necessary to advance the science of biotechnology. Biotechnology is now moving into the arena of providing products from the science, solving problems in the health care and agriculture industries. The more near-term products are certainly the high-profit therapeutics, which have focused on major elements such as cancer and cardiovascular disease. Products in the areas of neurology and brain biochemistry will be coming in the late 1990s and products in the autoimmune area and rational drug design also seem to be moving along rapidly. Areas in which products are needed are improved sample collection and noninvasive products. Biosensors seem closer to commercialization because of advances in microprocessing, microelectronics, and materials allowing for in vivo monitoring of critical drugs, blood gases, and intercellular monitoring.

Figure 5-3 shows a time line of business innovations that illustrates the industry's evolution. The 1990s appear to many industry watchers to be a period of consolidation. Many companies will not have the critical mass in terms of technology or management and will not be able to continue to sustain the funding required to give them enough time to reach the marketplace. At the same time, technology is advancing at an incredible rate, the rate-limiting factors today are not technology and science, but quality management and financing. Figure 5-4 shows a time line of technical innovations. For *Star Trek* fans, Dr. McCoy's tricorder is probably not too far off in our future. To gain the benefit of the advances in technology, mergers and joint ventures between venture-backed companies, which are currently in need of financing, will provide the synergy needed for each to succeed and optimize the core of their capital resource by eliminating duplications in overhead. Probably, each will fail, as they continue to compete.

A review of the biotechnology industry and marketplace shows that fundamental shifts have occurred over the past two decades, and as we enter the 1990s, it is expected that we will see another shift. In the 1970s

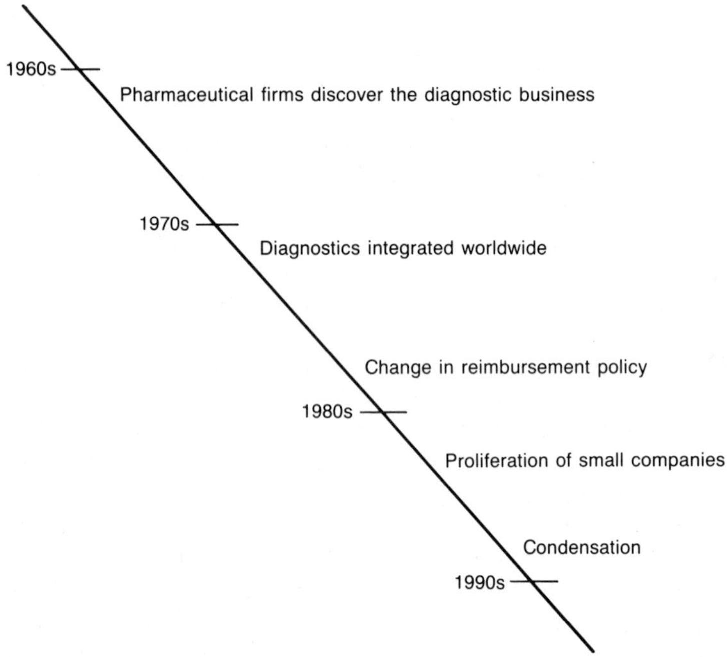

FIGURE 5-3 Time line of business innovations. The 1990s—a period of consolidation?

through the mid-1980s, biotechnology was driven by fundamental research and was academically driven, with few commercial products. The market estimates or potential really had not been validated. We were watching a phenomenon that was technology-driven, and investors were chasing deal after deal and driving up the prices unrealistically, based on the fundamentals of these early companies.

The period from the mid-1980s to the present has been a time of corporate partnering, some form of consolidation, and a rationalization of prices. It is also a period that saw the industry develop revenues from products and start to build a strong base of operations in the diagnostic, therapeutic, and scale-up technology market areas. Biotechnology companies' products are now more market- or application-driven.

Although we look at it as a new technology, in reality, biotechnology is an ancient industry, which has come from brewing and baking and has been renewed today by DNA identification and an understanding of biological mechanisms. It is a fundamental technology, which holds keys to our understanding of the immune system, genetic disorders, and ultimately, our ability to synthesize products that will treat various problems, primarily in agricultural, health care, and environmental areas.

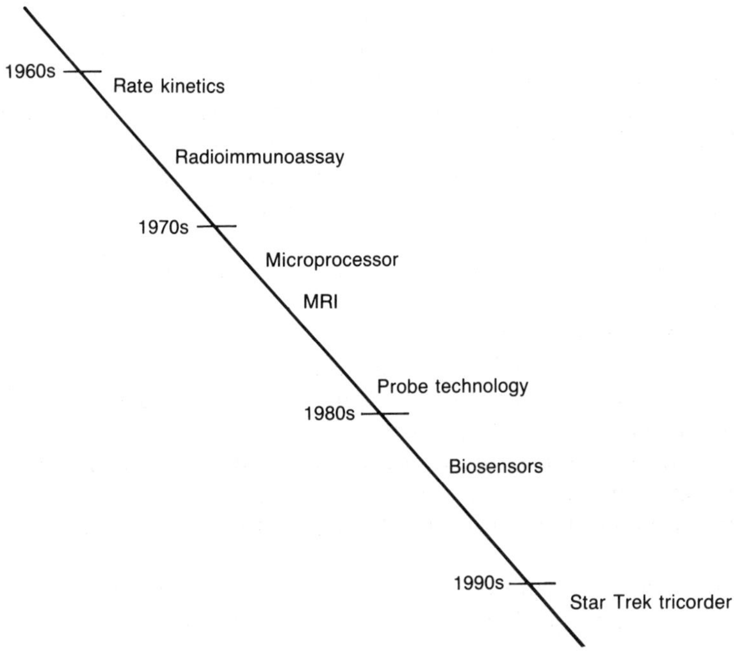

FIGURE 5-4 Time line of technical innovations. To gain the benefit of technologic advances, mergers and joint ventures between venture-backed companies currently in need of financing will provide synergy needed for each to succeed.

As a venture capitalist evaluates companies for potential new investments, he considers a number of specific criteria. From a market standpoint, the size and location of the market is an important factor: Are the markets large enough to support this company and a number of competitors, and are they accessible? Are there existing markets or are these markets that must be developed by the company through missionary selling? What are the resource requirements? Regulatory issues must be addressed to enter these markets. Are they small and focused to allow a start-up to build an expensive direct sales operation, or are they diffuse and large, where a more appropriate approach would be collaboration with a distribution partner? The products are evaluated. For instance, does the product the company is developing solve or create a problem or need? Is it a single product, or has the company developed a strategy that will allow development of a family of products, thus reducing risk and improving and enhancing the companies' opportunities to grow?

As you look for venture capital and try to match the profile of your company with venture investors, you should look for investors who have been active in your industry and have tracked it very closely. In the early period of your company's life, when your team is small, look for investors who have a strong network in your industry, those who can be an invaluable asset. Investors can offer essential additional services, such as helping to identify the potential additions to your management team. Corporate investors have their own limited partners or network. A fund has developed corporate relationships in which the corporations have specific strategic areas of interest that may cover the focus of your company. If your syndicate includes investors with corporate partners, they can become an asset in helping you from the introduction to their corporate partners to assisting you in negotiations.

As mentioned, trends in the venture community are shifting to later-stage investment. As your start-up begins to generate cash and you produce the image of a viable growing concern, you are increasing your strength from the standpoint of negotiation during fund-raising. The risk in your company is decreasing, and as a result, you will significantly increase the number of venture capital firms that will be interested in your firm.

Overall, the opportunity to raise venture capital for your company still exists. On average, $250 million was raised in each of the past few years.

You need to appreciate the dynamics of the financial market at the time of your fund-raising, as these will dictate how much leverage you have and will also assist you in determining which funds to approach. As you do, the funds have their own investors who are willing to take the risk if they believe the reward is commensurate with that risk. Overall, venture investors have become more realistic in understanding that the time horizons and capital requirements are greater than originally estimated in the early years. The willingness of these investors to support your company is contingent on you and your management showing an understanding of the requirements of building a company and financial strategy that will provide for the long term.

BIBLIOGRAPHY

Batey, G. E. 1990. *Entrepreneurship for the nineties.* Englewood Cliffs, N.J.: Prentice-Hall.

Genetic Engineering News, Volume 10, No. 1, January 1990.

Genetic Engineering News, Volume 10, No. 4, April 1990.

Genetic Engineering News, Volume 10, No. 5, May 1990.

Genetic Engineering News, Volume 10, No. 6, June 1990.

Genetic Engineering News, Volume 10, No. 7, July/August 1990.

Jones, L. 1990. *Trends in venture capital*. 1989 ed. Wellesley Hills, Mass.: Venture Economics.

Mintz, L. 1990. *Venture capital and technology: Investment trends, future opportunities*. Needham, Mass.: Venture Economics.

REFERENCES

Porter, M. E. 1980. *Competitive strategy: Techniques for analyzing industries and competitors*. New York: The Free Press.

Pratt S. E., and Morris, J. K. 1988. *Pratt's guide to venture capital sources*. 12th ed. Wellesley Hills, Mass.: Venture Economics.

CHAPTER 6

GROWING THE ORGANIZATION: PEOPLE-POWER FROM RESEARCH AND DEVELOPMENT START-UP TO MARKET-DRIVEN ENTERPRISE

Donald A. Lotufo

For the past ten to fifteen years a new science has been evolving from a number of learned academic institutions in North America as well as in Europe and Japan. This science, biotechnology, has been mounting a tenacious assault in a quest for the key to life as we know it. During this brief period, scientists have been able to extract genes from the heart of a living cell, dissect them to modify genetic information, and, finally, put them back into the same cell. As we know, genes hold the blueprint of life; they determine eye, hair, and skin color; they affect our longevity, personality, and intelligence; and, importantly, they govern our lifelong health and susceptibility to disease.

Let there be no doubt that this unique technology has proven that it will make it possible for this and future generations to modify humankind, and in the minds of some, tamper with the core of existence. Many learned individuals predict that biotechnology will be many times more advanced and complex in its technological breakthroughs than computer and information technology, which has had such a profound impact on our lives.

Through genetic engineering, scientists have already created new life forms. Gene splicing has proved to be a science that will provide a new green revolution: plants that produce their own nutrition; crops that grow where they have never grown before; and vegetables that will taste better, grow more rapidly, and resist disease. Biotechnology organizations today are developing hormones that will enable livestock to grow fatter and faster and, for those that produce a product such as milk, will result in a manifold increase of significantly more nutritious products.

Although biotechnology is still in its infancy both in academia and within biotechnology companies, the science is concentrating on a potpourri of endeavors. A direction that a majority of biotechnology organizations have taken concerns the field of pharmaceutical and health care products. In the minds of many, genetic engineering promises an abundance of novel therapeutic and antiinfective agents and vaccines that will one day vanquish many of this universe's scourges. Even as you read this, exciting progress is being made in the area of cancer therapy. Today there are close to one dozen new oncogenes that have been developed during the past five years. It does not take Nostradamus to predict that this century's remaining decade will undoubtedly bring more inventions of disease prevention and control as compared with any other ten-year period in our history.

Many experts estimate that the aggregate worldwide market for biologically designed commercial products may well reach $70 to $100 billion in a span of less than 20 years. Anticancer drugs alone are predicted to achieve $10 to $15 billion worldwide by the year 2000.

Ever since James Watson and Francis Crick found the secret of DNA's double helix more than three decades ago, scientists have worked in a supercharged environment of discovery. Consider that it has only been about fifteen years since Paul Berg of Stanford University astounded the scientific community by splicing genes and earning the Nobel Prize. Many of my friends and associates in the pharmaceutical and biotechnology industries continue to express amazement at what was perceived to be the impossible, not ten years ago but just one or two years past. Although I am not a scientist, having been in this "catbird seat" for the past ten years and viewing the developments in biotechnology, I, like many others, am convinced that we have barely scratched the surface.

Obviously, considerable money will be made in the process of transforming science and biotechnology into viable commercial products and services. During the 1990s, the biotechnology business will have a significant impact on new chemicals of various forms as compared with the total worldwide size of the conventional chemical industry during the 1970s and 1980s. Over the past few years, venture capitalists, private investors, and established corporations have been tripping over themselves rushing to put up more than $5 billion to stimulate more than 300 start-up biotechnology ventures. As there has been over the past ten years, there will of course be a continuing shake-out among the biotechnology community. However, very few investors are ready to write off this industry. In fact, many of the largest and most renowned biotechnology companies are partially owned today by chemical, food, or pharmaceutical giants or have close working relationships with many Fortune 500 corporations as well as European and Japanese conglomerates.

My introductory remarks in this particular chapter are simply provided to offer a platform for the views that will follow. Having spent an entire working lifetime both as a human resources practitioner and, for the past ten years, as a human resources consultant dealing with biotechnology companies, I am convinced that this industry will either survive or atrophy at some point, predicated on two obvious factors: the first is capital, and the second is people. In combination they represent a "hand-in-glove" fit. Out of synchrony, millions of dollars of investments and a multitude of careers will be affected.

As it has been with other technology-based new science industries, the centerpiece issue is promoting creativity, invention, direction, and ultimately, commercial by-products. As many architects of industry acknowledge, these goals are easily stated, but elusive to implement. Much of the following section of this chapter represents my personal views and expressions. Although I have made no attempt to arrange them chronologically, the anecdotes represent the human resource dynamics that I have observed within the biotechnology industry.

HUNTING FOR THE PREY—1980

Back in 1980, shortly after leaving Bristol-Myers (now Bristol-Myers-Squibb) and entering the field of executive search consulting, one of my first biotechnology clients was the Biogen Corporation (located in Cambridge, Massachusetts). Two of my colleagues from Bristol-Myers, Bob Fildes and Mark Skaletsky, had joined the newly established U.S. operation for Biogen, and I was assisting them in identifying candidates for several positions. I vividly recall that most of these positions targeted candidates already with pharmaceutical and health care organizations. More importantly, I remember the difficulties I encountered in getting candidates to even listen to my pitch. After all, biotechnology was still in its infancy, untested, and, as a new age science, far from predictable. One had to be a risk-taker with near suicidal tendencies to leave the bastion of an established corporation to join an untested biological company with limited resources and an unproven science.

During the past near ten years, I have continued to provide executive and management search and other human resources consulting services to a number of biotechnology organizations, and the difficulties I encountered initially have diminished appreciably. Today, we find that many of the scientific, medical, research and development, marketing, business development, and administrative support positions within biotechnology organizations large and small are in the hands of individuals who have had previous careers with pharmaceutical and health care, chemical, energy, and other industrial organizations as well as with prominent consulting firms, academic institutions, and governmental agencies. Clearly, the desirability and legitimacy of biotechnology as a career pursuit has continued to grow in a dynamic fashion over the past decade.

The evolution of biotechnology resembles what has been faced by other start-up and new technologies (i.e., aerospace, computers, and semiconductors). Many of the charter members of biotechnology organizations were and, to a lesser degree today, are academicians sometimes supported and influenced by investors who have taken an active role in the operations of the business. Although this statement can be readily contested, it has only been in recent years that biotechnology has proven to be an effective science—converting bench-scale laboratory activities into real commercial products. No doubt there will be readers of this chapter who will be part of the biotechnology industry who will disagree with this statement. However, if one reviews the total value of goods and services produced worldwide today and compares that number with the goods and services produced through biotechnology, the latter figure would represent a speck on the head of a pin.

IMMERSION OF BIOTECHNOLOGY PILGRIMS

Human resources management and planning has become an increasingly visible and urgent concern in biotechnology corporations, both large and small. Among the factors heightening interest is the expanding awareness that employees are a unique resource, not merely a cost of doing business. Therefore, human resources merit the same close attention, conservation, and improvement as any other vital resource. Employees are also increasingly seen as a mirror of an ultimate customer base. Understanding the internal environment necessitates understanding the external environment and vice versa. The expanding, unstoppable diversity emerging in the biotechnology work force, which includes the increased numbers of women in business, changing family structures, utility of individuals entering the biotechnology field from various industries, academic institutions, teaching hospitals, and governmental agencies, further challenges the manner in which human resources are managed. This issue is further affected by an increasing awareness of pay for effective performance, equity in the business venture, benefits, perks, amenities, and numerous other undefined direct and indirect incentives.

Again and again, I hear from my clients and associates in the biotechnology community about their growing concern in regard to the need to introduce further controls on day-to-day business operations without impairing creativity. This issue is far from exclusive to technology-based organizations. There is almost an axiom that the larger and older an organization becomes, the greater the need for controls of one form or another, a situation that can often have a debilitating impact on independent thinking and creativity, whether in the research and development or the finance department. To some measure, this is further influenced by the regulatory impact on biotechnology organizations. I vividly recall that several of my clients in the early 1980s who had never been part of a drug or food clearance process were amazed by the arduous requirements imposed by the Food and Drug Administration and other regulatory agencies in order to get product approval. This response was particularly true of many esteemed academicians who had taken on key roles with start-up biotechnology organizations.

The executives of fast-growing biotechnology companies are a rare breed in my opinion. They are sometimes described as the "racehorses of industry." They are also a bunch of lonely, terrified individuals, and most of them know that every day of their lives, everything they have in the world is pledged and on the line. Often they face a world of problems that are somewhat different from individuals holding the same titles with Fortune 500 companies. How do you keep an entrepreneurial spirit when the number of employees quadruples? Where do you get the money to expand when your concern is meeting payroll six months into the future? How do you find a professional manager who will not burden his subordinates with procedures that are not required by the entire organization?

The investment community that has provided much of the funding for the biotechnology industry has recognized that growing fast does not necessarily ensure success. The adage "work fast, die young" continues to be a big threat. Look at some of the high-flying computer companies that have been hitting the wall. If we were to track the Fortune 500 for the first five or ten years that they made this list, you would find a high failure rate, a high attrition rate, and many organizations that actually shrank in size.

LEADERSHIP—CLASSICAL OR EXPERIMENTAL?

Organization analyzers and social scientists suggest that the old, tired ideas about leadership are a thing of the past. I share that view and believe that the more successful organizations in the 1990s will be learning organizations, constantly adaptive enterprises with employees free to think for themselves, identify their own problems and opportunities, and go after them directly. In these organizations, particularly those that are technology-based, the leaders will ensure that everyone has the resources and power to make swift day-to-day decisions. In reality, the leaders, whether they have the title *managing director, chief executive officer, chief operating officer,* or *executive vice president,* must become the best learners of all.

As we have seen over the past ten years, several biotechnology organizations have come and gone. Some have been absorbed by Fortune 500 companies, in part or in full; others have become captive research houses for larger organizations; and many have failed due to under-capitalization, unfocused science, and, yes, ineffective management.

I hold the view that the individuals holding the reins in biotechnology organizations during the next twenty years will be somewhat different from the focused specialist in existence today. The functional leader of the 1990s will be a person who takes a broader view, does not harbor as many fears, and has an insatiable appetite to explore the uncharted. That person will know how to delegate authority rather than give orders, how to negotiate more skillfully, and how to manage and intervene even where he or she does not have control. Most of tomorrow's leaders within the biotechnology industry will have a more precise vision, which is subject to change and yet can be measured.

As it will affect every industry in the 1990s, globalization will mature from a buzzword to a pervasive reality. The emerging economic advancements of the Asia–Pacific countries and both the opportunities and upheavals resulting from Europe 1992 clearly mean that biotechnology companies will have to meet global requirements. There always has and probably always will be a controversy concerning the suit of armor, so to speak, that a corporate leader should have. Although there is also a substantial amount of hot air emerging from so-called head-hunters in my industry about the ideal qualifications a person should have, let there be no doubt that a manager by whatever title will in the future be called on to embrace more forcefully the novel and to respond effectively and quickly to the unexpected.

Today in businesses worldwide, leaders are changing with increasing frequency; hence the players must also change. Measuring performance to the third decimal is becoming much more common, even in the laboratory. If one follows various technical publications tracking the biotechnology industry, it is apparent that a great number of management changes have occurred in this industry over the past two or three years. Certainly, part of this turnover is due to impatient investors as well as to promises from biotechnology leaders that have been off the mark. The fact that a senior official is a Nobel laureate in some area of science is not an exclusive license to practice in this business. Today, the requirement is delivery of a product, not simply promises.

The manager of the future will be someone who has the ability to manage in new and changing situations, is able to respond effectively to ambiguity, and is able to coordinate a complexity of issues effectively. This will be further complicated by flatter organizations. For the most part, all industrial organizations will be obliged to move away from the hierarchical model organization that was adapted from the military and will die by the same route as the Winchester rifle. The modern ideal in organizational design will be flat, and many companies have already been obliged to cut layers of management between the chief executive and the laboratory section head. This means the span of control will be broadened, with the numbers of people reporting to each manager expanding dramatically. Also, leaders of the future will have to know how to manage employees with a diversity of backgrounds. Some of this need has been particularly evident within the biotechnology organizations that we all know.

PASSAGES OF REALITY

Since a number of the preceding passages have been somewhat abstract and represent highly personal views shared by some but not by all, let me now get to some of the pragmatic aspects of the bioscience industry. Several of the bromides that follow concern subjects and issues that I have faced during the years that I have been in executive search consulting. They are, however, real, not imagined, and by themselves are not intended to be criticisms, though they do focus on the realities of new industries and new businesses and the human resources issues that can often impede progress.

I am a firm believer that what Vince Lombardi called "basic football" is somewhat analogous to the manner in which you recruit, train, motivate, and compensate people for their work. Namely, you have to communicate with them and tell them what they should know about the business, treat them fairly and honestly, compensate them equitably, and provide for reasonable growth and promotion. If we accept these premises, there will be a further appreciation for certain of the stories that follow.

Some seven years ago Bob Fildes went out to the Bay Area of California to join Cetus Corporation as Chief Executive Officer. At that time Cetus was a "charter member" of the biotechnology industry. On joining Cetus, Bob found that over 90% of the employee population was dedicated to research and development. After Bob and I had several meetings, he provided me with an organizational table that allowed for the introduction of marketing, business development, legal, financial, scientific, medical, and other newly established units intended to move Cetus from a research and development "boutique" into more of a wall-to-wall commercial enterprise. I recall that a groundswell of concern emanated from the research and development organization. Up to that point, the research and development organizations were "kings of the hill." They had free rein, and the research and development programs in place had not been as fastidiously measured for their commercial worth as, perhaps, they might have been. Nonetheless, Dr. Fildes had to review carefully and quickly the research and development programs that were potentially commercially viable and those that were not. Unfortunately, a number of research programs had to be aborted. Unanimity of opinion on programs that were abolished could not be reached. Schisms therefore developed.

One further complication occurred, in that Cetus was looking almost exclusively to the pharmaceutical and health care industry for senior executives. A vast majority of the senior scientists in the research and development organization were individuals with limited, if any, long-standing industrial experience. Hence, there was an absence of synergy between newly appointed officials and members of the research and development organization. The research and development people no longer viewed themselves as the "centerpiece" of the organization. As a result, Bob Fildes and his associates spent considerable time initiating various information and communication forums to permit the research and development organization to interact more effectively with the commercial and administrative groups. Eventually, Cetus was able to overcome a number of major obstacles. It was a laborious and time-consuming process, but it was essential that it be put in place.

If there is a message to this story, it is simply that a chief executive officer cannot readily guide an organization through its various stages of growth without taking away or requiring that originating groups perceive that new business units may be sharing in what were perceived as their inviolable accountabilities and responsibilities. Furthermore, it may even appear that the original organization and members of a business enterprise may find themselves taking a back seat to units that become the centerpiece. One good example of this is the positioning of the regulatory affairs official in recent years; another is the accountability that has been given to the patent attorney's office with biotechnology companies in view of the ever-expanding cross-patent and other legal issues that this industry has faced.

Several years ago, one of my biotechnology clients, engaged in developing cancer products, retained me to recruit a senior marketing person with experience in cancer therapeutic products. I did identify a gentlemen by the name of Rick Maradie, who was then the marketing director for the largest-selling cancer drug, Adriamycin (doxorubicin). He had also established the first U.S. sales organization dedicated exclusively to oncology. There is no doubt in my mind that Maradie would never have considered the opportunity with my client had it not been for the significant stock equity that was offered to him. This was one of the key perks in attracting this individual, along with many other individuals from established pharmaceutical houses to start-up, limited finance, and still speculative biotechnology firms. I know for a fact that Maradie could readily have joined a number of other established drug firms that were aggressively moving into the oncology market. Today, Maradie tells me he does not regret his decision, and has been involved in other biotechnology firms himself.

I can think of many other illustrations of unique human resources issues that the biotechnology industry has faced, but one in particular has a message. A few years ago the chairman of the board of one of the preeminent United States–based biotechnology companies asked me to meet and interview a woman who had been one of the first female product managers with Procter & Gamble Company. He had met this woman through his church. She had recently married and relocated from Cincinnati, Ohio, to the West Coast and was interested in reestablishing herself with a West Coast corporation. She had studied the biotechnology industry and was intrigued by it as a potential employer. I had been asked to meet her to discuss establishing a formal market research function in the company. I discussed at some length her interest in such an opportunity. After several meetings, it became increasingly apparent to me that she would have extreme difficulty in adjusting to the culture and personality of my client. After having spent some eighteen years with Procter & Gamble, a sophisticated, research-based organization, this woman readily admitted that the somewhat unstructured environmental climate of a biotechnology organization would be difficult for her to adjust to. This issue was further exacerbated by the absence of definitive outside research consulting services relative to the biotechnology industry. Hence, where she would be operating somewhat autonomously, initially with limited staff, her role would be further complicated by an absence of research materials similar to the credible Nielsen Reports.

This illustration, by itself, is not intended to show the absence of sophistication within the biotechnology industry, dating back some five years, or the absence of appropriate skills on the part of this former Procter & Gamble product manager. It does show that quite often individuals who are selected for biotechnology organizations must be capable of operating in a less-structured environment. I continue to hear from friends I worked with at Bristol-Myers-Squibb who have joined biotechnology organizations. They report that one of the issues that perplexed them in some cases, was the absence of clearly defined "charts of account." Whether they are research and development scientists or marketing and business development specialists, they have found that biotechnology organizations often do not have the "support systems" that prevail within established pharmaceutical, health care, chemical, and other industrial organizations. For some, those who prefer a more laissez-faire environment, this situation is attractive. Certainly, in my experience, many of the candidates that I have interviewed for my biotechnology clients have given credence to the absence of constrictions, which, in their opinion, often prevail within the older, larger, and structured corporations. The old adage "different strokes for different folks" is applicable here.

Another parable as it relates to this industry comes to mind quite vividly. Some years back a longtime friend, Bob Havranek, who was then Senior Vice President of Operations with Cetus Corporation, had implemented GLPs (Good Laboratory Practices) and subsequently GMPs (Good Manufacturing Practice) at Cetus. At the time of introduction of these required federally legislated policies, Cetus' research and development organizations consisted of some 350 to 400 scientists and support staff. The criticism that emanated from the laboratory following the introduction of these policies could have readily been heard throughout the Bay Area of California. Quite naturally, up until that point, scientists had kept their own laboratory logs; some were most comprehensive and exhaustive and in other instances pertinent data applicable to research and development programs were confined to nonstructured three-by-five binders. In effect, there had not been any need up until then to establish uniform laboratory documented procedures. Although I believe Cetus management did an outstanding job in preparing the laboratory staff for these Good Laboratory Practices, it is also my opinion that this policy decision had further restricted the otherwise emancipated research and development laboratory. Please keep in mind that a large percentage of the laboratory people had never worked in an industrial setting where uniform laboratory documentation procedures had been in place. Could this policy have been implemented differently? I do not believe so. Should the laboratory personnel have been informed that this was a federally imposed program, not exclusively a determination of Cetus management? It was. Regardless, it is another example of the evolutionary process that biotechnology organizations have been going through, and the impact it has on what I perceive to be its second most important asset beyond capital: people. As I have noted before, stimulating the creativity of scientific and research personnel in a biological organization without atrophying the organization is far from simple. It is the same platform that has been faced by Dave Packard at Hewlett–Packard, Edward Land at Polaroid, and Henry Ford at Ford Motor Company. Few have been able to master the process of running an organization without impairing its research or commercial purpose and at the same time permitting the human resources of the organization to "do its thing."

A TENET

There is neither a pat answer nor alchemy to effective human resources management. One major shortcoming on the part of many executives-in-charge, however, is failure to pay attention to what makes people tick today in the workplace, based on the industry, culture, demographic distribution, turnover, growth or stagnated state, and leadership. What "turns on" motivation today may well "turn it off" tomorrow. Keeping creativity fueled is not easy, but if you do not give it priority of the highest order, atrophy can readily set in.

CHAPTER 7

HOPE IN THE VIAL: THE CLONING AND EXPRESSION OF RECOMBINANT PRODUCTS

Debra A. Peattie

Gone are the days, typified by the years 1977–1978 in Cambridge, Massachusetts, when scientists quarrelled among themselves and the general public read horror stories in the newspapers about the potential dangers of using genetically engineered bacteria or animal cells to produce recombinant gene products. By the 1990s, recombinant DNA technology has taken its place as a cornerstone of the biomedical industry and has propelled many of its early proponents into positions of wealth and power.

This chapter focuses on two fundamental aspects of biotechnology—the cloning and expression of recombinant gene products. Each topic is considered in broad terms, with appropriate references provided for the more detailed needs of some readers. The next two sections review several terms and concepts essential to the jargon of cloning and expression. A discussion of specific eukaryotic (e.g., plants, animals, and humans) expression cloning vehicles follows, and, finally, future generations of recombinant proteins and nonrecombinant drugs are considered. There are also tables of key terms and technical innovations in biotechnology included, and there is a brief section on "How to Interpret Vector Constructs."

NUTS AND BOLTS OF CLONING DNA

Jargon: the basics

Far from the science fiction concept of human replicons, clones are collections of identical cells or molecules that originated from a single ancestral cell or molecule. In the vernacular of molecular biology, and pertinent for our considerations here, a clone is a piece of DNA that has been specifically selected and that can be maintained and propagated within a cloning vector. Cloning vectors most frequently derive from plasmids or phage, can be tailored to meet specific needs, and serve as vehicles for manipulating cloned DNA fragments. DNA libraries and the polymerase chain reaction (PCR) provide the primary means of cloning with today's technologies. DNA libraries, collections of cloned DNA fragments from an organism, can be constructed with genomic DNA from the organism or with complementary DNA (cDNA) that has been synthesized *in vitro* from the messenger RNAs (mRNAs) of the organism. Like good lending libraries, the best DNA libraries are the most complete ones—they offer access to all genes or gene transcripts from an organism. Most cloning in biotechnology focuses on cDNAs rather than on genes because cDNAs do not have introns. Introns, commonly found in the DNA of higher eukaryotes, are portions of genes that are transcribed into mRNA but are removed from the messenger prior to translation into protein. This property distinguishes them from exons, the portions of genes that contain protein-coding information. Most gene products relevant to commercial biotechnology derive from genes of higher eukaryotes; therefore, this chapter focuses on the construction and screening of cDNA libraries, on the use of the PCR reaction to amplify cloned cDNAs, and on the *in vitro* expression of cloned cDNAs.

Bookmobiles: cDNA Libraries and Cloning Vehicles

cDNA libraries provide a powerful way to isolate, amplify, and maintain cDNA sequences. Two fundamental considerations to be made in constructing a DNA library involve the type of vector to use and the type of output information desired. The vector type can be either plasmid, phage, or cosmid, while the output information—essentially, the molecules to be screened—can be DNA, RNA, or protein. Plasmids, circular and autonomously self-replicating extrachromosomal DNA elements found within bacterial cells, and bacteriophage (hereafter called phage), viruses that infect bacteria, possess three essential features for cloning vehicles: (1) a replication site, (2) one or more genes that can provide a means of selection, and (3) one or more cloning sites where foreign DNA can be inserted without disrupting an essential function. Phage vectors offer *in vitro* the extremely efficient and reproducible DNA packaging that they demonstrate *in vivo* and have become such powerful cloning vehicles that even teleologic skeptics now confess that viruses might exist for a reason. Engineered from naturally occurring plasmids in the bacterium *Escherichia coli*, artificial plasmids are necessities of recombinant DNA technology. Plasmid copy number is defined as "high" or "low" and can be based on the number of chromosomes per cell or directly on the number of cells. In general, a plasmid that exists in 20 or more copies per cell is considered to be a high copy number plasmid, whereas one that exists in fewer than 20 copies is considered a low copy number plasmid. Some plasmids can continue to replicate after the bacterial cell has ceased to grow and can accumulate up to a thousand copies per host chromosome. After construction, plasmids are introduced into bacteria by transformation.

Like plasmids, phage offer several technical benefits as cloning vehicles. Constructs of the bacteriophage lambda, which constitute the bulk of phage cloning vehicles, multiply rapidly in *E. coli*, a bacterial strain grown easily in the laboratory. In addition, large DNA fragments are maintained stably in lambda vectors, recombinants can be recovered by *in vitro* packaging of the phage DNA, and the recombinant phage can be amplified, screened, and stored easily [see Hohn 1979; Hohn and Murray 1977; Kaiser and Murray 1985]. Several types of selection [see Lech et al. 1989] allow recombinant phage to be distinguished and selected from nonrecombinant phage [see Huynh et al. 1985], and the approximate burst size of 10^6 phage per plaque provides an abundant quantity of the cloned DNA.

Cosmid vectors, designed in name and form to capitalize on the advantages of both plasmids and phage, are plasmids that contain the cohesive sites (the *cos* sites) of lambda phage. The *cos* sites allow the DNA to be cut and packaged into phage heads, and cosmid vectors are particularly useful for cloning large pieces of DNA. This quality is not of primary importance in the construction of cDNA libraries, since cDNA fragments are usually less than 4 kb (kilobases) in length. As a result, cosmid libraries commonly are used to construct genomic libraries, usually from organisms with large (greater than 10^7 bp [base pairs]) genomes.

It's Up to You: Expression and Nonexpression Libraries

When constructing DNA libraries, the scientist must decide whether to construct an "expression" library or a "nonexpression" library. An expression library consists of DNA (usually cDNA) cloned into a vector that allows the inserted DNA to be translated into protein rather than merely replicated as DNA. Conversely, a nonexpression library consists of DNA cloned into a vector that ensures perpetuation by replication but does not generate proteins from the inserted DNA. Expression libraries are used most frequently for isolating clones via the functional properties of the recombinant proteins. Functional screening protocols can rely on immobilized hybridization of individual phage plaques to probes such as antibodies or DNA fragments, or they can utilize enzymatic and/or biological assays to detect secreted recombinant products. There can be caveats to selecting clones from expression libraries, however. A significant potential problem, and one that should be taken seriously, is that the antigenic epitope encoded by an expression clone may react with the antibodies used for screening while not actually encoding the protein of interest. One scientist I know recently spent many months isolating and sequencing seven clones isolated by antibodies from a *Schistosoma mansoni* cDNA library constructed in λgt11, a popular lambda phage expression vector [Huynh et al. 1985; Snyder et al., 1989; Young and Davis 1983a]. All of these clones, each isolated with polyclonal antibodies made against a purified glycolytic enzyme from *S. mansoni*, turned out to encode paramyosin, an abundant cytoskeletal protein previously identified, cloned, and sequenced from *S. mansoni* DNA libraries. Other such stories abound. Unfortunately, they are not to be found in the scientific literature but are usually heard over drinks at a scientific meeting. Such incidents notwithstanding, however, expression libraries provide a powerful means of obtaining specific clones [see Young and Davis 1983b], particularly when cloning a cDNA of unknown sequence that encodes a protein of known function.

Nonexpression DNA libraries are screened with nucleic acid probes. Such probes can be heterologous DNAs or RNAs (e.g., a rat cDNA or cDNA transcript to screen a human library), a homologous probe (e.g., a truncated cDNA or RNA transcript to screen for a longer cDNA), or a synthetic oligomer (e.g., a DNA or RNA oligonucleotide constructed with knowledge of the partial or complete protein sequence).

Making Mountains out of Molehills: Polymerase Chain Reaction

Otherwise known as PCR, the polymerase chain reaction is a popular and powerful technique for enriching specific DNA sequences. Based on repeated cycles of annealing, synthesis, and denaturation, DNA sequences bounded by short, sequence-specific primers can be amplified by a factor of 10^9 or 10^{10} over a period of a few hours in the presence of a heat-stabile DNA polymerase from *Thermus aquaticus*, a heat-tolerant marine bacterium [Innis et al. 1990; Mullis and Faloona 1989; Saiki et al. 1985, 1988]. DNA sequences amplified by PCR can be cloned and sequenced; alternatively, they can be sequenced directly via the dideoxy chain-termination technique [Sanger et al. 1977], chemical sequencing [Maxam and Gilbert 1980], or "genomic sequencing" [Church and Gilbert 1984; Ohara et al. 1989]. Sequences can be amplified by PCR from genomic DNA, cDNA, total or poly(A)+ RNA, or from DNA cloned in a vector (e.g., complete DNA libraries). The polymerase chain reaction offers an alternative to screening DNA libraries for specific sequences because it accomplishes *in vitro* what the phage or plasmid does *in vivo*, i.e., DNA amplification. Indeed, the amplification factor is large enough to permit radioactive detection (e.g., by Southern analysis) of a single-copy target sequence in the presence of a 10^{13}-fold excess of nontarget DNA.

As with many things too good to be true, however, there are some disadvantages to PCR cloning. One drawback is that this technology does not offer the ability to clone very large DNA fragments. At present, accurate, reproducible PCR amplification of DNA fragments greater than 5 kb in size is often difficult and unpredictable; thus, library construction as a means to clone large DNA fragments is unlikely to be replaced by this technique. Another disadvantage is that *Taq* DNA polymerase, which has no proofreading functions, produces single-base substitution errors once every 9000 nucleotides and produces frameshift errors once every 41,000 nucleotides [Tindall and Kunkel 1988]. This error rate is approximately four times greater than that found with the Klenow fragment of *E. coli* DNA polymerase [Sambrook, Fritsch, and Maniatis, 1989]. The implications of each limitation are straightforward: standard cloning techniques are probably more useful than PCR amplification for direct cloning of cDNAs or genes larger than 5 kb, and several independent clones of the same target sequence should be obtained and sequenced from independent amplification reactions if DNA sequence *per se* is crucial. A third limitation of PCR is that two regions of DNA sequence must be known in order to construct two primers that will hybridize to opposite strands of the template DNA with their 3′ ends facing each other. If only a single, relatively short (less than 50 bp) region of DNA sequence is known within the target clone, then cloning by hybridization with a nucleic acid oligomer is in order.

The short length of PCR-generated products and the rather high misincorporation rate of the reaction notwithstanding, the PCR technique can be used to generate enough cDNA to construct a library from minute amounts of mRNA. Once first-strand cDNA has been synthesized with reverse transcriptase, using oligo (dT) or short primers of random sequence as primers, PCR can be used to amplify these strands if a known DNA sequence is added to the 3′ terminus of the first strand [Belavsky, Vinogradova, and Rajewsky, 1989]. In practical terms, this means that a cDNA library can be constructed from the mRNA of as few as one or two mammalian cells. Obviously, such libraries would not be likely to contain cDNAs greater than approximately 5 kb in length, and it would be crucial to compare DNA sequences from several independent clones before assuming any one of them to be correct.

Like many techniques, PCR sometimes requires just the right "touch" to work. Certain variables, including primer length, primer sequence, primer concentration, annealing temperature, extension temperature, and salt concentration should be explored when using PCR to search for and amplify a specific DNA sequence. For example, cloning a gene from *Plasmodium falciparum* (the human malaria parasite) that is related to the mammalian multiple drug resistance (mdr) gene and is perhaps involved with drug resistance in certain malarial strains [Wilson et al. 1989] was facilitated by beginner's luck and the "right touch." Having devised two DNA primers based on conserved protein sequences in the human and mouse mdr genes and having synthesized relevant the primers based on the codon bias of *Plasmodium* [Saul and Battistutta, 1988], the workers at a nearby university repeatedly tried various conditions to amplify a homologous sequence from *P. falciparum* DNA. Finding that none of these conditions worked, they sidelined the project but stored the primers in the freezer. Several months later a rotating graduate student entered the laboratory and was given the primers and *P. falciparum* DNA as a negative control for the PCR reaction. One try and one day later, the student visualized two specifically amplified bands of DNA on an agarose gel, and cloning a *P. falciparum* gene related to the mammalian mdr gene had begun.

MAKING YOUR CLONES TALK: EXPRESSION OF RECOMBINANT PROTEINS

Expression Vectors

Once a protein-encoding DNA sequence has been cloned, the next step, often nontrivial, is to produce or "express" the encoded protein—preferably in abundant amounts. This is accomplished by inserting the cloned DNA fragment into an efficient expression vector. Such vectors usually are different from those used to construct expression libraries and often must be engineered individually for the recombinant product at hand. There are many different types of expression vectors, but the fundamental choice is between prokaryotic expression vectors and eukaryotic expression vectors. Prokaryotes, single-celled organisms without nuclei or other organelles, have a comparatively straightforward means of producing proteins from messenger RNAs, whereas eukaryotes, single-celled or multicellular organisms with nuclei and other cellular organelles, possess a more intricate system of protein expression. Theoretically speaking, it is possible to express any DNA sequence, regardless of its origin, in either a prokaryotic or eukaryotic vector. In practical terms, however, not all eukaryotic proteins can be expressed efficiently in prokaryotes. Eukaryotic proteins often have specific folding requirements that cannot be met in bacterial expression vectors. In addition, posttranslational modifications such as glycosylation, phosphorylation, gamma-carboxylation, beta-hydroxylation, disulfide bond formation, and/or proteolytic cleavage are processes common in eukaryotes that do not occur in bacteria [Harris 1983; Sambrook et al. 1989]. Despite this, prokaryotic expression vectors remain extremely popular and useful (they usually are far easier to manipulate than eukaryotic ones), and good prokaryotic vectors can express a foreign protein as 1 to 30% of the total cellular protein.

Facility of manipulation is outweighed, however, by the need for recombinant proteins to be biologically active. Most recombinant products targeted for the marketplace derive from higher eukaryotes such as humans, animals, or plants. This fact immediately focuses the issue mentioned above: prokaryotic expression vectors may be simpler to use than eukaryotic ones, but they also may be untenable for expression of biologically active eukaryotic proteins. Unfortunately, the ideal expression system for DNA—one that can mediate expression of cloned DNA from any species with high efficiency and simultaneously can ensure biological activity of the protein product—does not exist. Indeed, if there is any "rule" of expression constructs, it is that each protein product will bear a unique profile and will present a set of previously unencountered problems to the experimenter. Although general guidelines can be used, the expression system for any given protein will demand optimization on an individual basis. Three eukaryotic expression systems, each proven to produce high yields of functional recombinant protein products from eukaryotic sources, will be discussed briefly here: insect cells, yeast, and mammalian cells. The essential characteristics of achievable expression level and biological activity will be considered for each.

SPECIFIC EUKARYOTIC EXPRESSION CLONING VEHICLES

Insect Cells

A baculovirus expression vector system has been devised for the expression of foreign genes in Lepidopteran insect cells [Doerfler 1986; Granados and Federici 1986; Smith et al. 1983]. It is a helper-independent recombinant virus vector that has been used successfully to express genes from plants, mammals, bacteria, and viruses, often producing proteins that are immunogenically, antigenically, and functionally similar to the authentic gene products [Luckow and Summers 1988a]. An efficient expression system, baculoviral constructs have yielded fusion and nonfusion proteins at levels ranging from 1 to 500 mg per liter [Summers and Smith 1987]. The powerful promoter, both highly expressed and tightly regulated, in this expression system is a modified version of the *Autographa californica* nuclear polyhedrosis virus polyhedron promoter [Luckow and Summers 1988b]. The factors controlling expression efficiency in baculoviral constructs have not been well defined in molecular terms; it is clear, however, that the growth stage of the insect cells, the ratio of viral cells to insect cells, and the quality of the medium and serum components are crucial to achieve optimal expression [Summers and Smith 1987].

Yeast

Proteins secreted from eukaryotic cells usually possess a signal sequence—a hydrophobic amino terminus extension—that directs translocation of the protein into the endoplasmic reticulum. When using yeast as an expression vector, either the mammalian signal sequence or a yeast signal sequence can be used. *Saccharomyces cerevisiae* offers several advantages for large-scale cultures. It grows rapidly, doubling its population approximately every 90 minutes when grown in rich medium with a glucose carbon source, it has a well-characterized fermentation technology, and it has extensively characterized genetics [see Broach 1983]. Yeast usually produces heterologous proteins at lower levels relative to prokaryotic expression systems [Marino 1989], but it often produces recombinant proteins with biological activities that so far elude the prokaryotic constructs [see Kingsman et al. 1987].

Mammalian Vectors

Mammalian expression systems produce low yields of proteins relative to the other expression systems discussed here. Countering this disadvantage, however, is the fact that superior biological activity often is obtained with mammalian systems. Providing the natural environment of the eukaryotic cell for eukaryotic proteins has permitted significant advantages in the past few years. Mammalian expression systems can be divided into two types: (1) systems that use transfected DNA and (2) systems that use viral constructs.

Mammalian expression systems that use transfected DNA. Transfection-derived expression systems can be either transient or stable. Typical transient mammalian expression vectors contain eukaryotic control sequences attached to bacterial plasmids. *Cis*-acting control elements such as promoters and enhancers derived from animal viruses are cloned into plasmid DNA such that they mediate expression of the inserted DNA fragment. The fundamental expression "cassette" consists of a promoter to regulate RNA transcription, the inserted coding sequence to encode the eukaryotic protein, signals for transcription termination, and signals for polyadenylation. Optional additions to this cassette can include enhancer sequences and intervening sequences bordered by correct splice donor and acceptor sequences. The crucial prokaryotic elements that must be incorporated into the vector construct include an origin of replication that functions in *E. coli*, a selectable gene for antibiotic resistance, and unique restriction sites for insertion of the eukaryotic DNA. Other DNA elements can be added to facilitate identification of recombinant cells or to increase the level of expression of the cloned genes [see Sambrook and Gething 1988; Sambrook et al. 1989, section 16.8 ff].

Although transient expression systems commonly die within a few days after recombinant protein expression begins, they are easier to establish than stable expression systems and are extremely useful for several purposes. For example, they offer a means to screen secreted products from cDNA libraries, they offer a rapid way to verify that a selected cDNA clone encodes a protein with biological activity, and they provide a method to assay site-directed mutants for interesting properties. Once a transient system indicates that a recombinant protein has properties of interest, then a stable system (i.e., a permanent cell line) can be established [see Kaufman et al. 1985].

Mammalian expression systems that use viral constructs. Several eukaryotic viruses contain regulatory elements that can be placed in plasmid expression vectors for expression of eukaryotic proteins. Of such viruses, elements from simian virus 40 (SV40) [see Cone et al. 1987], bovine papillomavirus (BPV) [see Hsiung et al. 1984], and vaccinia virus [see Fuerst et al. 1987; Paoletti et al. 1984] are commonly used.

SECOND-GENERATION FORMS OF RECOMBINANT PROTEINS

Having made a recombinant protein that can be marketed and used as a drug, the next goal often is to make it again and make it better. For example, once recombinant tissue plasminogen activator (tPA) became available for use in humans, scientists immediately began to focus on second-generation tPA. This second-generation molecule—a modified form of tPA—would have the same dramatic thrombolytic effect on blood clots but would have a longer biological half-life than first-generation tPA [see Topol 1989]; in addition, it would be produced in a cost-effective manner [see Ratner 1990]. Other desirable attributes that might warrant refinement include improving drug delivery to target cells and tissues (e.g., by altering solubility properties or increasing protein affinity for target receptors) and decreasing or eliminating side effects of the drug [see McPherson and Livingston 1989; Sjodahl 1989]. Current recombinant products such as tPA, insulin, erythropoietin, human growth hormone, factor VIII, epidermal growth factor, and platelet-derived growth factor are likely to enjoy large markets and sales in the future; however, production of second-, third-, and fourth-generation constructs of these proteins—presumably better and cheaper than their forerunners—is probably only a matter of time. Clearly, the same forecast can be made for enzymes crucial to industrial processes.

BIOMOLECULAR TAILORING: PROTEIN ENGINEERING

Protein engineering is a powerful means of altering structural and functional properties of proteins. Simply speaking, the underlying concept of protein engineering is to identify regions of a protein critical for its function (e.g., residues involved in the binding of the protein to a receptor), to predict changes that might augment functionality of these crucial regions, and to incorporate these changes into a new version of the protein via recombinant DNA technology. Such changes, fundamental to an increased understanding of molecular recognition and biological catalysis, are often implemented by site-directed mutagenesis of the cloned DNA fragment encoding the protein. In essence, this type of mutagenesis involves altering specific nucleotides in the cloned DNA sequence in order to vary the amino acid sequence of the protein it encodes. Many physical characteristics of a protein can be addressed in this fashion. For example, protein engineering can address the importance of surface electrostatic charge in enzyme pockets, the functional significance of particular amino acid residues around an active site, or the relative impact of regions of the protein far removed from the active site [see Russell and Fersht 1987].

Changes in protein sequence induced by site-directed mutagenesis of DNA originally relied on the fact that amino acids in protein derive from nucleotide triplets—or codons—in the DNA and mRNA sequences. Because messenger RNA is synthesized faithfully from DNA within a cell, one only need alter individual nucleotides within a DNA sequence to induce corresponding changes in mRNA. For example, Table 7-1 shows that the mRNA triplet UCC encodes serine (Ser), one of the twenty natural amino acids. To change a UCC-encoded Ser residue in a recombinant pro-

TABLE 7-1 TABLE OF CODONS

Amino Acid	Abbreviation	Triplet(s)
Alanine	Ala, A	GCU GCA GCC GCG
Arginine	Arg, R	CGU CGG CGC AGA CGA AGG
Asparagine	Asn, N	AAU AAC
Aspartic acid	Asp, D	GAU GAC
Cysteine	Cys, C	UGU UGC
Glutamine	Gln, Q	CAA CAG
Glutamic acid	Glu, E	GAA GAG
Glycine	Gly, G	GGU GGA GGC GGG
Histidine	His, H	CAU CAC
Isoleucine	Ile, I	AUU AUA AUC
Leucine	Leu, L	CUU CUG CUC UUA CUA UUG
Lysine	Lys, K	AAA AAG
Methionine	Met, M	AUG
Phenylalanine	Phe, F	UUU **UUC**
Proline	Pro, P	CCU CCA CCC CCG
Serine	Ser, S	UCU UCG **UCC** AGU UCA AGC
Threonine	Thr, T	ACU ACA ACC ACG
Tryptophan	Trp, W	UGG
Tyrosine	Tyr, Y	UAU UAC
Valine	Val, V	GUU GUA GUC GUG

tein to phenylalanine (Phe) would require altering the first cytidine of the UCC triplet to a uridine so that the new codon, UUC, would encode Phe. If this change were made in the sixth triplet within a cloned DNA sequence, for example, then the sixth amino acid of the resultant protein would be Phe instead of Ser.

A recent biosynthetic method now makes it possible to incorporate unnatural amino acids into proteins [Noren et al. 1989]. By increasing the repertoire of amino acid residues available in protein constructs, novel electronic and steric properties can be integrated into protein structures. This ability to refine and extend protein design should provide new insights into protein structure as it relates to function and could provide avenues to create novel antibodies, enzymes, and receptors with properties unattainable with the natural amino acids. In addition to site-directed mutagenesis and protein constructs with novel amino acids, chimeric molecules provide another way of altering therapeutic molecules. Immunotoxins—molecules that combine the specificity of monoclonal antibodies with the destructive qualities of toxin molecules—are examples of such present-day "magic bullets" [see Blattler et al. 1989; Frankel 1988; Pastan et al. 1986]. These chimeras are seen by many as powerful drugs of the future to combat cancer, the idea being to direct a powerful toxin to cancerous cells via the monoclonal antibody portion of the molecule.

What does the future hold for protein engineering as a means of drug design? This question, or perhaps especially its answer, is of great interest to both scientists and financial investors alike. Because the limits of protein engineering technology will likely result from our predictive limitations rather than our experimental ones (i.e., predicting the best ways to alter a protein will be much more difficult than actually altering it), this question is tough to answer. One intriguing possibility is that engineered proteins will be chimeras of protein/nonprotein. For example, target specific proteins covalently attached to synthetic drugs might prove beneficial if solubility and delivery problems could be overcome. Similarly, proteins designed to target cell nuclei might provide a way to deliver specific DNA constructs to a cell nucleus known to be deficient or defective in a certain gene. Finally, since certain RNA constructs can function as enzymes [Cech 1989; Kruger et al. 1982], one can envision protein/RNA chimeras in which a targetting protein moiety delivers an RNA enzyme to its site of action.

Is there an alternative to engineered proteins as significant drugs of the future? Clearly, the path of major biotechnology companies has been forged by marketing expressed proteins. As this chapter has described, the strategy of isolating, sequencing, and expressing cDNAs that encode medically relevant proteins is a powerful one now rooted in durable yet still-emerging technology. Even assuming that problems of physiologic ac-

TABLE 7-2 KEY TECHNICAL INNOVATIONS IN BIOTECHNOLOGY (DNA MANIPULATIONS)

Innovation	Date	Reference	Impact
Site-specific DNA restriction enzymes discovered	1969	Smith & Wilcox, *J. Mol. Biol.* 51:379 Arber & Linn, *Annu. Rev. Biochem.* 38:46	DNA molecules can be cleaved at defined sites
Reverse transcriptase discovered	1970	Baltimore, *Nature* 226:1209 Temin & Mizutani, *Nature* 226:1211	RNA can serve as a coding template for DNA synthesis; the sequences of mRNAs can be produced and studied as DNAs
Restriction enzymes used to map a DNA molecule	1971	Danna & Nathans, *Proc. Natl. Acad. Sci. U.S.A.* 68:2913	DNA molecules can be charted and distinguished from each other
Splicing of DNA molecules from different sources to create functional plasmid	1973	Cohen et al., *Proc. Natl. Acad. Sci. U.S.A.* 70:3240	DNAs from different species can be used to encode function
Chemically synthesized gene expressed in *E. coli*	1977	Itakura et al., *Science* 198:1056	DNA made *in vitro* can be expressed as though it had been made *in vivo*. Genes from higher organisms can be manipulated in lower organisms.
DNA sequencing technology described	1977	Maxam & Gilbert, *Proc. Natl. Acad. Sci. U.S.A.* 74:560 Sanger et al., *Proc. Natl. Acad. Sci. U.S.A.* 74:5463	Every nucleotide within a DNA molecule can be identified according to its position within the molecule

TABLE 7-2 ***(continued)***

Innovation	Date	Reference	Impact
Automation of DNA synthesis	1983	Hunkapiller et al., *Nature* 310:105	DNA molecules can be constructed rapidly and inexpensively
Polymerase chain reaction (PCR) technology developed	1985	Saiki et al., *Science* 230:1350	Specific DNA sequences can be identified, amplified, and isolated from minute quantities of DNA

TABLE 7-3 KEY TERMS IN BIOTECHNOLOGY

Term	Definition
Amino acids	Building blocks of proteins
cDNA	DNA synthesized from an RNA template; usually contains no introns
Clone	(1) set of cells or molecules derived from a single parent; (2) specific piece of DNA in a vector
Codon	Nucleotide triplets in DNA and mRNA; encode specific amino acids in protein synthesis
DNA	Deoxyribonucleic acid; comprises genes; composed of adenine (A), cytosine (C), guanine (G), and thymine (T) attached to a deoxyribose-phosphate backbone
Eukaryote	Nucleated organism
Exon	Portion of a gene that is transcribed into RNA; translated into protein if the RNA is mRNA
Expression library	Library that permits cloned DNA to be translated into proteins
Gene	Stretch of DNA that encodes one or more RNA transcripts; the RNA, as mRNA, is usually translated into proteins
Intron	Portion of a gene that is transcribed into RNA; not translated into protein if the RNA is mRNA; removed by splicing
Library	Set of cloned DNA or cDNA fragments
Messenger RNA	mRNA; RNA that encodes protein
Nucleotides	Building blocks of DNA and RNA

TABLE 7-3 *(continued)*

Term	Definition
Nonexpression library	Library that does not permit cloned DNA to be translated into proteins
PCR	Polymerase chain reaction; method of amplifying specific DNA molecules *in vitro*
Phage	Bacteriophage; bacterial virus; useful cloning vector
Plasmid	Common cloning vector; circular, autonomously replicating DNA elements
Prokaryote	Anucleated organism
Promoter	Region of DNA associated with a gene and involved in the binding of RNA polymerase for transcription initiation
Restriction enzyme	An enzyme that cleaves double-stranded DNA at or near a specific sequence recognition site; nomenclature indicates the organism from which the enzyme was purified and the number of restriction enzymes preceding it from that organism, e.g., *Eco*RI was the first restriction enzyme purified from *Escherichia coli*
RNA	Ribonucleic acid; composed of adenine (A), cytosine (C), guanine (G), and uracil (U) attached to a ribose-phosphate backbone
Transformation	Stable introduction of foreign DNA into a cell or organism
Vector	Constructed DNA vehicle for manipulating DNA

tivity/delivery and difficulties in manufacturing can be contained by improved generations of recombinant DNA drugs in the future, one might ask if there are unlimited proteins of medical relevance—suited to the constraints of biotechnology—remaining to be discovered. Proponents of synthetic drugs—e.g., pharmaceutical companies—will argue that the answer to this question is no, that marketable biotechnology drugs are limited in scope, leaving the greatest future strides in drug design and development to be made by organic chemists in conjunction with molecular modellers and molecular biologists.

Proponents of biotechnology will argue that the answer to this question is yes, that the horizons open to recombinant drugs are just emerging. Only time and hard work will demonstrate how best to approach the challenge of designing hope in the vial.

HOW TO INTERPRET VECTOR CONSTRUCTS

General Concept

Maps of vector constructs are represented circularly or linearly. They provide key pieces of information about the vector such as sites where foreign DNA fragments can be cloned, selectable markers that can be used to identify cells transformed with the vector, selections that will distinguish the parent vector from the recombinant vector, and sites of transcription initiation. For example, let us consider the maps of the small *E. coli* plasmid cloning vector pUC19 [Yanisch-Perron et al. 1985] and the larger *E. coli* lambda cloning vector λgt11 [Young and Davis 1983a].

Salient Features of the pUC19 Map

Restriction sites. The restriction enzymes listed on the map of pUC19 (Fig. 7-1) will cleave the vector at the positions indicated by numbers; the numbers themselves refer to specific nucleotides within the vector sequence. For example, the restriction endonuclease *Aat*II cleaves at nucleotide 2617 in between the *Nde*I site at 183 and the *Xmn*I site at 2294.

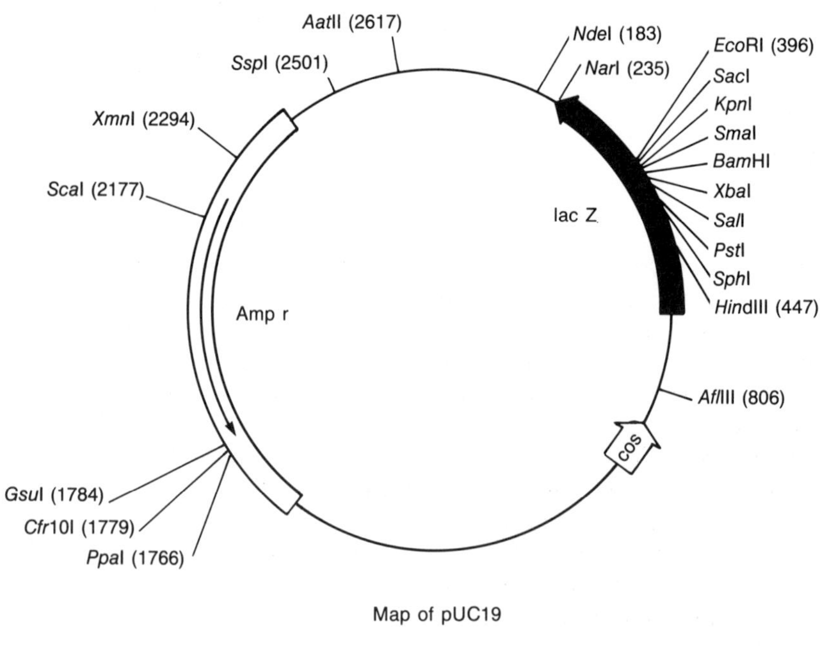

FIGURE 7-1 Map of pUC19.

The entire pUC19 plasmid is 2686 bp in length and is represented as a circle since the plasmid exists and replicates as a circle *in vivo*.

Amp^r. Gene encoding resistance to the antibiotic ampicillin; cells transformed with pUC19 will grow on ampicillin. The plasmid need not have DNA inserted in the polylinker to confer ampicillin resistance. The arrow indicates the direction in which the ampicillin gene is transcribed.

lacZ. Gene involved in lactose metabolism that permits chromogenic selection of cells transformed with plasmid constructs containing DNA inserted in the polylinker.

***Eco*RI (396)–*Hin*dIII (447).** Polylinker cloning sites that can be used as insertion sites for DNA fragments. These cloning sites are not present anywhere else in the pUC19 molecule. Note that DNA inserted in these sites will disrupt the integrity of the lacZ gene; this will be reflected in the chromogenic color selection mentioned above.

Ori. The origin of DNA replication; the arrow indicates the direction in which replication proceeds.

Conclusions to Be Drawn from This Map

1. DNA fragments can be cloned into sites generated by *Eco*RI, *Sac*I, *Kpn*I, *Sma*I, *Bam*HI, *Xba*I, *Sal*I, *Pst*I, *Sph*I, or *Hin*dIII.

2. Cells transformed with this plasmid can be grown in liquid media or on solid media that contain the antibiotic ampicillin.

3. Cells transformed by pUC19 constructs with DNA inserted in the polylinker can be identified by growth in the presence of ampicillin and can be identified chromogenically if IPTG (isopropyl-β-thiogalacto-pyranoside) and x-gal (5-bromo-4-chloro-3-indolyl-β-D-galactoside) are present.

4. The vector is approximately 2600 bp in length.

5. The restriction enzymes *Nde*I, *Nar*I, *Afl*III, *Ppa*I, *Cfr10*I, *Gsu*I, *Sca*I, *Xmn*I, *Ssp*I, and *Aat*II, each cut once within the plasmid.

Salient Features of the λgt11 Map (Fig. 7-2)

Restriction sites. As in the pUC19 map, the restriction enzymes listed here will cleave the vector at the positions indicated by numbers, and the numbers themselves refer to specific nucleotides within the vector se-

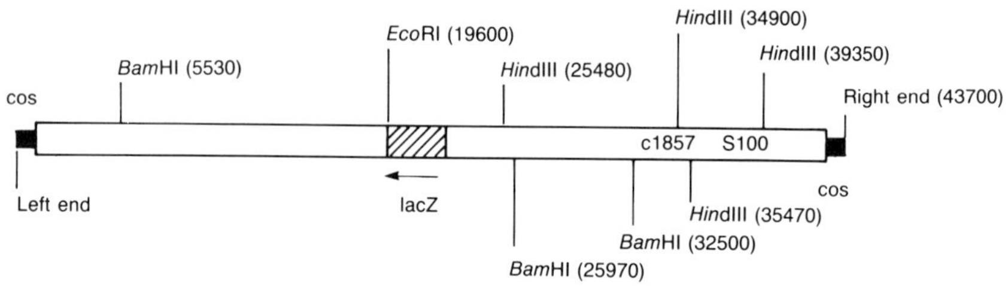

FIGURE 7-2 Map of λgt11.

quence. Thus, the restriction endonuclease *Eco*RI cleaves at nucleotide 19,600 in between the *Bam*HI site at 5530 and the *Hin*dIII site at 25,480. The entire λgt11 molecule is approximately 46 kb pairs in length. Although usually represented in a linear fashion, lambda has "cohesive" ends, which, when base paired, allow the molecule to exist as a circle.

cos. Indicates the cohesive left and right ends.

lacZ. Gene involved in lactose metabolism that permits chromogenic selection of cells transformed with plasmid constructs containing DNA inserted in the polylinker. The arrow indicates the direction of transcription.

Ori. The origin of DNA replication; the arrow indicates the direction in which replication proceeds.

cI857. A gene that encodes a temperature-sensitive repressor molecule; the repressor is involved in the processes of lysogeny (cells are infected with the bacteriophage but do not burst) and lysis (infected cells burst).

S100. A mutation in the S lysis gene.

Conclusions to Be Drawn from This Map

1. DNA fragments can be cloned into the *Eco*RI site located in the 3′ end of the lacZ gene.

2. Cells infected with recombinant λgt11 constructs can be identified chromogenically if IPTG (isopropyl-β-thiogalacto-pyranoside) and x-gal (5-bromo-4-chloro-3-indolyl-β-D-galactoside) are present.

3. The vector is 43,700 bp in length.

4. Infected cells can be induced to lyse (thereby releasing great numbers of phage) by increasing the temperature of the culture medium.

REFERENCES

Belavsky, A., Vinogradova, T., and Rajewsky, K. 1989. PCR-based cDNA library construction: general cDNA libraries at the level of a few cells. *Nucleic Acids Res.* 17:2919.

Blattler, W. A., Lambert, J. M., and Goldmacher, V. S. 1989. Realizing the full potential of immotoxins. *Cancer Cells.* 1:50.

Broach, J. R. 1983. Construction of high copy yeast vectors using 2 μm circle sequences. *Methods Enzymol.* 101:307.

Cech, T. R. 1989. RNA as an enzyme. *Biochem. Int.* 18:7.

Church, G. M., and Gilbert, W. 1984. Genomic sequencing. *Proc. Natl. Acad. Sci. U.S.A.* 81:1991-1995.

Cone, R. D., Weber-Benarous, A., Baorto, D., and Mulligan, R. C. 1987. Regulated expression of a complete human beta-globin gene encoded by a transmissible retrovirus vector. *Mol. Cell. Biol.* 7:887.

Doerfler, W. 1986. Expression of the *Autographa californica* nuclear polyhedrosis virus genome in insect cells: homologous viral and heterologous vertebrate genes—the baculovirus vector system. In *The molecular biology of baculoviruses, current topics in microbiology and immunology,* Vol. 131, edited by W. Doerfler and P. Bohm, 51. New York: Springer-Verlag.

Fuerst, T. R., Earl, P. L., and Moss, B. 1987. Use of a hybrid vaccinia virus-T7 RNA polymerase system for expression of target genes. *Mol. Cell. Biol.* 7:2538.

Granados, R. R., and Federici, B. A. 1986. *The biology of baculoviruses, vol. 1: Biological properties and molecular biology.* Boca Raton, Fla: CRC Press.

Harris T. J. R. 1983. Expression of eukaryotic genes in *E. coli*. In *Genetic engineering,* edited by R. Williamson, 127-185. New York: Academic Press.

Hohn, B. 1979. In vitro packing of λ and cosmid DNA. *Methods Enzymol.* 68:299.

Hohn, B., and Murray K. 1977. Packaging recombinant DNA molecules into bacteriophage particles *in vitro*. *Proc. Natl. Acad. Sci. U.S.A.* 74:3259.

Hsiung, N., Fitts, R., Wilson, S., Milne, A., and Hamer, D. 1984. Efficient production of hepatitis B surface antigen using a bovine papilloma virus-metallothionein vector. *J. Mol. Appl. Genet.* 2:497.

Huynh, T. V., Young, R. A., and Davis, R. W. 1985. Constructing and screening cDNA libraries in λgt10 and λgt11. In *DNA cloning: A practical approach,* vol. 1, edited by D. M. Glover, 49-78. Washington, D.C.: IRL Press.

Innis, M. A., Gelfand, D. H., Sninsky, J. J., and White, T. J., eds. 1990. *PCR protocols: A guide to methods and applications.* New York: Academic Press.

Kaiser, K., and Murray, N. E. 1985. The use of phage lambda replacement vectors in the construction of representative genomic DNA libraries. In *DNA cloning: A practical approach,* vol. 1. edited by D. M. Glover, 1. Washington, D.C.: IRL Press.

Kaufman, R. J., Wasley, L. C., Spiliotes, A. J., Grossels, S. O., Latt, S. A., Larsen, G. R., and Kay, R. M. 1985. Coamplification and coexpression of human tissue-type plasimogen activator and murine dihydrofolate eductase sequences in Chinese hamster ovary cells. *Mol. Cell. Biol.* 5:1750.

Kingsman, S. M., Kingsman, A. J., and Mellor, J. 1987. The production of mammalian proteins in *Saccharomyces cerevisiae. Trends Biotechnol.* 5:53.

Kruger, K., Grabowski, P. J., Zaug, A. J., Sands, J., Gottschling, D. E., and Cech, T. R. 1982. Self-splicing RNA: autoexcision and autocyclization of the ribosomal RNA intervening sequence of tetrahymena. *Cell* 31:147.

Lech, K., Brent, R., and Irwin, N. 1989. Lambda as a cloning vector. In *Current protocols in molecular biology,* edited by F. M. Ausubel, R. Brent, R. E. Kingston, D. D. Moore, J. G. Seidmen, J. A. Smith, and K. Struhl, 1.10.1-1.10.11. New York: Wiley.

Luckow, V. A., and Summers, M. D. 1988a. Trends in the development of baculovirus expression vectors. Biotechnology 6:47.

Luckow, V. A., and Summers, M. D. 1988b. Signals important for high-level expression of foreign genes in *Autographa californica* nuclear polyhedrosis virus expression vectors. *Virology* 167:56.

Marino, M. H. 1989. Expression systems for heterologous protein production. *Biopharm.* 10:18.

Maxam, A. M., and Gilbert, W. 1980. Sequencing end-labeled DNA with base-specific chemical cleavages. *Methods Enzymol.* 65:499.

McPherson, J. M., and Livingston, D. J. 1989. Protein engineering: new approaches to improved therapeutic proteins, Part 1. *Pharm. Technol.* 13:22.

Mullis, K. B., and Faloona, F. A. 1989. Specific synthesis of DNA *in vitro* via a polymerase-catalyzed chain reaction. In *Recombinant DNA methodology*, edited by R. Wu, L. Grossman, and K. Moldave, 189. New York: Academic Press.

Noren, C. J., Anthony-Cahill, S. J., Griffith, M. C., and Schultz, P. G. 1989. A general method for site-specific incorporation of unnatural amino acids into proteins. *Science* 244:182.

Ohara, O., Dorit, R. L., and Gilbert, W. 1989. One-sided polymerase chain reaction: the amplification of cDNA. *Proc. Natl. Acad. Sci. U.S.A.* 86:5673.

Paoletti, E., Lipinskas, B. R., Samsonoff, C., Mercer, S., and Panicali, D. 1984. Construction of live vaccines using genetically engineered poxviruses: Biological activity of vaccinia virus recombinants expressing the hepatitis B virus surface antigen and the herpes simplex virus glycoprotein D. *Proc. Natl. Acad. Sci. U.S.A.* 81:193.

Ratner, M. 1990. T-PA trials, tribulations, and litigation. Biotechnology 8:385.

Russell, A. J., and Fersht, A. R. 1987. Rational modification of enzyme catalysis by engineering surface charge. *Nature* 328:496.

Saiki, R. K., Scharf, S., Faloona, F., et al. 1985. Enzymatic amplification of beta-globin genomic sequences and restriction site analysis for diagnosis of sickle cell anemia. *Science* 230:1350.

Saiki, R. K., Gelfand, D. H., Stoffel, S., Scharf, S. J., Higuchi, R. et al. 1988. Primer-directed enzymatic amplification of DNA with a thermostable DNA polymerase. *Science* 239:487.

Sambrook, J., Fritsch, E. F., and Maniatis, T. 1989. Expression of cloned genes in cultured mammalian cells. p. 16.3. *Molecular cloning: A laboratory manual.* Cold Spring Harbor, N.Y.: Cold Spring Harbor Laboratory Press.

Sambrook, J., and Gething, M.-J. 1988. Vectors for high level expression of proteins in mammalian cells. *Bethesda Res. Lab. Life Technol. Focus* 10:1.

Sanger, F., Nicklen, S., and Coulson, A. R. 1977. DNA sequencing with chain terminating inhibitors. *Proc. Natl. Acad. Sci. U.S.A.* 74:5463.

Saul, A., and Battistutta, D. 1988. Codon usage in *Plasmodium falciparum. Mol. Biochem. Parasitol.* 27:35.

Sjodahl, J. 1989. Protein engineering in Japan: towards atomic biology. *Tib Tech.* 7:144.

Smith, G. E., Summers, M. D., and Fraser, M. J. 1983. Production of human beta interferon in insect cells infected with a baculovirus expression vector. Mol. Cell. Biol. 3:2156.

Snyder, M., Elledge, S., Sweetser, D., Young, R. A., and Davis, R. W. 1989. Lambda gt11: gene isolation with antibody probes and other applications. In *Recombinant DNA methodology*, edited by R. Wu, L. Grossman, and K. Moldave, 309. New York: Academic Press.

Summers, M. D., and Smith, G. E. 1987. A Manual of Methods for Baculovirus Vectors and Insect Cell Culture Procedures. *Texas Agric. Exp. Stn. Bull.* 1555:5.

Tindall, K. R., and Kunkel, T. A. 1988. Fidelity of DNA synthesis by the *Thermus aquaticus* DNA polymerase. *Biochemistry* 27:6008.

Topol, E. J. 1989. Recombinant tissue plasminogen activator: implications in therapy. *Semin. Hematol.* 26(3 Suppl. 3):25.

Wilson, C. M., Serrano, A. E., Wasley, A., Bogenschutz, M. P., Shankar, A. H., and Wirth, D. F. 1989. Amplification of a gene related to mammalian mdr genes in drug-resistant *Plasmodium falciparum. Science,* 244:1184.

Yanish-Perron, C., Vieira, J. 1985. Improved M13 phage cloning vectors and host strains: nucleotide sequences of the M13mp18 and pUC19 vectors. Gene 33:103.

Young, R. A., and Davis, R. W. 1983a. Efficient isolation of genes using antibody probes. Proc. Natl. Acad. Sci. U.S.A. 80:1194.

Young, R. A., and Davis, R. W. 1983b. Yeast RNA polymerase II genes: isolation with antibody probes. *Science* 222:778.

CHAPTER 8

SCIENCE, TECHNOLOGY, AND POLITICS: REAL FACTORS IN SELECTING A MAMMALIAN CELL BIOREACTOR

Randall G. Rupp and Melvin S. Oka

During the past few years, therapeutic proteins have been produced from cultured mammalian cells. These proteins are generally very high value-added products which are usually administered in relatively small doses. For these proteins to remain in the highly competitive and rapidly changing market (and not to be replaced with other, less expensive treatments) their costs must be reduced. The final product price is tightly linked to the cost-to-produce (CTP) these proteins. The most fundamental way to decrease the CTP is to increase the culture yields while decreasing the time in culture and the quantity of raw materials consumed. This chapter will discuss some of the considerations that we think are most important in developing strategies to address this problem and, based on these considerations, what the most successful manufacturing strategy is likely to be at this point. When the goal is to obtain product, either for research or production, a conventional stirred tank, run in a batch mode, with adequate controls on temperature, pH, agitation and oxygenation, is the "best" and most *expeditious* method of culture. It is necessary to discuss briefly and to review some of the major culture strategies employed and some general details on bioreactor operations before entering into a discussion concerning our stated position. Space prevents an exhaustive discussion or even description of all the variations of reactors that have been developed to culture cells. For the purpose of this chapter, we will consider two basic *operational* strategies (e.g., batch and perfusion) illustrated in Figure 8-1 while only touching on actual reactor design (e.g., hardware) as shown in Figure 8-2. For a very readable review of this area, which covers reactors in more detail, the reader is referred to a recent review [Griffiths 1988]. We also wish to apologize if your favorite reactor is not mentioned.

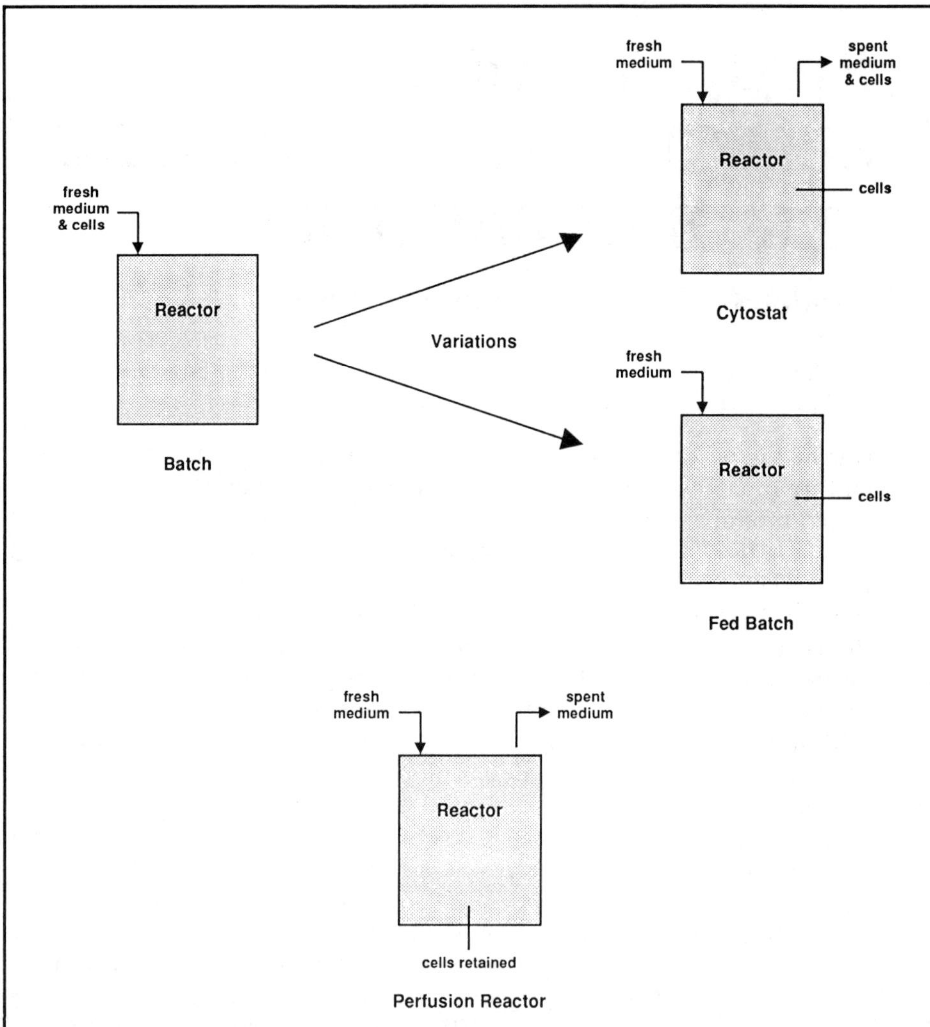

FIGURE 8-1 Strategies for operating bioreactors.

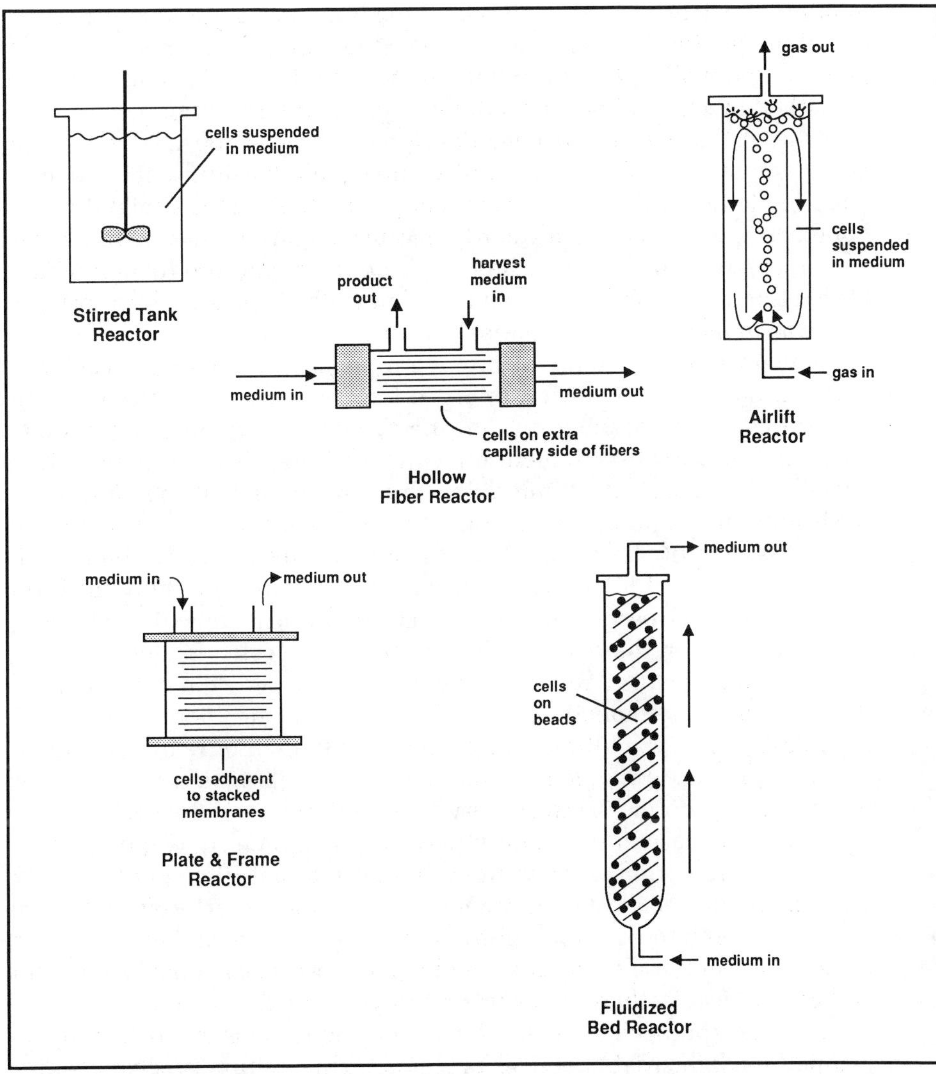

FIGURE 8-2 Basic bioreactor types.

BATCH AND PERFUSION CULTURE

Historically, cell culture is performed in sterile chambers into which cells and nutrient mixtures are introduced. Oxygen is provided and carbon dioxide is removed. In some systems, such as spinner flasks and roller bottles, some form of agitation is used to enhance circulation of medium to allow efficient delivery and use of medium and oxygen, and removal of carbon dioxide. The chamber content is maintained at physiologic temperature, normally 37°C, during the period of culture. The end of culture is marked by removal of the medium or cells for harvesting.

One of the first approaches developed to lower the CTP was to increase the number of cells per unit volume (cell density) with no special attention to increasing specific cellular productivity. This straightforward approach was developed primarily because engineers and nonbiologists were made responsible for scale-up development and production. Out of these efforts were derived a number of novel bioreactors whose primary goal was to increase cell densities per unit volume.

One operational technique applied in many of these novel reactors is medium perfusion. Perfusion refers to the feeding of the cultures by infusing fresh nutrient medium into the reactor during culture, with concomitant removal of spent medium while retaining the cells in the culture chamber. The basic idea underlying this strategy is that the addition of fresh nutrients supports more cells, thereby resulting in higher product titer. This can only be accomplished in a scalable system if a filter mechanism is either within or outside the reactor to retain cells, and this filtration system must permit the free exchange of nutrients and waste products in and out of the reactor while retaining the cells. Under this mode of operation, reactors can attain high cell densities. If the rates at which the medium can be added and removed are adequate, and oxygenation is not rate-limiting, the ultimate cell density is limited only by the amount of physical space provided for the cells. These perfusion reactors have become very popular because very high cell densities can be achieved with some increase in the amount of cellular product. Among these types of reactors are stirred tanks with on-line or internal filters; hollow fiber, static-maintenance, plate-and-frame reactors; and other reactors that retain cells within the reactor while continuously adding and removing medium. Many of these reactors are problematic because a reliable, efficient, and easy-to-use method of perfusion has yet to be developed.

Another approach is the fluidized bed reactor. This was originally developed for industrial chemical processes. This culture process requires cells be cultured on the surface of a solid substrate (usually a particle in the form of a bead), which is suspended in a stream of nutrient medium. The reactor is configured in such a way that the medium flow is rising while the bead is settling (due to gravity). The particles in the bed reach an equilibrium state in which the forces moving the particle downward are countered by the fluid force moving upward. The beads therefore never settle to the bottom of the chamber but remain suspended in the liquid in a "fluidized" state. The putative advantage to this type of reactor is that high medium-exchange rates, which are usually higher than other types of perfusion reactors, can support very high cell densities. Furthermore, both adherent and nonadherent cell types are theoretically capable of being cultured in this system. However, these reactors are very expensive to operate because of the cost of the hardware, particle substrate, and other components. Furthermore, operators must be highly skilled and dedicated, often with a great deal of experience and a Bachelor's or higher level degree in biological or engineering sciences. This type of reactor is particularly susceptible to bed expansion as a result of changes in the specific gravity of the beads due to cell growth. The change in specific gravity of the beads alters the equilibrium within the reactor chamber. These changes can ultimately cause the loss of significant amounts of the cell bed because the beads become "lighter" and are lost from the top of the reactor through the outlet tube. This loss of cell mass will ultimately adversely affect reactor productivity (e.g., loss of cells means less productive capacity).

The stirred tank reactor operated in batch mode is the simplest reactor to operate. With this operation, one places the cells and their growth medium in a vessel and permits them to grow to a specified cell density. When the culture is terminated, the cells and their medium can then be separated and the intracellular or extracellular product obtained. A modification to this method is to add medium to the initial culture at some point during the growth period. This is known as a fed-batch reactor. These same reactors can be fed continuously with fresh medium if cells and medium are drawn off at the same rate that fresh medium is provided. This technique is known as a chemostat (first developed in microbial systems to study microbial metabolism). With reactors operated in either mode, the ultimate cell density is determined by the balance of nutrients and waste products in the medium, the amount of surface area provided (in the case of anchorage-dependent cells), and doubling time of the cells.

CRITICAL CRITERIA

None of the reactors discussed above or those that we are aware of are optimal for *all* cell types. Several novel reactors are effective and efficient for specific cells and products. With this in mind, the following is a list of some of the critical questions that should be answered before adopting a cell culture technology:

- How versatile and generally applicable is the technology?
- What is the cost to build the reactors and the facility to house them?
- What is the maintenance cost for the reactors?
- How sophisticated do the operators need to be?

- How does one define a "batch," especially if the reactor is in continuous operation?
- How much time will elapse before becoming operational?
- How patient is management, and what risk levels are they comfortable with?
- What are the effects of delays in production on ultimate goals?

Versatility of Technology

Ideally, the technology of choice should handle a variety of cell types. More specifically, an ideal technology should support both adherent (those that require a solid substratum on which to grow) and nonadherent (those that can grow suspended in the medium) cell lines, the key advantage is that a highly versatile system allows for the manufacture of a variety of products using different cells without the need for building a manufacturing plant around every cell type and product that comes along. Most therapeutic proteins are currently being produced from recombinant Chinese hamster ovary cells (CHO) or, in the case of antibodies, from murine hybridomas. Both of these cell types will grow in suspension culture. (CHO cells can be grown as either anchorage-dependent or independent cells.) The expansion (culturing and subculturing cells until sufficient numbers are obtained to inoculate a production vessel) of suspension cultures is considerably easier than expanding anchorage-dependent cultures. However, if anchorage-dependent cells are used, a surface must be provided on which they can attach and grow. There are a number of substrates that can be used to provide the necessary surface. By allowing anchorage-dependent cells to attach to these surfaces and then suspending the culture in growth medium, one can achieve effective cell growth and high cell densities. Stirred tank bioreactors can therefore be used for anchorage-independent and anchorage-dependent cells—provided a suitable substrate for the anchorage-dependent cells can be identified. It should be added that air lift reactors can also be used for both types of cells. It is a little more difficult to grow anchorage-dependent cells in an air lift reactor, however.

Alternative reactors, e.g., hollow fiber and plate-and-frame are more limited in the kinds of cells that can be used successfully. These types of reactors, depending on the type of material in contact with the cells and the configuration of the "growth chamber" can support adherent or nonadherent cell types. Their major limitations are in scale-up. It is difficult to subculture cells from them, and because of the difficulty in producing them, it is difficult to obtain reactors in sufficient size to use them in large-scale production. Another significant problem that affects these kinds of reactors is the non-homogeneity of the culture environment. It is possible, even probable, that nutrient concentration gradients in the reactor chamber will create gradients of cells in various states of starvation. If the immediate environment or nutritional state of the cells can affect the quality of the product (i.e., alterations in glycosylation, disulfide bonds, etc), a reactor with a nonhomogeneous chamber environment might yield a nonhomogeneous product. This is problematic because the Food and Drug Administration wants to be reassured that the process is reproducible and because replicating identical nonhomogeneous reactor conditions from run to run is rather difficult.

Cost of the Reactors and the Facility to House Them

Alternative bioreactors have originated from a desire to develop a proprietary technology that could be exploited for commercial needs. These reactors without doubt have some utility for certain cell lines and products. In most cases, however, these reactors with their associated support equipment cost more to construct (or to buy) than does a stirred tank reactor that accommodates the same number of cells. In addition, these reactors may require somewhat unconventional facilities in which to operate, such as very large warm rooms, very large autoclaves, high throughput washroom support, and intense engineering support services. Also, in the past, some vendors have demanded royalties on final sales for the use of their proprietary technology. Of course, this directly affects the bottom line.

Maintenance Costs for the Reactors

Hollow fiber, plate-and-frame, perfusion-cytostat, and perfusion-maintenance reactors all require some elaborate mechanism to contain the cells and allow unimpeded flow of nutrients. The problem is that at this time there is no adequate way to do this. As cells grow and die, cell debris causes fouling of the materials used to contain the cells. This slows the flow of essential medium. Without sufficient flow of fresh nutrients to the cells and removal of exhausted medium from the cells even more cells will die. To avoid this "death spiral" it is necessary to change the clogged filters frequently or to transfer the culture to another reactor. Either way, this is an expensive way to maintain a high density culture.

At these high cell densities, there is little room for error in the culture. In addition to increased production of the desired protein, catabolites such as lactic acid (causing low pH) and specific cellular inhibitors are increased. Also at high cell densities oxygen can be an important rate-limiting nutrient. This increase in catabolites, along with the necessity for adequate oxygenation, mandates that medium flow and dissolved oxygen be rigorously maintained. Even minutes without adequate control on these parameters can cause a loss of control of the culture. To maintain this control requires sophisticated, expensive equipment.

Furthermore, early failure of a perfusion reactor generally costs more than failures at the same point in a batch run. Many of the perfusion-type reactors depend on long operations to reduce the overall cost of material. Initial costs are usually high because these novel reactors usually employ expensive hardware and can use expensive cell substrates (i.e., beads) as part of the system. Because failure rates must be kept especially low in perfusion cultures, operators must be trained to a very sophisticated level.

Sophistication of the Operators

Animal cell culture is less of an art than it used to be, but there is still some art involved and the skill of the operator is essential to success. Appropriate reactors with adequate control and monitoring of important physiologic parameters can minimize the amount of "green thumbism" necessary to culture cells reproducibly. One can obtain data on agitation rates, temperature, pH, dissolved oxygen and other parameters. These data can be used to establish operational parameters that can be measured and monitored. It is imperative (for a therapeutic product) that these operational conditions be reproducible within limits and that control of the production process is maximized. Setting and adhering to these limits is often difficult if not impossible with some of the novel bioreactors that are being used. Operations on some reactors can be so complex that advanced engineering degrees may be necessary or teams of operators might be required to perform routine functions. These reactors are very labor-intensive, requiring constant attention and manipulation by trained personnel. Such labor-intensive activities over very long periods (e.g., extended runs in perfusion reactors) can be demoralizing unless the reactor facility is adequately staffed.

On the other hand, the use of a stirred tank run in a batch or fed-batch mode can permit the definition of a production cycle that can be reproduced from run to run. The protocols defining this stirred batch process can be written and followed logically. The operation and maintenance of the equipment can be defined and protocols clearly established. Since these operations are relatively straightforward, substantially less training and expertise is necessary to operate this kind of reactor as compared with a perfusion reactor (e.g., a fluidized bed reactor). In addition, operating many of the novel bioreactors within specified limits is either difficult or the established limits are so broad that they are meaningless.

How Does One Define a "Batch"?

The production of a biological agent to be used as a pharmaceutical requires strict adherence to production and quality assurance protocols. A major tenet is that every component that becomes a part of or comes in contact with the final product must be well characterized. Also, there must be a well-defined set of production parameters to which each "batch" must conform. If one is using a mammalian cell culture technology that employs continuous perfusion to permit the culture to be maintained for weeks or even months, it is unclear what defines a batch or a batch end point. One can define the batch as a certain quantity of crude product that is used to begin a particular purification. However, it is difficult in this scenario to keep appropriate records of what raw materials went into the particular purification batch. For this reason, it is much easier to control the entire production process if one uses a batch process that has a well-defined beginning and end to the culture period. Cells are inoculated into small vessels using well-defined components (flasks, bottles, media) and can continue to be expanded and ultimately cultured in the production vessel to yield product. All of the components used during the production process can be clearly and well defined. The records for each particular batch can be compiled and maintained as a discrete package of information that can be thoroughly scrutinized.

How Much Elapsed Time Before Becoming Operational?

Generally, if not always, there is intense pressure to obtain the protein of interest either for research, clinical trials, or production. The more novel and/or complex the culture system the more time it takes to begin to manufacture the product. The time it takes to install and begin operations for cell culture reactors at a small, research scale is about the same for any of the various reactors. Small hollow fiber, air lift, plate-and-frame, and stirred tank systems are readily available for this scale. However, as one needs to increase the scope of the operation to provide more product, it is necessary either to use multiple small reactors or to begin using larger reactors. The use of multiple small reactors to produce product is inefficient and can be cost prohibitive. Multiplicity, while a shortcut, is usually not a good alternative to a large-scale manufacturing process.

A subtle, though significant, advantage to working in batch culture technologies is simply that both laboratory and engineering personnel are familiar with the concepts and techniques involved. This means that laboratory biologists and biochemical engineers can speak with each other using concepts familiar to both backgrounds. The laboratory biologist usually works in batch culture and understands the biology of the system fairly well. The biochemical engineer, especially one experienced in the microbial fermentation area, has a vast literature to work from that deals with scaling batch reactors. This can facilitate the exchange of information between these two groups, which results in steady developmental progress.

Developing an optimized perfusion process of any kind takes much longer than optimizing a batch process. Batch culture processes usually run up to a week. Perfusion cultures run for longer periods, sometimes as much as six months. Running repeat and replicate experiments to optimize operating conditions and, more importantly, to develop optimized medium and environmental conditions (for optimal product secretion) will take an extremely long time. In other words, more experiments to optimize the system can be run using a batch culture than a perfusion culture system in the same amount of development time. Therefore, the reactor most easily and expeditiously scaled from bench to production is the stirred tank, of which there are many examples in the vaccine production area and recombinant protein production area.

How Patient Is Management, and What Risk Levels Are They Comfortable With?

Many decisions in biotechnology are driven by factors other than scientific and technical ones—especially in start-up companies, in which funding for promising ideas or products is of utmost importance. Under these stressful conditions, the management personnel of these companies, who may or may not be scientists, often have difficulty understanding the amount of time necessary to develop the project. Typically, when a product is being readied for advanced research or clinical trials, upper management tends to ask, ad nauseam, "Where's my stuff," meaning, "When will there be sufficient protein for further testing?" Availability of product for research or clinical trials is always rate-limiting, at least at some point in the development of the project. Under these conditions, choosing a novel bioreactor that has not been extensively investigated may lead to research on the bioreactor rather than on the production of material for research and clinical trials. The scale-up of novel bioreactors, while attractive from the standpoint of generating new data for publication, requires extensive development time and research without assurance of success. A stirred tank reactor using a batch process is in our opinion the most likely choice under these conditions. With the stirred-tank system there is much more certainty of success, and the number of references to their use is much more extensive than with other bioreactors. Making upper management comfortable in these stressful times is not an unimportant task. Do not minimize this fact.

What Are the Real Effects of Delays?

Once a product derived from mammalian cells is determined to be of interest, it is important to begin clinical testing as expeditiously as possible. For one reason, it serves no useful purpose to delay finding out whether the product has therapeutic benefit. Second, delays in getting a useful product on the market decrease the income from the sales of the product. For example, if the real market for a biological agent is $365 million a year, each day lost in getting to market costs $1 million in revenues. This is a result of loss of market share and perhaps competitive position. It is usually not the case in biotechnology that your product is the only one in development to address a particular market. The apocryphal quote attributed to General Bedford Forrest is applicable to this situation. He is supposed to have said "Git thar furst with the mostest." Words to live by in the highly competitive biotechnology marketplace. It is therefore important to use technologies that are well understood and simple.

SUMMARY

We hope this discussion of the choice of bioreactors has encompassed the areas that are most important in making the correct decision. As in practically all areas of science, there is probably no one, correct answer to any given situation. Therefore, when it comes to selecting a bioreactor for a manufacturing process, the ultimate decision may be made on the basis of availability rather than on the type of reactor. Time is often the reason for making key decisions. Basing a decision on time factors is almost always wrong, however. "Quick and dirty" solutions to projects, especially in advanced research or clinical trials, do not exist. The importance of selecting a cell culture technology that can product research, clinical trial, and production material cannot be overestimated. Unless the goal of the company is to develop novel bioreactors, it is essential that a technology is adopted that permits an adequate, predictable supply of protein product—even if this means using a reactor that is somewhat suboptimal.

In our opinion, whether one is involved with a large, well-funded company or a start-up company with few revenues, it is important to begin with a stirred tank bioreactor operated in a batch mode. These reactors are well understood, there is a wealth of information in the literature on their operation, and the major manufacturers of bioreactors have the most experience with the stirred tank. In addition, small bench-scale reactors are available for research. The data obtained from these is readily transferable to larger reactors for scale-up. We concede that the cell densities in a stirred tank bioreactor run in a batch mode may not be as high as in alternative reactors discussed above. However, for all of the reasons discussed, they clearly are the vessel of choice for production of animal cell–derived protein products for many biotechnology companies.

So, after years of trying to solve the problem of the high expense of animal cell culture using alternative bioreactors and improved engineering principles, we are left with the conclusion that these efforts have contributed little to reducing product costs. It is clear to us that the major emphasis on driving costs down should be placed at the cell itself. That is, the largest gains will come from increasing the specific productivity (the amount of product per cell per time). For a recent review of this area, the reader is referred to Oka and Rupp, 1990. Therefore, rather than spending money on alternative bioreactors, it could be better spent on improving the cell's protein synthetic machinery. Increases in specific productivity have a direct impact on the cost-to-produce. If the specific productivity increases 30%, a nearly comparable decrease in the cost-to-produce is seen. There are many areas of cell biology and molecular genetics that have the potential of bringing biotechnological products to the market at more competitive prices than we see today.

REFERENCES

Griffiths, J. B. 1988. Overview of cell culture systems and their scale-up. In *Animal cell biotechnology*, vol. 3, edited by R. E. Speir, and J. B. Griffiths, 179-220. New York: Academic Press.

Oka, M. S., and Rupp, R. 1990. Large scale animal cell culture: A biological perspective. In *Large scale cell culture technology*, edited by A. S. Lubiniecki, 71-93. New York: Marcel Dekker.

CHAPTER 9

THE PROTECTION OF INTELLECTUAL PROPERTY IN BIOTECHNOLOGY

Mark A. Hofer

Like an advertising campaign or a highly trained salesforce, patents are merely another marketing tool with which to exert leverage over competitors. In the extremes, they may be skillfully obtained and used effectively to promote and protect a product, or caution may be thrown to the winds and patents may be ignored or not procured at all. Particularly in biotechnology, patents are proving to be especially potent weapons and—in the view of both venture capitalists and investment bankers who deal with publicly oriented financing—essential assets. Before we can learn how patents may be effectively used, a short course on what patents are is in order. However, this course is not intended to be comprehensive, as there already are a number of readily available, excellent treatises on the subject [see, for example, Rosenberg 1989; Cooper 1987] and any attempt to do more than synopsize the relevant features would be beyond the scope of this chapter.

A patent is merely a monopoly right, granted by the issuing government, for a specified term of years, to exclude others from making, using, or selling the claimed invention. In the United States, the term is generally 17 years from the date of issue, although the term may be shortened by virtue of a terminal disclaimer (to be found on the face of the issued patent), in order to prevent an unconstitutional extension of the monopoly of overlapping subject matter from an earlier issued patent, or extended, in order to compensate for time "lost" while the invention was awaiting Food and Drug Administration approval. In Japan, the term is generally 20 years from the date of filing but no more than 15 years from the date the patent is published for opposition. A "European" patent, issued by the European Patent Office, must be converted into national patents in each of the participating countries in which coverage is desired. The national patents will have terms based on the laws of the individual countries, generally from about 15 to 20 years.

This monopoly right is granted to the patentee (who may be under a contractual or other obligation to assign his rights to another, such as his employer) in exchange for the revelation and teaching in the patent specification of an invention provided by the patentee, provided that the invention meets certain statutory criteria. In the United States, the patent must possess novelty and utility, and the claimed invention must be unobvious when assessed against the prior art. It must also comply with complete description requirements to enable another to make and use the invention, and it must disclose the best method contemplated by the inventor at the time of filing. If the written description cannot enable one of ordinary skill in the art, then a deposit (in an approved depository) of enabling materials is required. Foreign patent requirements include absolute novelty (any publication anywhere prior to filing can destroy patentability), utility, and nonobviousness. At present, the Europeans do not permit patents on animals (such protection is not provided by statute, and the issue is also clouded with arguments of a public policy nature) nor do they have the best-mode requirement of the United States. The astute reader will recognize the implications of absolute novelty—namely, the patent is awarded to the first to file. Although such does not support the laudable goals of the U.S. system intended to reward the first-to-invent, it does serve the public by clarifying who possesses the superior rights—an interest definitely not served by the U.S. system, which shrouds the issue with the mystical cloak of interference practice. There is also an enablement teaching requirement in the foreign arenas, although it generally is less demanding than that of the United States and sometimes also demonstrates surprising subtleties. Canadian courts (as exemplified by *Pioneer Hi-Bred Ltd.* v. *the Commissioner of Patents* [6/22/89] regarding a new soybean variety made by artificial cross breeding but cultivated by natural means) made a peculiar distinction in this regard. One must be enabled to duplicate the inventor's work as opposed to making and using the invention discovered by the inventors.

Thus, upon issuance of the patent in the United States, or publication of the patent application filed in foreign countries,[1] the public is benefited

1. Typically, when attempting to procure patent coverage, a U.S. inventor will first file in the United States and then, within one year of the filing or priority date, will file in one or more foreign countries, generally using the Patent Cooperation Treaty or the European Patent Convention, or both, claiming the benefit of the priority date. Pursuant to foreign practice, the foreign application is then typically published within six months after its

by learning of the invention, thereby serving the constitutional goals of advancing the arts and sciences. At that point, the patentee can wield his patent (or collect damages from the foreign publication date) within the territory of the nationality granting the patent, to enjoin others from practicing his invention.

It is vital to note, however, that being awarded a patent does not imply or mean that the patentee has the right to practice his own invention. This limitation occurs because there may be other patentees possessing issued patents to broader inventions. Such other patentees also have the right to exclude others from practicing their patented invention and thus can prevent the patentee from practicing his invention if by doing so he uses the broader, patented invention. Stated alternatively, patentability does not mean freedom from infringement.

An example may serve to illustrate this apparent incongruity. U.S. Pat. No, 4,766,075 (assigned to Genentech[2]) possesses claims to a DNA isolate (thereby distinguishing it from the nonisolated gene existing in nature) that encodes human tissue plasminogen activator (tPA). However, in order to practice this invention (e.g., produce tPA by transforming mammalian cells with vectors or plasmids comprising the claimed DNA), it is probable that you will need to employ the famous patented inventions of Cohen and Boyer, assigned to Stanford University. Fortunately, Stanford University has chosen to make these pioneering patented inventions available to the public under nonexclusive, royalty-bearing licenses rather than to prevent everyone from making, using, or selling the invention, as is their right. Naturally, they retain that right against those who choose to benefit from the invention, but who fail to obtain a license and thus pay for the privilege of protection from suit. Because the Cohen–Boyer inventions were supported in part by government funding, the U.S. government retains certain march-in rights, and Stanford University must comply with its obligation to capitalize efficiently on the invention to ensure that the public obtains the greatest benefit from its tax dollars. Thus, Stanford University would most likely have been unable to justify preventing everyone but themselves from practicing the invention. Similarly, I believe Stanford would have found the granting of an exclusive license equally unjustifiable.

The analysis does not stop there, however. Other pertinent patents may deal with the use of the particular mammalian cells chosen to be

---

filing. Thus, by watching the foreign patent application publications, one can surmise the nature of a competitor's position, albeit 18 months in arrears.

2. Genentech has sued Burroughs Wellcome and Genetics Institute in the Federal District Court for the District of Delaware for infringement of this patent. As of this writing, the case is still in discovery. The United Kingdom equivalent of this patent was held invalid for overbreadth.

transformed, components of the vectors or plasmids that may be necessary (e.g., promoters, enhancers, origins of replication) to enable production and/or secretion of the desired protein, methods of culturing the cells, the media used to support the cells, methods of purifying the protein, or even the protein itself. Obviously, if the inventions of all such patents must be used because alternative approaches are impractical or uneconomic, and assuming that the license to each patent carries with it some royalty burden, then you can easily find yourself owing more royalties than the eventual product can generate after cost of goods and sales are deducted from revenues. Thus, in the process of identifying a new business and focusing on a product, the biotechnology entrepreneur is wisely counseled to spend considerable time with patent counsel and to conduct extensive patent searches in order to understand the nature of patent rights possessed by or pending to others. Only through this process can the entrepreneur understand the likely patent-related ramifications of selecting one development route over another or aiming his scientific team toward any particular product. Obviously, the choices of direction should be heavily weighed in favor of the areas that are not seriously impeded by the patents of others and, even more preferably, toward the direction in which there may be opportunities for the fledgling company to obtain its own patents.

Up to this point, we have assumed that all patents encountered have been valid and enforceable. The Court of Appeals for the Federal Circuit (CAFC),[3] by the delivery of innumerable rulings holding patents valid rather than invalid, has greatly strengthened the presumption of validity that accompanies an issued patent, particularly in the arena of obviousness (35 U.S.C. 103).

Although the U.S. Supreme Court in, *Graham* v. *John Deere,* espoused the standards for determining obviousness (standards involve the scope and content of the prior art, the ordinary skill in the art at the time of filing, and the differences between the prior art and the invention as a whole), such standards can be easily met by a patentee armed with only

3. The CAFC is the federal appeals court to which all appeals from the various federal district courts are directed. This court is the "patent court," which came into being in October 1982 in order to standardize patent law by using judges who have scientific and patent-related backgrounds. The jurisdiction of this court is broader than that of the Court of Customs and Patent Appeals, which it replaces, in that the CAFC has exclusive subject matter jurisdiction over all patent issue-related appeals, thus eliminating the federal circuit appeals courts as an alternative forum of appeal from the federal district court. Unfortunately, it is my opinion that the CAFC has, in its zealousness to protect patents, strained the fundamentals of patent law in so many ways that the only safe prediction with respect to patents and this pro-patent court is that they will find some way to find the patent in question (or at least one claim thereof) valid. In effect, the long-standing desire to have a dedicated patent court has been fulfilled, to the chagrin and curses of many defendants.

the flimsiest of arguments. A glaring case in point is *Hybritech* v. *Monoclonal Antibodies.* In *Hybritech,* Greene was awarded a patent claiming the so-called sandwich immunoassay technique that employed two monoclonal antibodies for detecting analytes. The assay was identical to the well-known sandwich techniques used with polyclonal antibodies. Polyclonals are derived from animals that have been immunized with the antigen for which the antibodies are desired. The worst polyclonal preparations include antibodies reactive for the antigen of interest as well as antibodies of widely diverse specificity, which may react with totally undesired antigens. Conversely, the best polyclonal antibody preparations contain few if any cross-reacting antibodies. Often the polyclonal preparations are purified to eliminate undesired antibodies until essentially only one species of antibody remains. Scientists searched for years to find a more efficient method of making such purified antibodies and were rewarded in 1975 by a short, but explosively important publication by Kohler and Milstein, which set forth the procedure for making monoclonal antibodies, i.e., antibodies of only one specificity. Greene, using the Kohler-Milstein procedure, simply substituted the monoclonal antibodies for the polyclonal antibodies previously used in the sandwich immunoassay approach and claimed the thusly modified assay. In effect, Greene repatented the old approach using a more highly purified ingredient to do no more than what was previously accomplished. If ever there was an "obvious" invention, this was it. Every patent lawyer I have ever discussed this case with feels likewise. Unfortunately, there were a few patent lawyers in Washington (those sitting on the CAFC hearing Hybritech's appeal of the lower court's finding of invalidity due to obviousness among other grounds) who felt otherwise. Given this decision, it has become very difficult to justify an opinion of obviousness, particularly in genetic engineering.

Only in the Patent Office during prosecution does the patentee encounter any real need to present a serious nonobvious argument, since every examiner issues at least one obviousness rejection as standard practice. Unfortunately for the public at large, the examiners (particularly in the biotechnical arts in which the turnover is great and the government has difficulty competing against the demand for patent attorneys from the private sector) are often inexperienced and generally so overburdened[4]

4. Serial No. 100-59 of the Hearing before the Subcommittee on Regulation and Business Opportunities of the Committee on Small Business, House of Representatives, One Hundredth Congress, Second Session, Washington D.C., March 29, 1988, reported that in 1987, 1400 examiners were responsible for examining 126,597 cases filed in that year alone. The number of cases represented an increase of over 5000 cases from the all-time high of the previous year. Of these applications, nearly 5000 were in the field of biotechnology, creating a backlog of 7000 cases. If those in the printing process, in reexamination, and those yet to be identified and allocated within the biotechnology groups are

that they either adopt unreasonable intransigent positions or are unable or do not have the time to reflect reasonably on the patentee's nonobviousness arguments and, accordingly, are overly solicitous in their acceptance thereof. As a result, one may experience widely divergent approaches and philosophies by different examiners even though the general nature of the subject matter they are examining remains substantially similar. The unfortunate effect of this is to exacerbate the unpredictability of the scope of patent coverage to be obtained. The wise entrepreneur will factor this into his plans as he sets out to build a business. Similarly, he should develop a contingency plan in the event his competitor unexpectedly receives a windfall in the nature of extensive patent coverage due to the issuance of a patent with claims having a far broader scope than could have been reasonably expected after examination of his published patent application. Such a plan could be predicated upon a change in direction to avoid infringement (if possible), an attempt to respect the patent by licensing or otherwise joining forces with the competitor, or on opposition of the patent via defending an infringement action on grounds of invalidity or bringing a declaratory judgment action, or moving his activities to a territory devoid of such patent impediments.

In general, most biotechnology managers complain that it always seems their competitors have the luck of drawing the "easy and reasonable" examiners, whereas they are constantly saddled with the "unreasonable sticklers." In actuality, everyone experiences much the same luck, only their perceptions differ because their individual focuses do. As expected, however, the competitor's focus tends to change to become more closely aligned with the just issued "blockbuster" patent his attorney was able to extract out of the Patent Office. Such changes, and the accompanying press releases, only exacerbate the feeling that the other guy has all the luck.

While the predictability of the application of the obviousness standard within the Patent Office has suffered in recent years, such has not been the case with the more objective criteria of 35 U.S.C. 102. This section of the patent statute contains many hurdles and pitfalls; however, we will examine only subsections a and b, which deal with publications and "on-sale bars." Simply, you may obtain a patent if the invention claimed has not been published, on sale, or in public use for more than one year prior to the filing date. Thus, you can still obtain patent coverage after

---

counted, the number, as reported by *Messrs.* Quigg and Tegtmeyer, is closer to 11,000. All of these were to be handled by 42 examiners. Eleven of these examiners were expected to leave and be replaced. Given the complexity of the examining task, the need for years of experience in order to become accomplished, any manager will recognize that a personnel turnover of this proportion cannot lead to consistent and appropriate examination.

public disclosure of the invention provided that you file the pertinent patent application (which may be simply the published paper with one or more claims attached) before the one-year grace period expires. The occurrence of publication prior to filing will, however, have destroyed absolute novelty and thus the ability to procure any foreign patent coverage. This substantial distinction between U.S. and foreign law is one of the issues to be resolved in the ongoing worldwide patent harmonization efforts.

I believe it likely that this issue will be resolved in favor of a compromise such as that presently practiced by the Japanese—that is, the inventor may file a patent application on a published invention provided it is his own publication and he files within six months of the publication date. This would seem to serve public policy in that it provides the unsophisticated inventor with a limited opportunity to rectify a mistake while eliminating the right to patents by inventors who have slept on (or actively hidden) their inventions but who are suddenly prodded to life by seeing their invention duplicated and published by others who are benefiting society by their timely disclosure.

One of the first things an enlightened entrepreneur will do is implement a publication review procedure to prevent unanticipated or untimely publication of potentially patentable inventions prior to the filing of a pertinent patent application. The generation of an Idea Disclosure, Invention Memorandum or similar document will not save this loss. The purpose of such documents is to create admissible evidence demonstrating and dating conception, the first act of invention. (The second act is reduction to practice, which may be either actual or constructive in nature. At the time an embodiment of the invention, working in its intended environment occurs, actual reduction to practice is said to take place. If, before this event occurs a patent application claiming the invention is filed [including the requisite enabling description], a constructive reduction to practice is said to have occurred.) The date of conception could eventually prove to be vital in the event a pertinent patent application is filed on the invention and the Patent Office discovers another contemporary patent or patent application claiming overlapping subject matter.

In the United States, this situation leads to a Patent Office–instituted proceeding called an *interference*, whose purpose is to decide who is the first inventor of the overlapping subject matter. This procedure is intended to satisfy the requirement under U.S. law that patents be awarded to the first inventor—a laudable goal. This is in contrast to the law virtually elsewhere in the world which provides that the first to file, and thus teach the public, is awarded the patent. Obviously, this is another serious impediment to the harmonization efforts and is not readily subject to a compromise position. Due in part to the extremely burdensome and expensive nature of interferences, generally inventors who are better fi-

TABLE 9-1 INTELLECTUAL PROPERTY POINTS-OF-INTEREST TIME LINE

1790	First U.S. patent granted for a process for making potash
1870	Congress passes an Act codifying the patent laws
1873	U.S. PTO grants Louis Pasteur a patent for a method of "preparation of yeast free of microbial contamination," which includes a claim to purified yeast
1916	Federal Court rules that a guiding principle in determining infringement involves Doctrine of Equivalents (*Telescope Cot Bed* v. *Gould Medal Camp*)
1930	U.S. Congress passes the Plant Patent Act; PTO authorized to grant patents on asexually reproduced (grafted) plants
1946	Trademark Act promulgated
1949	PTO begins recommending the deposit of microorganisms
1952	U.S. Congress passes the new patent statute (still in use today)
1963	Obviousness of a compound depends on similarity of the compound to all the characteristics of the prior art compounds, not just the formula (In re: *Papesch*)
1966	U.S. Supreme Court mandates the three-part test of obviousness in *Graham* v. *John Deere*
1970	Congress passes Plant Variety Protection Act; Patent Office empowered to grant patents on sexually reproduced plants
1970	Court of Appeals for the Federal Circuit (CAFC) rules that in certain circumstances words may be inadequate to enable and a deposit must be made (In re: *Argoudelis*)
1977	European Patent Convention enters into force with thirteen signatories
1978	Patent Cooperation Treaty enters into force with forty signatories
1980	U.S. Supreme Court rules that a live genetically engineered microorganism constitutes patentable subject matter within the meaning of section 101 (*Diamond* v. *Chakrabarty*)
1981	U.S. Supreme Court rules that laws of nature, physical phenomenon, mathematical expression, and scientific truths are not patentable (*Diamond* v. *Diehr*)
1985	CAFC rules that under certain circumstances a laboratory may serve as its own depository until issuance (In re: *Lundak*)
1985	CAFC rules that section 101 of the Patent Act can also cover plant material (Ex parte *Hibberd*)
1987	Board of Appeals in the U.S. PTO rules pleuroploidy shellfish is patentable (Ex parte *Allen*)
1988	U.S. PTO issues first patent on a transgenic animal (the "Harvard Mouse")
1988	Process Patent Amendments Act promulgated

TABLE 9-1 ***(continued)***

1988	U.K. court overturns patentability of Genentech's recombinant tPA patent on the basis that it is overbroad
1989	Revisions to U.S. Trademark Act of 1946 go into effect
1990	U.S. District Court in a jury verdict upholds patentability of Genentech's tPA patents

nanced win. Since those same well-heeled inventors were presumably able to avail themselves of counsel earlier and more readily, they are also more likely to have filed earlier than the "basement" inventor and, hence, win again. Since an earlier filing benefits the public at an earlier date, and since predictability (i.e., it is relatively easy to discern overlapping subject matter and, on the face of published patents or applications, determine chronological priority) also benefits the public (by clarifying who owns the rights and thus limiting imprudent commercial development), in my opinion the U.S. public would be well served by adoption of the first-to-file system prevalent everywhere else in the world, along with banishment of interference practice, which generally benefits only the lawyers.

Anomalous results may be discovered where different patents and patent owners control the same technology in different parts of the world. For example, on January 2, 1985 (the priority date) Orion filed in Sweden a patent application for coverage on a one-step DNA hybridization assay using a solid phase. Amgen and Abbott, in a joint venture, filed in the United States on substantially the same subject matter but did so approximately six months later (June 13, 1985). Orion did not file in the United States until approximately one year after its priority date or, another six months after the Amgen–Abbott filing. Because in absolute terms, Orion had the first filing, foreign patents (based on the first-to-file system) should only issue to Orion; to date, this has been the case. Meanwhile, in the United States, because the Amgen–Abbott patent was physically filed in the U.S. PTO first, it issued first (first come, first served). However, it also claims overlapping subject matter with the later physically filed (but entitled to the earlier priority date) Orion application. Theoretically, therefore, an interference should be declared between the patent and the pending application.[5] In the interim (and potentially after the interference as well), the technology is controlled in the United States by the Amgen–

5. The situation is even more complex, because Orion will be restricted to proving its acts of invention which took place in the United States. Those acts are therefore likely to be limited to the earlier of the priority date or the date of importation of the Orion technology into the United States. One can only surmise when the Amgen–Abbott joint venture made their invention. Such proofs would only be disclosed in the interference.

Abbott joint venture and in the rest of the world (to the extent patents have issued) by Orion. It would seem that a situation such as this, which offends common sense, should never occur and, indeed, would not occur if the United States adopted the first-to-file system prevalent in the rest of the world. Until such time, however, the biotechnology entrepreneur should do his best to obviate such a possibility with regard to his own inventions and maximize the chances of securing worldwide coverage. This is accomplished by filing the pertinent patent applications as early as possible, thereby behaving as if the entire world were operating under the first-to-file system. Some tempering of when this occurs will take place on consideration of the ability to satisfy the statutory enablement requirement. Filing too early runs the risk of failing to enable. Although an enablement failure that is subsequently discovered by the patentee can be cured by another filing, even if this filing is characterized as a CIP (continuation-in-part), it cannot properly benefit from the earlier filing since that first application is not entitled to its filing date because of its enablement deficiency and thus its incompleteness at the time of its filing.

So far, we have not addressed the subject matter on which claims can be predicated and patent coverage obtained.[6] The patent statute (35 U.S.C. 101) provides that one may obtain patent coverage, assuming all other elements are met, on articles of manufacture, processes, and compositions of matter. On first inspection of these clauses, it is not apparent whether living organisms are included within section 101. Fortunately, this issue has already been addressed by the U.S. Supreme Court in *Diamond* v. *Chakrabarty*, wherein the court determined that microorganisms produced by genetic engineering are not excluded from patent protection. The decision in that case makes it clear that whether or not the invention comprises living matter is not the pertinent issue, rather it is whether the living matter is the result of human intervention. Applying such a test to host cells transformed with DNA such that the transformed cells possess characteristics apart from untransformed cells (e.g., they produce a new protein), would clearly result in a finding that such cells fall within patentable subject matter, again assuming all other conditions of patentability are met.

---

6. Remember, the monopoly right to exclude others provided by patents is measured by the claims, not by the scope of the specification. The specification is provided to teach the public how to make and use the invention, and thus it provides understanding regarding the nomenclature employed in the claims. When determining infringement, patent counsel will first look to the claims to determine whether there is literal infringement, and if not, whether there is infringement under the doctrine of equivalents (taking into account file history estoppel, state of the art, pioneering nature of the invention, and other relevant aspects). In a nutshell, the doctrine of equivalents allows a patentee to capture within the scope of his claims a pirate who would make use of the patentee's contribution but avoid infringement by making a small, relatively insignificant change to escape the literal language of the claims.

The analysis is more complex with respect to DNA (deoxyribonucleic acid), the "stuff of life" first described in a meaningful way by Watson and Crick in 1953. This high molecular weight organic polymer is really a *sui generis* molecule unlike all other chemicals. Although it is composed of only four types of monomeric building blocks (bases comprising adenine [A], which pairs with thymine [T], and cytosine [C], which pairs with guanine [G]; each pairing is referred to as a base pair), it is unlike other polymers in that the placement or order of the bases contains critical information decipherable by the cellular machinery. With the DNA instructions, the cell can synthesize a wide range of molecules, in a timely fashion, in order to metabolize, grow, and reproduce. On the other hand, a single point mutation in the millions of base pairs typically found in the chromosomes of a cell can have disastrous lethal consequences or no effect whatsoever. It has now become commonplace to identify and isolate a sequence of DNA—either genomic or cDNA—that encodes a desired protein, ligate it with suitable regulatory elements, and insert it into suitable mammalian host cells, which, under suitable culture conditions, produce the desired protein. This process is well known, as are the techniques for identifying the DNA encoding the protein. Only the actual base pair, or DNA sequence itself is the new information uncovered by practicing what are now conventional techniques. From the DNA sequence, the complete amino acid sequence of the protein can be deduced, and this too may represent new information.

Simply because it is new, is it patentable? If the actions of the Patent Office are any indication, the answer appears to be a clear yes.

Should it be patentable? In general, the mere application of skilled work and diligence, however time consuming or expensive, is insufficient to warrant patentability. One must return to the statute and apply the tests of novelty, nonobviousness, etc. Specifically with respect to DNA sequences, the English Court of Appeal has ruled on the issue in a case involving DNA encoding tPA. In *Genentech, Inc.* v. *Wellcome Foundation Ltd.* the court ruled that all of the claims issued to Genentech in U.K. patent 2,119,804 B should be revoked. An excellent short review article about this was written by Thurston [1989]. (A similar court case is now pending in the U.S. District Court of Delaware.) That English court found that Genentech did not invent tPA; indeed, it had been described in the literature many years earlier. Accordingly, those claims directed to recombinant tPA were found to be "unsound." On the other hand, tPA produced by recombinant means did not previously exist; its utility was unquestioned. Assuming it was unobvious, one could question the court's finding that recombinant tPA was also "unsound." Still, Genentech's admitted goal was to produce a tPA that was substantially identical to natural tPA, presumably so that it could be administered therapeutically to achieve substantially the same effects as the natural tPA (albeit at greater levels of effectiveness because of its presence at concentrations greater than those normally encountered). By all accounts, Genentech has been successful in this regard. It would thus seem that the court properly concluded that the distinguishing word "recombinant" that Genentech had put in their claims was not a limitation sufficient to distinguish the claims and the tPA over the known art.

The U.K. court further dispensed with claims directed to (i) recombinant cloning vectors comprising tPA encoding DNA and (ii) replicable expression vectors capable of expressing tPA in a transformed cell on the basis that a person ordinarily skilled in the art could not have created all of the products specified by these claims without performing additional inventive acts. The court also summarily dismissed the process claims, worded in broad functional terms, as lacking an inventive step. Only one process claim, namely claim 19, found any acceptance with the court. Number 19 claimed the process described in the application—admittedly a process no one would follow after being provided with the goal of the described process—the sequence itself.

How such DNA claims will ultimately fare elsewhere in the world remains the subject of intense speculation and expensive litigation. There are those who would argue that no human DNA sequence should be patentable, because to provide otherwise is to take the first step down the slippery slope toward patenting a human. They raise questions: If one gene is found patentable, what about two genes? A partial chromosome? An entire chromosome? A cDNA version of a chromosome? Proponents of this rationale point out that the DNA is known to exist; only the minute details concerning its sequence remain unknown, and their discovery is a matter completely within the ken of those of ordinary skill in the art. While the first such discovered DNA sequence might be patentable under this theory, all subsequent DNA sequences would be rendered obvious since one need merely add the appropriate probes (the design and selection of which have been the subject of innumerable articles) to the "hopper," turn the crank to screen a genomic or cDNA library for evidence of hybridization, identify positives, and sequence and assemble the full gene.

From a public policy perspective, the arguments against patentability of DNA sequences are even more compelling. This derives in large part from the peculiar nature of the DNA molecule itself. Intuitively it would seem that one should not be allowed to prevent all others (by patenting) from making, using, or selling a piece of DNA, which is a product of nature and present within all of us. Furthermore, patenting the DNA also effectively gives one the right to full control of the protein encoded by that DNA, since transformation of host cells with the DNA comprises the only known practical means for making the protein. Thus, possessing a

DNA claim provides one with control over the DNA and the encoded protein, both of which exist in nature, albeit not in purified form, in addition to control over host cells transformed thereby, admittedly a construction of man. Such DNA claims can thus effectively control the field of biotechnology with respect to the encoded protein.

The patenting of DNA purified from nature extends naturally from the traditional patenting of microorganisms purified from nature. However, the isolation of a new mineral from the earth would not be patentable. It would not seem logical that the distinction of life should imbue a newly purified substance with patentability. The pro-patent proponent would argue that the granting of patent monopolies on DNA sequences is warranted from a public policy perspective because without the promise of such a reward, researchers (such as those in the biotechnology industry) would not undertake the expensive and time-consuming cloning efforts. However, that same argument has been used to justify Orphan Drug Status (which entails a seven year approval monopoly from the Food and Drug Administration [FDA]) for human growth hormone, insulin, and erythropoietin, among many other multimillion dollar a year drugs. I believe that with respect to all of these drugs as well as those for which approval is still pending, there would be, in the absence of patent and FDA monopoly grants, many biotechnology companies competing in the same market and thus providing the public the choice and benefits of competition. The markets are too large to ignore if there is an avenue of participation available.

Of greater disadvantage to the public is that the granting of a claim to a DNA sequence (generally in terms of XYZ sequence, its allelic variations, and any sequence that hybridizes thereto) also puts a cloud over the freedom to develop second-generation molecules. Often, substantial changes in a molecule's activity, half-life or other characteristic may be made by comparatively insubstantial changes to the native DNA sequence. Yet, the second-generation DNA sequence may well hybridize to the native DNA sequence and thus literally infringe such a claim. As a result, a potentially superior molecule may also be kept off the market, to the detriment of the public. To think that the original entrepreneur will be willing to license a competitor whose second-generation molecule would completely replace the original molecule fails to acknowledge the nature of the entrepreneur. It should be remembered that this person is typically endowed with a large ego (generally necessary for survival, especially in a start-up industry) and that he has preached consistently to shareholders that their company "owns" a particular molecule and hence the relevant market. To believe that such a person would give up his 100% share of the market reflects incredible naiveté. A court, however, is not so egocentrically shackled and may well find it inequitable to the public to curtail the sales of a second-generation molecule not otherwise available. In such an instance, one might well expect the court to "force" the patent owner to accept a royalty-bearing licensing arrangement.

Given the fact that the Patent Office has issued patents containing claims to DNA sequences and the pro-patent nature of the CAFC, one can also expect the district courts, when faced with the prospect of enforcing such controversial patent rights, to take a Solomonic approach, letting plaintiffs and defendants both "win a little." (Of course, recognize that to be held to infringe a single valid claim constitutes a complete loss for the defendant regardless of how successful he has been in invalidating the remaining claims.) The case of *Amgen* v. *Genetics Institute and Chugai* points this out with remarkable clarity.

Amgen sued for infringement of its patent claims to DNA (encoding erythropoietin, EPO) and host cells while Genetics Institute countersued on its purified EPO patent. The claims in the Amgen patent which claimed, in functional terms, DNA encoding EPO and the host cell claims on which they were dependent were invalidated on the basis that they were overbroad and beyond the support of the specification that had only a specific DNA sequence (inexplicably, some claims that were also dependent on the invalidated claims and that suffered the same deficiency were not invalidated). The remainder were held valid, and since there was no noninfringement argument presented, also infringed. The claims in the Genetics Institute patent that used the term *about* in describing the minimum specific activity were invalidated (except, inexplicably, for one claim). The remainder were held valid and infringed. In effect, the magistrate, specially appointed to hear this case (indeed it was the last federal case she heard, as she was sworn in as a Massachusetts state court judge twenty minutes after rendering her written opinion), left Amgen and Genetics Institute in the same position (discounted the expenditure by each of many millions of dollars in attorneys' fees) as when litigation began more than two years earlier; Amgen could use the DNA to grow transformed host cells to make EPO but could not have purified EPO without infringing Genetics Institute's patent; and Genetics Institute could have purified EPO but could not make it using cells transformed with DNA encoding EPO without infringing Amgen's patent. Thus, the 184 page decision maintained the stalemate situation that analysts had long recognized and expected to be resolved by cross-licensing. There was, however, a bright note from Genetics Institute's perspective, because the court specifically held that Chugai, Genetics Institute's partner, was free to continue importing EPO into the United States without infringing Amgen's patent. As a result, Genetics Institute is likely to win an injunction against Amgen once their own EPO product, produced in Japan and imported into the United States, is approved by the FDA, an event that is imminent. Whereas Genetics Institute had previously been willing to negotiate with Amgen, they now seem to have the upper hand and may well drive a harder bargain than might have been driven earlier had Amgen not, in the eyes of the marketplace, backed themselves into the corner of having to be king of the EPO mountain. This situation may well change once pending legislation, discussed below, makes importation by Chugai of EPO made from cells transformed by patented DNA an act of infringement.

In the federal court proceedings, Amgen again attempted, as it had in earlier International Trade Commission (ITC) proceedings, to convince the court that claims to cells were tantamount to claims to processes, because the cells were really "manufacturing sites" for producing the EPO molecule. Remember that importation into the United States of a product produced abroad by a process patented in the United States is an infringement in the U.S. Like the ITC, the Massachusetts federal district court was unimpressed with the argument. After all, cells are tangible things and processes are not. Hence, Chugai's importation of EPO, albeit manufactured by host cells transformed by patented DNA encoding EPO, was not held to be an infringement. To rectify this, Amgen has cleverly been lobbying for the enactment of pending legislation that would result in (1) extension of 337(a) of Tariff Act of 1930 (19 U.S.C. 1337 (a) (1) (B)) and (2) amendment of 35 U.S.C. 154 and 271 to (i) include the right in an ITC proceeding to exclude the importation of a product made or produced by use of an essential biotechnological material (e.g., DNA) covered by a valid and enforceable U.S. patent and (ii) to make such importation an act of infringement under the federal patent statute, which finds its remedy in federal court. Such legislation would obviate the existence of a foreign haven for producing products for importation into the United States. Effectively, claims to DNA would gain in their level of importance. Regardless of whether these legislative changes are enacted, it clearly makes sense to obtain patents, especially to DNA sequences, whenever possible. Better to have a patent in hand and question its appropriateness privately (of course) than be forced to examine the other person's patent and have to question its appropriateness publicly.

Copyrights, trade secrets, and trademarks also have their place in protecting biotechnical intellectual property. However, in the limited space left, I will only focus on the latter because of recent changes that may affect the entrepreneur's timing of securing trademark protection. (The Trademark Revision Act of 1988 took effect on November 16, 1989, for registrations and renewals after such date.) Previously, actual use of the mark in interstate commerce was required prior to the filing of an application to register such a mark on the Principal Register. Registration was of twenty years' duration and thereafter was renewable on a showing of token use.

The new revisions allow for the filing of an application on a bona fide allegation of intent to use. Before issuance of the registration and within six months after the date of the application, however, the applicant must make actual commercial use of the mark. The six month grace period may be extended for another six months on submission of a written request and payment of a fee. Thereafter, up to four additional six-month periods may be requested on a showing of good cause (as yet undefined). In order to help clear the Register of unused marks and make them available to the public, the new registration is effective for only ten years and renewable on a demonstration of actual use; token use is no longer sufficient. Priority is now awarded to the earlier of the application filing date or the date of first use. The revision does not, however, alter the requirement of actual use prior to registration on the Supplemental Register, and, accordingly, for those marks that would be descriptive or are surnames it remains advisable to establish a bona fide commercial use before filing. Still, the changes are particularly helpful in instances when one is presented with a number of possible choices, which will only get narrowed down to the final choice on final marketing information or clearance checks. Thus, a working reserve of "marks" can be created as a particular project approaches its commercial phase.

Hopefully, this chapter has given you a sense of the dynamic character of intellectual property, particularly in the biotechnology arena. No chapter this short can hope to be comprehensive, however. Given the space constraints, I have attempted to mix conventional garden variety wisdom with insights into the limitations of such understanding in its application to biotechnology. The opinions espoused are my own and do not necessarily reflect those of my employer, Genzyme Corporation. Although the opinion of counsel should be sought as early as possible prior to initiating any new project, company, or other enterprise in order to appreciate the specific role of intellectual property, you need not have the benefit of such counsel to recognize that patents are likely to have an important role in the success of your endeavor.

Postscript: In the interim since this chapter was written, the Delaware jury found valid the Genentech tPA patent in the *Genentech v. Burroughs-Wellcome* case; the CAFC ruled on the appeal in the *Amgen v. Chugai Pharmaceutical and Genetics Institute* case invalidating G.I.'s purified EPO patent for failure to enable and generally upholding the validity of Amgen's EPO encoding DNA patent. The CAFC ruling thereby vindicates Amgen's "king of the mountain" position.

REFERENCES

Amgen v. Genetics Institute and Chugai, U.S.P.Q. (Dist. Ct. Ma., December 11, 1989).

Cooper, I. P. 1987. *Biotechnology and the law.* New York: Clark Boardman.

Diamond v. Chakrabarty, 206 U.S.P.Q. 193 (1980).

Genentech Inc. v. Wellcome Foundation Ltd., 14 U.S.P.Q. 2D 1363 (1990).

Graham v. John Deere, 148 U.S.P.Q. 459 (U.S.S.Ct., 1966).

Hearing before the Subcommittee on Regulation and Business Opportunities of the Committee on Small Business, House of Representatives, 100th Congress, 2nd Session, Washington, D.C., March 29, 1988. Serial no. 100-59.

Hybritech v. Monoclonal Antibodies, 802 F.2d 1367, 231 U.S.P.Q. 81 (CAFC, 1986).

Kohler, G., and Milstein, C. 1975. Continuous cultures of fused cells secreting antibody of predefined specificity. *Nature* 256:495.

Pioneer Hi-Bred Ltd. v. the Commissioner of Patents [Canada, June 22, 1989].

Rosenberg, P. D. 1980. *Patent law fundamentals.* New York: Clark Boardman.

Thurston, J. P. 1989. 66 Case Comment 2EIPR.

Watson, J. D., and Crick, F. H. C. 1953. Molecular structure of nucleic acids; a structure for deoxyribose nucleic acid. *Nature* 171:737-738.

CHAPTER 10

IMPACT OF GOVERNMENT REGULATION ON COMMERCIAL BIOTECHNOLOGY

David J. Glass

For much of commercial biotechnology's history, the need to comply with government health and safety regulations has been a major factor affecting the development of products. To some, it may seem paradoxical that a technology based on the exploitation of nature's own molecules and processes should become subject to excessive regulation and widespread public concerns over safety. Yet, this should not be surprising when you consider that biotechnology is the first major new technological innovation to be commercialized since the potential health and safety risks of older industries became evident, often through well-publicized disasters. The general public is no longer willing to accept blindly government or industry assurances that any new technology is safe, particularly when their own neighborhoods are involved, and most people prefer to believe that any technology is dangerous unless "proven" to be 100% safe. It was in this context that the biotechnology industry matured, and for this reason, government regulation has played a prominent role in that maturation.

Many industries or technologies are, of course, subject to significant amounts of health and safety regulation, generally at the time products are to be introduced into the marketplace. Indeed, the industry most affected by the biotechnology revolution, pharmaceuticals, is one of the most heavily regulated industries in the United States. What is significant about biotechnology regulation is that substantial amounts of federal and local oversight apply to the use of the technology at the research stage and that a broad regulatory scheme for both research and commercialization has evolved in the absence of any proven risks associated with the regulated technologies. The latter phenomenon is unique to biotechnology, and is clearly an outcome of the public's uneasiness over technology in the 1970s and 1980s. Although biotechnology regulation is becoming more firmly rooted in scientific principles and less in fear of the unknown, the field is also subject to broader trends that will affect the way technology-based products are regulated in the 1990s.

This chapter will review the history of biotechnology regulation and summarize its status at present. Particular emphasis will be placed on how regulations currently affect the commercialization of biotechnology products and how particular regulations create opportunities for market exclusivity that can be used to a company's benefit. Trends and important issues being considered for the future will also be discussed. This chapter will mostly concentrate on the United States, but will also briefly mention recent international developments.

HISTORICAL PERSPECTIVE

Early Years: Health and Safety Concerns over Laboratory Research

Ironically, much of the history of biotechnology regulation owes itself to the decision of the pioneering scientists of recombinant DNA technology to take unprecedented steps to examine publicly the safety of the technology at the very outset of its use in research laboratories. This action, widely praised at the time, has lately been criticized in hindsight for misleadingly guiding the tone of all future public policy discussions toward an assumption of risk.

The early days of the public debate over the safety of recombinant DNA research have been well documented [Krimsky 1982]. The announcement that scientists had invented the tools to combine genetic material from different sources resulted in enormous public interest and generated significant discussion on moral, ethical, and safety questions. In response to this debate, leading scientists in the field called for a voluntary moratorium on certain types of recombinant DNA (rDNA) experiments, while also convening a scientific meeting to discuss and evaluate what was known about potential health and safety effects of the technology. This meeting, held at the Asilomar Conference Center in California in 1975, recommended that research be allowed to proceed as long as appropriate safety guidelines were developed.

The task of developing such guidelines fell to the National Institutes of Health (NIH), which promulgated the first version of "Guidelines for

Research Involving Recombinant DNA Molecules" in June 1976 (41 *Federal Register* 27902). The Guidelines attempted to classify all categories of possible rDNA experiments on the basis of potential risk. Certain experiments, judged the riskiest, were prohibited. For those that would be allowed to proceed, the Guidelines specified applicable procedures and safeguards under which they could take place. To do this, the Guidelines defined four levels of "physical containment," specifying both the required features of laboratory design and the standards for laboratory practice at each containment level. The lowest level of containment, P1 (later called BL1, for "biosafety level one"), was reserved for the experiments thought to be the safest, while riskier experiments were assigned to the higher-numbered categories.

The Guidelines also empowered an administrative office within the NIH (the Office of Recombinant DNA Activities) and an outside scientific advisory committee, the Recombinant DNA Advisory Committee (RAC), giving them the responsibility to oversee compliance. RAC approval was needed for certain categories of experiments, and the RAC was given the power to recommend changes to the Guidelines themselves. The Guidelines required that any institution conducting rDNA experiments establish an institutional biosafety committee (IBC), including members from outside the institution, to oversee compliance and review rDNA experiments.

The NIH Guidelines, however, did not have the power of full-blown regulation. They were mandatory only for experiments conducted at institutions receiving funding from NIH or another government funding body. The penalty for noncompliance was loss of federal funding: a severe penalty for a scientist, but far short of a criminal or civil penalty under the law. The NIH suggested that institutions not receiving federal funding comply voluntarily with the Guidelines.

The fact that the Guidelines were not mandatory was disturbing to some. Congress briefly considered passing legislation making compliance mandatory, but ultimately took no action [Krimsky 1982]. This led to the passage in the late 1970s of several local ordinances and two state laws that required compliance within those jurisdictions. The most famous of these was adopted in Cambridge, Massachusetts, in 1977, after a frenzied series of City Council debates. These first ordinances did little more than mandate compliance with the Guidelines, but, beginning in 1981, a number of communities began adopting ordinances requiring institutions to obtain permits for rDNA research and larger-scale manufacturing. These laws were designed to give local governments better control over biotechnology research in the industrial laboratories that were by then beginning to appear. Cambridge was again the first to pass such a law, and several other towns and cities, all in Massachusetts, followed suit throughout the

1980s [Krimsky, Baeck, and Bolduc 1982]. Today, about twenty communities and four states have such laws on the books.

The NIH Guidelines remain in effect to this day, with the most recent version published at 51 *Federal Register* 16958–16985. They have been progressively relaxed as more experience has been gained with rDNA techniques, and as specific risk assessment experiments have shown initial concerns to be largely unfounded. Today, very few classes of experiments require the approval of an IBC, much less RAC approval, and the bulk of rDNA experiments are conducted at safety level one or lower. In fact, it was estimated several years ago that 80 to 90% of the rDNA experiments conducted across the country are considered so safe as to be exempt from the Guidelines [Milewski 1981].

Later Years: Environmental Concerns Lead to Commercial Regulation

The combination of voluntary research guidelines and a patchwork of local ordinances served as the only regulation directed specifically at biotechnology for some time. Commercial firms using biotechnology to develop new drugs and diagnostics were approaching the Food and Drug Administration (FDA) about the regulations for such products. FDA responded by issuing a "Talk Paper" in January 1983, stating that the agency would use its established procedures for these types of products, but that rDNA-derived products would require new approvals, even if the product was identical to a previously approved product. Aside from requesting information about the genetic changes introduced into the production organism, FDA was creating no new requirements for rDNA.

However, beginning in 1983, academic and commercial biotechnology research programs began to move out of the laboratory and into the environment. This development triggered a concern over possible environmental effects of biotechnology, while also exposing the fact that the federal government had done very little thinking about regulating the many areas of impending commercial uses of biotechnology.

Concerns first arose in the area of agricultural applications of rDNA technology. Agricultural products developed by any technology, and tested in laboratories or greenhouses, must also be tested in the field under conditions approximating actual farming use, before they can be approved and purchased by farmers. The first such field tests involving products produced by rDNA began to be proposed in 1982 and 1983, and the outcome of these proposals showed that guidelines adopted by a biomedical institute for indoor laboratory research were not well suited to cover outdoor agricultural experimentation.

The first few of these proposals all arose from academic research laboratories, which were subject to the NIH Guidelines. So-called deliberate releases to the environment were among the experiments originally prohibited under the Guidelines, but which were later put into a class in which they could be allowed on approval by both the RAC and the director of the NIH. In 1982 and 1983, three experiments involving genetically engineered plants were approved by the NIH, and in 1983, the NIH issued an approval for the first field test of an engineered microorganism. This was a proposed field test by professors at the University of California at Berkeley of an organism later dubbed the *Ice Minus* bacterium. It was a variant of a naturally occurring microbe that has a crucial role in ice nucleation on plant leaves: the natural microorganism contains a protein that promotes ice crystal formation, while in the engineered variant, the gene encoding this protein has been deleted. The Berkeley professors wanted to field test the variant's ability to allow potato plants to withstand freezing temperatures a few degrees lower than normally possible.

The NIH granted approval for this experiment in June 1983, following review by the RAC over the course of several months. Shortly before the experiment was to begin, it was stopped by a lawsuit filed by a coalition of activist groups led by author Jeremy Rifkin, alleging that the NIH had not followed proper federal procedures for assessing environmental risks in approving the experiment. Rifkin alleged that the NIH should have prepared an environmental assessment (EA) under the terms of the National Environmental Policy Act, and made that assessment available for public comment before issuing a final approval. Rifkin eventually won the suit (although a final decision took several years), and the NIH was forced to prepare environmental assessments for any deliberate-release experiments it intended to approve. [See Krimsky and Plough 1988, for a detailed history of the *Ice Minus* case.]

Although the court case took years to resolve, its impact on commercial biotechnology was more immediate. Many companies, particularly those developing products outside the pharmaceutical area, came to realize that the quasiregulatory structure under the voluntary NIH Guidelines was not well suited for the regulation of commercial products. This was particularly true if commercial deliberate-release experiments were to require the lengthy EA process over and above the already-long NIH review process. The Rifkin lawsuit opened the very real possibility that proposed field tests would be continually delayed by these time-consuming procedures, the need for public comment periods, and the ever-present threat of litigation by activists opposed to the technology.

At the same time, the federal government was beginning to realize that it needed a regulatory process more stringent than the NIH Guidelines in order to oversee the growing biotechnology industry. As noted above, the FDA was basically conducting "business as usual" with biotechnology products, but other agencies began contemplating how they might use their existing statutes and regulations to apply to biotechnology. The most prominent (and to the industry, most startling) of these was the Environmental Protection Agency (EPA), which, in the summer of 1983 announced that it was considering regulations under one of the laws it administers that could have potentially covered most, if not all, aspects of commercial biotechnology, even those that were already regulated elsewhere in the federal government. The law to be used was the Toxic Substances Control Act (TSCA), which empowers the EPA to be notified of any "new chemical substance" intended for commercial manufacture. EPA's 1983 announcement was that, by defining new microorganisms to be "new chemical substances," it could require premanufacture notification of any genetically engineered organism to be used in commerce [Budiansky 1983].

This announcement was quite controversial. First, many questioned the decision to call an organism a "chemical substance." Second, there was concern within the industrial community that the EPA would impose strict regulations on biotechnology activities, even on those activities already regulated by other federal agencies.

In the light of these developments, the Reagan administration decided to resolve the broader issues by convening an interagency committee to develop a unified regulatory approach to biotechnology. Critics have alleged that this action was taken to "rein in" the EPA and prevent it from imposing sweeping regulations. Although there may be some truth to this allegation, the fact that the commercial use of biotechnology stood to be affected by the regulatory programs of several federal agencies argued strongly for an interagency approach.

Thus, in April 1984, such a committee was formed as a Working Group of the Cabinet Council on Natural Resources and the Environment. The Working Group included representatives from seventeen federal agencies and cabinet departments. Its task was to "study and coordinate the government's regulatory policy for [biotechnology] products," and it published the results of its work in a Federal Register notice on the last day of 1984 (49 *Federal Register* 50856-50907). This notice, entitled "Proposal for a Coordinated Framework for Regulation of Biotechnology," had three components. First, it published a "matrix" of existing laws and regulations applicable to biotechnology products. Second, the FDA, the EPA, and the U.S. Department of Agriculture (USDA) each stated its policy with regard to biotechnology products. Third, the notice proposed a mechanism for the oversight of biotechnology regulation by two interagency committees: one to oversee scientific issues and one to cover policy issues. A public comment period on this notice followed.

The regulatory matrix was useful in that it established what would become a cornerstone of federal regulatory policy: that existing laws and regulations would suffice for regulation of biotechnology products. The agency policy statements were, in all cases, preliminary: the FDA and the USDA made no new proposals beyond their existing regulations, while the EPA put forward an initial proposal to apply TSCA and the pesticide law FIFRA (Federal Insecticide, Fungicide, and Rodenticide Act) to regulate biotechnology products. The EPA statement was controversial, primarily because the EPA was proposing a "process-based" approach, choosing to regulate only those organisms produced by rDNA and related techniques. The general oversight proposal was also controversial in that it contemplated creation of a body, the Biotechnology Science Board, that many felt would have had the undesirable power to question or overrule scientific decisions made by individual regulatory agencies.

This *Federal Register* notice generated a great number of comments from the public, enough that the Working Group eventually decided to rethink some of its basic assumptions. Its next public pronouncement was not until November 1985 (50 *Federal Register* 47174), when the Biotechnology Science Coordinating Committee (BSCC) was created as the new interagency committee. Established under the provisions of the Federal Coordinating Council for Science, Engineering, and Technology, the BSCC consisted of senior policy officials from agencies primarily involved in oversight of biotechnology research and commercialization. The BSCC assumed responsibility for continuing development of the coordinated framework, and published an updated version on June 26, 1986 (51 *Federal Register* 23202-23393). This notice provides the basis for current regulation of biotechnology, and will be described in more detail in the next section. The BSCC is still in existence, although its intended coordinating role never fully materialized, as each agency separately developed its own policies and guidelines. Recently, however, it has been at the center of a controversial attempt to create a common definition that all agencies could use to define the scope of organisms subject to regulatory oversight [Fox 1990] (and see below).

In fact, other coordinating mechanisms have arisen within specific federal agencies. In 1986, the USDA created the Office of Agricultural Biotechnology (OAB), to coordinate that department's biotechnology activities. The OAB is housed in the USDA's research branch, where it has overseen activities such as the Agricultural Biotechnology Research Advisory Committee and other programs related to agricultural research and field testing [Jones 1988]. However, the OAB has not been effective in relieving the political difficulties between the USDA's research branch and its regulatory divisions. More recently, the FDA established an Office of Biotechnology as an ombudsman for biotechnology companies going through the agency's regulatory process and as a central coordinating mechanism for that agency's biotechnology-related activities [*Biotechnology Newswatch* 1990a].

The coordinated framework largely endorses a status quo regulatory structure for the bulk of biotechnology-derived products. The entire interagency effort, while directed at the universe of biotechnology products, was in actuality driven by the need to regulate adequately a small number of products that were judged not to be covered by existing regulations: living organisms intended for use in the environment. However, because the government chose to do so by creating a regulatory structure for all biotechnology products, it perpetuated the public perception that biotechnology was something dangerous, in dire need of regulation. The ramifications are being felt to this day.

THE COORDINATED FRAMEWORK FOR BIOTECHNOLOGY REGULATION

Philosophy Behind the Coordinated Framework

The philosophy behind federal regulation under the coordinated framework is that biotechnology products can be regulated on a product-by-product basis, under existing statutory authority. In spite of charges by critics that this philosophy is politically motivated, this policy is sensible because biotechnology is a tool being applied to many industries, resulting in products of many different types. It would not make sense to abandon the product-specific regulatory structure that has existed for decades and create a new system for a small class of dissimilar products whose only similarity is their method of manufacture.

Hence, biotechnology products are regulated under the scheme shown in Table 10-1, which follows from the product categories traditionally regulated by three federal agencies, the FDA, the EPA and the USDA. The FDA, with existing authority over human drugs and biologics, animal drugs, medical devices, and human and animal foods and food additives, regulates the lion's share of biotechnology products, all under existing statutory power. The EPA uses existing authority to regulate new chemicals produced by biotechnology (under TSCA), as well as pesticides derived from or containing microorganisms (under FIFRA). Products regu-

TABLE 10-1 PRODUCT-SPECIFIC REGULATION OF BIOTECHNOLOGY

Food and Drug Administration
Human drugs
Human biologics
Medical devices
Animal drugs
Human foods and food additives
Animal feeds
U.S. Department of Agriculture
Animal biologics
Animal pathogens
Plant pests
Transgenic plants
Engineered agricultural microorganisms
Environmental Protection Agency
Pesticides
New chemicals
Miscellaneous environmental uses of microorganisms

lated under traditional USDA authority are primarily animal biologics and certain microbes derived from plant or animal pathogens.

The coordinated framework created a small number of new regulatory categories, in which existing laws could be used in new ways (preferably under new rules), to apply to biotechnology products. These types of products were primarily those engineered organisms intended for use in the environment. Thus, the USDA is using the Plant Pest Act to regulate environmental introductions of transgenic plants and most agricultural microorganisms. The EPA is using TSCA to oversee commercially funded deliberate releases of microorganisms not regulated elsewhere in the federal government. Outdoor experiments conducted by academic or government laboratories are to be regulated by the agency supervising or funding the work, such as the NIH, the USDA or the National Science Foundation. For example, the science branch of the USDA recently approved a proposal for outdoor growth of transgenic fish by a university research group (55 *Federal Register* 5752-5757), and a proposed agricultural field test by a University of Wisconsin researcher (55 *Federal Register* 29643-29645).

With very few exceptions, the policies being used to regulate these products are identical to those used for their nonbiotechnology counterparts. These policies are described in the sections below, which summarize regulatory policies and procedures for each important product area. Citations to individual regulations can be found in the "matrix" in the coordinated framework (49 *Federal Register* 50860-50877).

Human Drugs and Biologics

The FDA employs a three-tier system of testing to ensure that new drugs and biologics are safe and efficacious before being brought to market. When a new drug is identified in a research and development program, it must undergo preclinical testing in animals to assess safety and biological activity. This data is used to support an Investigational New Drug application (IND), which is submitted to the FDA in order to commence human clinical studies.

Clinical tests proceed in three phases. In Phase I, the drug is tested in a small number of healthy human volunteers, mostly to measure its safety, pharmacological effects, and how it is metabolized. Phase II, employs a larger group of patients actually suffering from the targeted disease and is designed to begin measuring the drug's effectiveness and to determine side effects. These tests are usually carried out at university or government medical centers. Finally, Phase III studies test the efficacy of the drug in 1000 to 3000 patients in many locations, and these trials must also show a low incidence of adverse effects.

After these trials, if the drug performs as expected, the sponsor must file a New Drug Application (NDA) with the FDA's Center for Drug Evaluation and Research, or in the case of a biologic, a product license application (PLA) with the Center for Biologics Evaluation and Research. These applications include all the data developed during clinical trials. FDA approval of an NDA or PLA gives the sponsor clearance to sell the drug for the indications approved. New drug applications are usually approved within about two years, and the entire regulatory process can take seven to ten years and cost up to $200 million, according to one recent estimate [*Biotechnology Newswatch* 1990b].

FDA drug regulation also extends to certain procedures and practices that must be followed in the conduct of clinical studies and in the manufacturing process itself. These procedures, known as Good Laboratory Practice and Good Manufacturing Practice, control almost every aspect of the drug development process, from research and development to clinical testing to pilot- and commercial-scale manufacture.

The FDA follows these standard procedures for the review of biotechnology-derived drugs and biologics, with certain additional data requirements for products of biotechnology. These requirements, largely relating to characterization or purity of the production organism, have been embodied in a series of "Points to Consider" documents (e.g., see 53 *Federal Register* 10948 for a notice of availability of one).

In spite of the length and costs of this process, 10 therapeutic products derived from biotechnology have been approved for sale by the FDA as of March 1990. Most notable among these are human insulin, tissue plasminogen activator (a clot-dissolving protein used as a heart attack treatment), and erythropoietin (a protein that boosts red blood cell counts, used for anemia and related diseases). Total U.S. sales in 1989 for all ten products have been estimated at $900 million [*Biotechnology Newswatch* 1990b]. It has recently been estimated that there are at least 104 other new biotechnology-derived drugs at various stages of the clinical research pipeline [see *Biotechnology Newswatch* 1990c for a report of a survey by the Pharmaceutical Manufacturers Association].

Animal Drugs and Biologics

Animal drugs are regulated by FDA's Center for Veterinarian Medicine quite similarly to the way in which human drugs are regulated. Sponsors must go through the Investigational New Animal Drug and New Animal Drug Application processes, which are analogous to the procedures for approval of human drugs. Although agency approval is not needed to commence clinical investigations of new animal drugs, the agency must approve the use of edible products derived from food-producing animals in which the drug has been tested. Approval of an animal drug can be just as time-consuming and expensive as approval of a human drug or biologic, particularly for drugs to be used in food-producing animals.

Animal biologics such as vaccines are regulated not by the FDA, but by the USDA's Animal and Plant Health Inspection Service, under the Virus, Serum, and Toxin Act. To obtain a product license for an animal biologic, the sponsor must prove purity, safety, potency, and efficacy of the product. USDA approval of these products is needed not only for commercial sale, but also for interstate transport.

The USDA and the FDA have entered into a memorandum of understanding to help ascertain under which law a given product might be regulated (47 *Federal Register* 26458). In one application of this memorandum, the FDA was given authority to regulate bovine interferon as a drug [Mackler and Lawrence 1983]. Animal therapeutic products that have been approved include vaccines for pseudorabies and scours. FDA approval is expected in 1991 for bovine somatotropin, a natural cow hormone used to improve milk production, which has attracted controversy in dairy states principally over its potential socioeconomic impacts.

Diagnostics

Diagnostics products for *in vitro* detection of human diseases or medical disorders are regulated as "medical devices" under the 1976 Medical Device Amendments to the Food, Drug, and Cosmetic Act. Devices are classified in one of three categories, depending on the risk of illness or injury associated with their use, with Class I being the lowest risk and Class III the highest. There are two basic schemes for approval of a medical device. For Class III devices, a premarket approval application (PMA) must be filed with FDA's Center for Devices and Radiologic Health, which will approve the application based on proof of safety and efficacy in clinical trials. Clinical trials are generally conducted under the terms of an investigational device exemption (IDE), which often requires approval by the FDA. The FDA has 180 days to review a PMA, but often extends this period by requesting more data. Approval of a device under the IDE–PMA procedures usually takes about two to three years.

An alternative route to approval is available for Class I or II devices that can be shown to be "substantially equivalent" to a device already on the market. This abbreviated procedure, called "510(k)" after the section of the act creating the procedure, often allows diagnostic products to come to market within the ninety-day statutory review period, after proof of equivalence to the existing device, which is usually shown by clinical trials.

Diagnostic tests are by far the largest category of biotechnology product approved by any agency. About 400 diagnostic tests for various human diseases or conditions, making use either of monoclonal antibodies or DNA probes, have been approved by the FDA to date [Miller, H.; personal communication].

Pesticides

Most of the genetically engineered microorganisms that will be used in the environment will be pesticides of some sort, and are therefore regulated by EPA's Office of Pesticide Programs, using existing regulations under the pesticide law FIFRA. All pesticide products, such as insecticides, herbicides, and fungicides, must be registered by the EPA before they can be sold commercially. However, it is necessary to conduct several years of field tests first to identify commercially viable pesticides, and then to gather the data needed to support EPA registration. For traditional chemical products, the EPA has realized the experimental nature of early, limited acreage field tests, by exempting them from regulatory scrutiny: experimental pesticides can normally be field tested at any level up to ten acres without EPA oversight. Only field-testing activities above this limit require companies to obtain an experimental use permit (EUP).

Under the EPA's biotechnology policy, however, the ten-acre exemption has been removed for a large class of genetically altered microbial pesticides, particularly including ones of recombinant origin. For such products, companies must notify the EPA of any proposed field test, regardless of size, and possibly obtain an EUP in order to conduct the test.

Although the EPA has circulated drafts of proposed rules that would narrow the scope of this oversight, it would still affect many biotechnology-derived microbial pesticides.

Because of a concern that genetically modified microorganisms might present unique issues of environmental behavior that are not seen with chemical products [see Alexander 1985], the EPA generally conducts risk assessments of even the smallest field tests of engineered microorganisms. Field-test proposers often submit detailed documents describing the genetic makeup of the organisms to be tested, laboratory and greenhouse studies addressing the potential environmental effects, the design of the field test, and measures to be taken to reduce possible dissemination of the organisms beyond the test site. It is often necessary to monitor the environmental behavior of the organisms during the field test by appropriate soil and air sampling.

After notification, certain small-scale tests of engineered microbes might be judged to need EUPs. Through 1989, twelve field tests of recombinant microbial or viral pesticides have been conducted under EUPs, and three such tests have proceeded without EUPs. Some thirty others, involving nonrecombinant alterations, have taken place under notifications alone [Betz 1990].

All larger-scale field tests, even for naturally occurring microbial pesticides, will require EUPs. To obtain these permits, it is usually necessary to conduct much of the "first tier" of testing that would later be required for product registration: product characterization, short-term toxicological testing, and a certain amount of environmental effects testing, including effects on nontarget organisms [Betz 1985]. In addition to permits issued for small-scale tests of engineered microbes, the EPA has awarded a number of EUPs for larger-scale testing of naturally occurring or mutant microbes and one recent EUP for a killed recombinant microbe. The review process is straightforward and requires about four to six months. Testing of genetically engineered microbial pesticides may also require a permit from the USDA, under the Plant Pest Act (see "Transgenic Plants," p. 183).

Registration of microbial pesticides for commercial sale follows existing EPA regulatory procedures [Betz 1985]. A complete set of the "Tier I" testing described above must accompany any request to register a pesticide, and depending on the results of these tests, additional testing in Tiers II, III, or IV may be needed. Generally speaking, natural microbial pesticides will not require additional testing, but some genetically engineered products might be judged to need some Tier II tests. Registration applications must also include proof of the product's efficacy. Most product registrations take at least six months from the time of submission.

Although microbial pesticides can be registered for substantially lower cost than chemical pesticides, the scope of required testing still results in large expenses. Estimates of the costs to register chemical products range from $10 million to $50 million, whereas it is possible to register a microbial agent for as little as a few hundred thousand dollars, or, at most, a few million dollars [see, for example, Rich 1988]. To date, no genetically engineered products have been registered, but about two dozen naturally occurring or conventionally modified microbes and viruses, in many different formulations, have been approved as pesticides by the federal government since the 1940s.

Chemicals and Miscellaneous Environmental Uses

The EPA is using TSCA to regulate the microbial production of certain chemicals or enzymes not regulated elsewhere in the government, as well as deliberate-release experiments not otherwise regulated. TSCA is a statute requiring manufacturers to notify the EPA's Office of Toxic Substances at least ninety days before commencing manufacture of any "new" chemical, i.e., one that is not already in commercial use. In the coordinated framework, the EPA ultimately limited its TSCA oversight to "intergeneric" organisms, deciding that such organisms, containing genetic material from more than one taxonomic genus, could be considered to meet TSCA's definition of "new." This definition may change, depending on the EPA's response to the "scope guidelines" (see below) in upcoming rule making.

EPA review of chemicals under TSCA is straightforward, and the vast majority of premanufacture notices (PMNs) are routinely approved within the ninety-day statutory review period. The EPA's use of TSCA to regulate contained biotechnology fermentations (e.g., for production of enzymes and chemicals) has been similar to this long standing practice. For commercial fermentation processes using intergeneric organisms, sponsors need to submit certain information pertaining to the genetic changes introduced, and the procedures that will be used to contain the organisms during fermentation (such as the NIH Guidelines). Because there is wide consensus that such indoor uses pose virtually no environmental risks, the EPA has routinely approved about a half dozen "contained use" PMNs between 1987 and the end of 1989.

To regulate deliberate releases, the EPA has removed TSCA's small-volume exemption for intergeneric microbes used in the environment. This exemption is usually available for certain activities, principally research and development, believed to pose extremely low environmental risks. In the coordinated framework, the EPA announced that, because of the potential for microorganisms to spread or multiply in the environment, it would request PMN reporting for all environmental uses of intergeneric organisms, even for research purposes, regardless of the scope.

The EPA's application of TSCA to environmental uses resembles its regulation of microbial pesticides. Companies have submitted similar documents describing the makeup of the organism, its potential environmental effects, the field-test layout, and a monitoring plan. Toxicological testing has not always been required. The EPA has generally completed risk assessments of field-test proposals within the ninety-day review period.

In order to approve a field test proposed in a PMN without also approving unlimited commercial manufacture, the EPA has been asking submitters to enter voluntarily into consent orders with the agency. These orders grant the company authority to conduct field trials under protocols expressly approved by the agency, while prohibiting further commercial activities. The EPA can usually conclude a consent order within about one additional month, resulting in an approval process of about four months or less. The EPA has approved six small-scale field tests of nonpesticidal agricultural microbes under TSCA through the 1990 growing season. It is presumed that such consent orders will be modified and eventually withdrawn to allow products to move into larger-scale testing and, ultimately, commercialization.

The use of genetically engineered organisms in waste cleanup, enhanced oil recovery, and other environmental applications would probably also be regulated under TSCA policies. To date, all applications in these fields have involved only nonrecombinant organisms, so that the biotechnology policies do not apply. Waste treatment applications may be subject to other EPA regulation under the hazardous waste laws.

Transgenic Plants

The use of genetic engineering to modify food-producing crop plants will be regulated at two stages of commercial activity: field testing and use in human or animal food.

Early field testing of transgenic plants is being regulated by the USDA, through their Animal and Plant Health Inspection Service (APHIS). In the coordinated framework, APHIS proposed new regulations under the Plant Pest Act that would allow it to review in advance proposed environmental uses or interstate shipments of genetically engineered plants or agricultural microorganisms. These regulations were finalized with some revisions in June 1987 (52 *Federal Register* 22892-22915). An institution wishing to conduct a field test of such organisms must consult APHIS and possibly apply for a permit to conduct the test. These applications include a description of the modifications made to the organism, data characterizing the stability of these changes, and a description of the proposed field test and the procedures to be used to "contain" the organisms in the test plot. Submitters must also assess potential environmental effects, such as the chance that the new plants might become weeds.

APHIS review of these field-test proposals has been straightforward, and is usually completed well within the 120 days allowed by the regulations. APHIS must also obtain the consent of the state where the test will take place, and this has not proved to be problematic. From 1986 to mid-July 1990, APHIS has approved almost 100 permits for field tests of transgenic plants (as well as eight for engineered microbes) in twenty-nine different states, encompassing twelve different plant species. All tests conducted to date have been small-scale, and although early tests with any given crop were conducted under certain constraints (e.g., prevention of flowering and pollen dissemination), these conditions have usually been removed or reduced in subsequent testing. This level of oversight is generally considered to be consistent with the lower level of perceived environmental risks associated with outdoor use of plants as opposed to microorganisms.

Later uses of transgenic food plants will likely encounter some regulatory oversight regarding their possible food use. This is a major emerging issue for the regulation of agricultural biotechnology, and will be discussed in more detail below. However, the following section will describe the general regulatory scheme for foods and food additives as it might apply to biotechnology products.

Foods and Food Additives

It appears likely that the FDA will become involved when biotechnology products intended for use in human or animal food begin to come to market. At this time, the scope of FDA regulation of such products is far from clear, but it will no doubt arise from existing authority under the Food, Drug, and Cosmetic Act. FDA's Center for Food Safety and Applied Nutrition is charged with preventing "adulterated" foods from being sold and is also responsible for approving the use of specific food additives. Companies wishing to market a product as a food additive must obtain FDA approval of a petition that includes information regarding the make-up of the additive and its safety in the proposed use, usually supported by toxicological testing. The agency also classifies certain foods and food additives as "generally recognized as safe" (GRAS). Although this category was created to cover food products with many years' historical records of safety, it is possible to seek GRAS status for new foods or ingredients. This can be done by submitting and obtaining FDA approval of a GRAS Affirmation Petition, but it is also possible for a company to decide unilaterally that a product is GRAS. Either action must be based on the appropriate publicly available and accepted data pertaining to safety.

In its policy statement in the coordinated framework, the FDA stated that it will review proposed food uses of biotechnology products on a case-by-case basis, but that it presumes that most recombinant DNA products will require agency review to be added either to the food additive or to the GRAS list. The FDA is already reviewing a number of food additive and GRAS petitions for food-processing enzymes produced through recombinant DNA, and in March 1990 it approved the first of these, to allow the use of recombinant-derived rennin in cheese processing (55 *Federal Register* 10932-10936). The FDA has not yet formally addressed the use of transgenic plants in human or animal food.

Miscellaneous Regulatory Issues

Laboratory regulation and worker safety. Biotechnology research is affected by a number of laboratory safety regulations, in addition to the NIH Guidelines. These include various safety standards applicable to many industries, such as radiation protection standards, hazardous waste disposal regulations, and federal and state right-to-know laws. More broadly, regulatory authority for worker safety rests with the Occupational Safety and Health Administration (OSHA), and similar state agencies. OSHA has made one policy statement regarding biotechnology (50 *Federal Register* 14468-69), which indicated that it would treat biotechnology manufacturing activities similarly to the way conventional processes are regulated. Stand-alone research biotechnology laboratories are regulated under a recently issued OSHA "Laboratory Standard" (55 *Federal Register* 3300-3335), which creates a "performance-based" approach to safety in all research laboratories.

Imports, exports, and interstate shipping. A variety of regulations affect the movement of biological materials between countries and between states. The U.S. Commerce Department's Bureau of Export Administration regulates exports of all commodities, and maintains regulations specific for biological materials, based on concern over spread of biological weapons. Regulations that took effect in early 1989 created a situation in which almost all exports of recombinant organisms would require a permit from the Commerce Department. In response to public comments that these regulations were overly broad, in March 1990 the department published revised rules requiring permits only for exports of potentially pathogenic organisms (55 *Federal Register* 11361-11362).

Imports into the United States might be regulated by the USDA's APHIS branch, which maintains regulations aimed at preventing the introduction into the United States of plant pests, noxious weeds, or animal pathogens. These regulations apply only to the specific species listed in the regulations. The same regulations affect interstate shipment, with consent of the agriculture department in the receiving state required for approval of the needed permits.

EMERGING ISSUES FOR THE 1990s

In spite of occasional allegations that the federal regulatory process is "confusing" or "in chaos" [e.g., Jaffe 1987 or Mellon 1989] the framework as described above is remarkably established and settled. However, in several particular areas, changes (often controversial) are likely to occur in the years to come. This section will highlight some issues likely to come into prominence during the first years of the 1990s.

Changes in the FDA Drug Approval Process

Recent developments relating to new therapies for life-threatening diseases are resulting in calls for changes to the long-established FDA procedures for clinical evaluation and regulatory approval of new drugs and biologics. Although not specific for biotechnology-derived drugs, these changes, if they come to pass, would profoundly affect the development of certain products of biotechnology.

Much of the impetus for change has come in relation to drugs being developed to treat the acquired immunodeficiency syndrome (AIDS). Because the disease is believed to be 100% fatal, activists on behalf of people with AIDS have argued that the FDA should abandon its usual concern that drugs must be proven safe before they can be marketed, since AIDS patients literally have nothing to lose in trying a drug that might produce side effects that would be unacceptable in other patient populations.

This concern led to the adoption by the FDA in 1987 of a new procedure known as the "treatment IND" (52 *Federal Register* 19466). Under a treatment IND, patients seriously ill can receive an experimental drug once it has completed Phase I and Phase II clinical testing. In these cases, Phase III testing would continue as planned, and data from the patients receiving the drug under the treatment IND might also be used along with the Phase III data. Several AIDS drugs have already been approved for use under treatment INDs, including azidothymidine (AZT) and dideoxyinosine (DDI).

The apparent success of the treatment IND process has led to calls for further streamlining of the drug approval process. One proposal, called the *parallel track*, would make experimental drugs available even earlier, after completion of Phase I trials. Again, this would apply only to seriously ill patients, for diseases such as AIDS or cancer, and the remainder of the approval process would proceed as usual.

These specific proposals come at a time when the FDA has been investigating ways to streamline the routine approval process for all drugs. Through measures like early consultations between the FDA and the manufacturer, final approval for certain AIDS and cancer drugs have been expedited, and a committee formed by then–Vice President Bush in 1988 has recently proposed further improvements. For this reason, the concepts of treatment INDs and parallel tracking are being considered for adoption for a wider range of drugs and diseases. Because of the considerable uncertainties in how such proposals would be implemented, and what their implications might be on drug manufacturers, these proposals have been controversial, and have not generally been supported by the pharmaceutical or biotechnology communities. See Ember [1989] for an overview of this area.

Orphan Drug Regulation

Federal legislation was adopted in 1983 to stimulate the development of therapeutic products for diseases with small patient populations, where the market size would ordinarily not justify the long and expensive drug approval process. The law allows companies to petition the FDA to have certain products designated as "orphan drugs" if the prospective market consisted of fewer than 200,000 patients. A company successfully gaining agency approval for a drug designated as an orphan is entitled to seven years marketing exclusivity for that product. Most observers consider this law to be a significant success, with 40 orphan drugs or biologics approved for sale and 256 other products designated as potential orphans, through the end of 1989, according to FDA's Office of Orphan Products Development.

In applying the law to biotechnology, however, problems have arisen. In several instances, two or more companies have been developing the same (or very similar) molecule for the same therapeutic purpose, and each has sought orphan status for the product. In some cases, it has been possible for more than one company to receive orphan status, because of minor differences in amino acid structure of the molecules, while other companies have been shut out. Other concerns with the law relate to disease indications, particularly AIDS, where the patient population is initially small but the potential exists for large markets in terms of revenues. This can be exacerbated by the ability to divide the total market for a given drug into small patient populations, by considering each disease indication separately (as allowed by law). Some observers believe that it violates the intent of the law to have large-market products qualify as orphans in this way.

A number of prominent biotechnology companies and their products have been in the center of these controversies, and there have been numerous proposals for legislative remedies to the perceived shortcomings of the law. In 1990, Congress passed legislation to amend the market exclusivity provisions of the Orphan Drug Act, but it was vetoed by President Bush. It is likely that orphan drugs will remain a contentious issue for some time.

Regulation of Environmental Applications of Biotechnology

The policies and regulations described above for the oversight of agricultural and environmental uses of biotechnology are far from settled, and remain a subject of controversy among industry, government, and public interest groups. Early biotechnology field tests attracted considerable public concern and opposition [Krimsky and Plough 1988; Van Brunt 1987]. Although many of these concerns have subsided [Sun 1988], many issues remain that could trigger future public debate over these technologies as companies move to large-scale testing and commercialization.

Scientifically, there is far less controversy over environmental safety issues for small-scale testing. Scientific monographs have supported the contention that small-scale field tests are not likely to pose unreasonable risks and that potential risks can be adequately assessed (Kelman et al. 1987; National Research Council 1989; Office of Technology Assessment 1988; Tiedje et al. 1989). Indeed, the results of the small-scale tests conducted to date confirm that predictions of ecologic behavior can be made accurately, and that such small-scale tests can be conducted safely [Baum 1989].

In spite of this, regulations for small-scale testing remain to be finalized. Much of the EPA's policies under TSCA have not yet been formally proposed as regulations, and institutions are being requested to comply with these policies voluntarily. The EPA has been attempting to draft proposed regulations to implement its policies for reviewing small-scale field tests under TSCA (and, to a lesser extent, FIFRA) since the publication of the coordinated framework in 1986. One well-publicized attempt in 1988 was aborted by the Office of Management and Budget because a "leaked" draft of the proposal attracted criticism from many parties. Instead of publishing that draft as a proposed rule, the EPA later informally made it available for public comment, and spent most of 1989 attempting to convert that public comment into a new regulatory proposal. Several provisions of that draft, notably its proposal to decentralize some field-test reviews by the use of local "Environmental Biosafety Committees," were highly contentious.

EPA rule making has been delayed by administration discussions on the appropriate scope of government oversight for small-scale field tests. In August 1990, after several months of interagency debate, the Office of Science and Technology Policy published a document defining organisms requiring federal oversight for a small-scale field research (55 *Federal Register* 31118-31121). Although these so-called scope principles are likely to be controversial, their publication will allow the EPA and other agencies to proceed with rule making. The EPA expects to circulate new draft regulations under both FIFRA and TSCA (sometime in 1991.)

A more far-reaching issue is the way in which both the EPA and the USDA will regulate field tests as companies move beyond small-scale experimentation toward large-scale use and commercialization. No large-scale field tests have yet been approved by either agency, and most observers agree that such tests will raise many new scientific issues. Although it is clear that environmental data generated during initial testing will be useful to support requests for broader usage, the procedures and public involvement in such decisions remains to be seen. Moreover, any regulatory scheme that is developed must accommodate product development procedures that are traditional for agricultural products. For example, new crop varieties are brought to market through a comprehensive process of breeding and field testing that takes several years and which might be conducted at research sites in numerous locations around the country. USDA regulation of transgenic plants must allow this nearly unrestricted field use once safety is established.

Finally, legislative solutions to the deliberate-release debate have occasionally been proposed. The latest, a bill to require permits for all outdoor use of genetically modified organisms, was introduced in the House in July 1990 by Rep. Robert Roe (D., N.J.). No action on this bill is expected before Congress adjourns for the year.

Developing a Regulatory Scheme for Food Biotechnology

As noted above, regulatory issues related to the human consumption of engineered crop plants and other biotechnology-produced foods will assume major importance for the agricultural biotechnology sector in the 1990s. In particular, a scheme must be adopted to allow new varieties of food plants to be introduced into the market without disrupting normal practices, while still ensuring adequate safety assessment. Of greater concern might be the public reaction to the marketing of genetically engineered crop plants for food use. Given the often-intense negative public reaction to the initial small-scale field tests of biotechnology products and to the presence of bovine somatotropin in milk products, it is quite likely that a similar reaction could arise when the first altered plants are proposed to be used in human foods.

Although the first sales of genetically engineered seeds for food crops are still some years away, a number of biotechnology and food processing companies formed the International Food Biotechnology Council (IFBC) in 1988 to carry out a study to define the scientific criteria that can be used to assess the safety of such uses. Through consultations and peer review from a broad range of outside reviewers, including government agencies in this country and abroad, the IFBC document is intended to promote a consensus as the basis for a sound regulatory policy for biotechnology's use in foods [Lindemann 1990]. Knight [1989] summarizes an early draft of the IFBC report and discusses some of the related issues regarding food biotechnology.

State Regulation of Biotechnology

Public concerns over the safety of biotechnology field testing, and the lingering perception that biotechnology is inherently dangerous, have led to a recent surge in interest in state biotechnology regulation [Crawford 1989; Davis 1989; Meeks 1988; Moreland 1987]. This trend, first evident in 1987, gained prominence in 1989 and will be a major issue in the 1990s, particularly for agricultural and environmental applications of biotechnology.

Although this development mirrors to some degree the history of the national debate over biotechnology, it is also consistent with a broader trend, in which citizens distrust federal regulatory agencies, whom they feel have failed to protect the public in specific well-publicized instances, and prefer to see additional oversight from state regulatory agencies. Moreover, the issue of biotechnology field testing has plugged into the same "not in my backyard" concern that has led to citizen protest against proposed sites for waste disposal and other facilities. Biotechnology's high visibility and the supposed scientific uncertainty about its risks have naturally created a demand for local oversight. Major areas of state legislative activity have been in consideration of bans or moratoriums on specific biotechnology products and in the regulation of field tests, the latter of which will be discussed here.

State governments have participated in the reviews of almost all federally approved biotechnology field tests since 1987, either informally or as required by federal regulations. Several states soon began to consider implementing state laws to formalize this process. Seven such laws have been passed since 1988. Hawaii, Wisconsin, and Illinois have adopted laws requiring that the state be notified of proposed biotechnology field tests; while in North Carolina and Minnesota, new laws now require institutions to obtain state permits for field tests. Florida has modified an existing law to cover some biotechnology products, and Oklahoma has created a permit scheme for organisms not regulated by a federal agency. Legislation relating to deliberate-release experiments has been considered in at least five other states over the past three years [Industrial Biotechnology Association 1989; Wisconsin Department of Natural Resources 1987], with other states contemplating regulatory approaches not requiring new laws. There has begun to be a significant amount of communication between the states regarding the need for and appropriate components of state regulation.

Most members of industry support some state involvement in the field-test review process, as a means of bringing local concerns to the attention of federal regulators and to reassure the public that regulators closer to home are involved in the review [Glass and Owens 1990; Strauss 1989]. Many feel that the proper state role is to complement but not duplicate the federal regulatory structure, since no state has the resources the federal government has to conduct reviews in such a rapidly evolving scientific field. An additional industry concern is that any state regulations adopted in the years to come should be as uniform as possible, so that companies wishing to sell products in all states are not subject to a patchwork of conflicting state regulations. Many see the attainment of these goals as a major challenge for the industry in the 1990s.

INTERNATIONAL REGULATION OF BIOTECHNOLOGY

Biotechnology will be commercially exploited all over the world, so its regulatory treatment in countries other than the United States is of concern to those in the industry. As is usually the case in many areas of regulation, the United States has a regulatory structure for biotechnology that is more complex and better developed than most other countries. However, recent years have shown increasing interest on the part of most other industrialized nations to adopt a biotechnology regulatory scheme.

Development of a regulatory framework in other countries has often followed the same historical pattern as that seen in the United States: early concerns over public health issues leading to controls over contained uses, followed by commercial regulation triggered largely by environmental concerns. For example, many countries have adopted laboratory and/or manufacturing guidelines similar to those of the U.S. NIH. The United Kingdom has maintained a committee, now called the Advisory Committee on Genetic Manipulation, that is analogous to the NIH RAC, and that administers that country's recombinant DNA guidelines. Other countries that have promulgated laboratory research guidelines similar to the NIH's include Japan, France, Australia, Ireland, Switzerland, the Netherlands, and New Zealand [Gibbs, Cooper, and Mackler 1987].

The adoption of guidelines for manufacturing have not always been noncontroversial. The most prominent case at the moment is West Germany, where political pressure from the Green Party is threatening the development of the biotechnology industry, even that sector dealing only with contained manufacturing. The German parliament has been considering legislation for some time that would require public hearings for construction of any new facility for research involving genetically engineered organisms. But before any such law could be adopted, a German court in November 1989 ordered Hoechst A.G. to stop construction of its planned plant for the manufacture of human insulin in the state of Hesse, because the country had no specific law authorizing such manufacturing facilities. Although a bill rectifying this situation was adopted in May 1990, a number of German companies are establishing research or production facilities in other countries, in large part due to the political situation at home.

The European Commission (EC) is attempting to standardize regulatory activities throughout its member states, and has adopted a directive covering contained uses of biotechnology. This directive is similar to the NIH Guidelines in establishing appropriate procedures, both for small and large scale, for the use of engineered organisms. The directive also classifies experiments on the basis of risk and establishes procedures under which member states must be notified or must actually approve certain biotechnology activities. Directives of the EC are laws that are binding on all member states, regarding the result to be achieved, but under which the method of carrying out the goal is left to each state. Thus, each country within the EC must adopt its own laws implementing this directive. Another international group that has been active in developing standardized biotechnology regulations is the Organization for Economic Cooperation and Development (OECD), which has developed guidelines for standard industrial good laboratory practice for small-scale and large-scale use of recombinant DNA organisms, and for field tests of modified organisms.

As in the United States, the issue of deliberate release has been controversial in many other countries. Some nations, such as Canada, France, Australia, Ireland, and the United Kingdom, have developed rational regulatory approaches that have allowed a number of biotechnology field tests to take place with a minimum of public concern. In some cases, such as in Canada and France, these are part of comprehensive government policies for regulation of all commercial biotechnology products.

However, countries such as Japan, Denmark, and West Germany have adopted regulations effectively banning or placing a moratorium on all environmental uses of biotechnology, although certain experiments may be approved on a case-by-case basis. Even the United Kingdom is considering tightening its approach to review of field-test proposals, by involving the Department of the Environment in the process. There has been considerable public concern over biotechnology field tests in several European countries, particularly ones in which the Green Party is strong. The Greens have also been effective at promoting discussion of the "fourth criterion," according to which the socioeconomic effects of products must be evaluated before approval for sale.

The EC has adopted a second directive aimed at outdoor uses of genetically manipulated organisms. Because of the profound philosophical disagreements between member states, and the strength of the Greens in EC politics, this directive has proven to be quite controversial. The version adopted in March 1990 requires each member state to approve research and development field tests within its jurisdiction, and to notify the other member states of such actions. For commercial products, however, a decision by one member state to approve marketing would be binding throughout the community, but could be challenged in some circumstances by another state. Industrial groups have some concerns over the latter provision.

In spite of this uncertainty, over 100 biotechnology field tests have taken place outside the United States, in countries such as the United Kingdom, France, Belgium, Spain, the Netherlands, New Zealand, Canada, and Mexico, and tests have been approved for 1990 in West Germany and Denmark [GBF 1990].

This brief discussion is meant only to portray that regulation of biotechnology has followed the same patterns in other countries as it has in the United States. Gibbs et al. (1987) provides a more detailed review of biotechnology regulations in many countries. However, it is primarily the industrialized nations that have begun to address regulatory issues, and such discussions have not yet begun in most of the Third World.

CONCLUSIONS AND STRATEGIC CONSIDERATIONS

Although biotechnology is subject to significant government regulation, this regulation does not present a barrier to commercialization that is unique to biotechnology. In most countries, biotechnology products will likely be subject to the same regulations as those used for similar traditional products. Since many of these products are already subject to significant health and safety regulation, the use of biotechnology in product manufacture will not create substantial new burdens.

However, regulatory issues remain critical in the planning and execution of a successful biotechnology business plan. In contrast to other fields of high technology, the lengthy regulatory testing and review periods in biotechnology are often the major time-limiting factors determining speed to market entry. The impact of regulations in the health care field is clear from the time it has taken even the more successful biotechnology companies to bring products to market. In agriculture, short delays in field-test or marketing approvals could mean the loss of a growing season, delaying commercialization by a full year or more. Because biotechnology regulations are still evolving (particularly in agriculture), the first "pioneers" to go through the process have often encountered additional problems, leading to longer delays.

Therefore, it is crucial for regulatory programs to be planned and managed carefully, to avoid lost time in bringing products to market. Fortunately, the "roadmaps" for successful product approvals are becoming more clear. All the federal agencies encourage companies, especially inexperienced start-ups, to consult with them as early as possible, to ensure that testing data and other information is developed according to procedures acceptable to the agency and to be sure that companies understand all requirements and time frames. Start-up companies might wish to make use of a regulatory consultant, or take advantage of seminars, workshops, or the networking available at trade associations to obtain the necessary expertise. Early planning of the regulatory process is needed to assess time lines and costs adequately, and to ensure that early laboratory research complies with agency regulations such as "good laboratory practice" guidelines.

More importantly, perhaps, the reality of long regulatory time frames has forced more recent biotechnology start-ups (and their investors) to be more pragmatic about the time and money needed to bring regulated products to market, and the ability of a start-up to become a stand-alone company, particularly in the pharmaceutical industry. It is not uncommon to find newer biotechnology companies adopting business plans that include early development of products for less-regulated markets (such as the research market), as a means of generating short-term revenue and profitability to sustain the longer-term development of regulated products. Transgenic Sciences, Inc., provides one example of this strategy. In agriculture, some firms have avoided more rigorous regulatory and public scrutiny by developing pesticide products based on naturally occurring or nonrecombinant microorganisms (e.g., Ecogen, Inc.), or by using killed recombinant bacteria (e.g., Mycogen Corporation) [Fritz 1989]. In any case, early optimism has long since disappeared that biotechnology products would receive favorable regulatory treatment because of their human or "natural" origin.

However, certain factors inherent in the regulatory process create opportunities for companies to enhance their competitive position and actually use regulations to their advantage. In certain regulatory areas, particularly drugs, biologics and pesticides, product approvals are given specifically to a single manufacturer, creating a period of market exclusivity that can give a company a substantial competitive edge. This is often cited as a situation that can be exploited by large companies at the expense of smaller companies less able to afford or manage the regulatory process; however, it magnifies the value to a company of any size of being "first through the gate."

Other opportunities are created as changing regulations or scientific knowledge force older products off the market. Restrictions on marketing human blood–derived products (such as human growth hormone), for fear of contamination with human immunodeficiency virus (HIV) or other viruses, have opened the market for replacement products of recombinant origin. Of course, the need to monitor the blood supply for HIV has itself created an enormous market for diagnostic tests. In agriculture, health concerns are resulting in the cancellation of many pesticide registrations and an increased call to reduce chemical inputs in agriculture. These trends may create market opportunities for microbial pesticides, insect-resistant plants, etc., and might also create entirely new businesses such as testing of environmental pollutants and bioremediation.

There are also instances in which the regulatory system explicitly creates or enhances market exclusivity. One example is the seven-year exclusivity available for the first company to gain FDA approval for an orphan drug. Another example, the Drug Price Competition and Patent Term Restoration Act of 1984, grants extended patent terms in partial compensation for the length of time required to obtain regulatory approval for drugs and biologics. Similar legislation has been proposed for animal drugs, and has been considered for other product types as well. Such built-in routes to market exclusivity can be made part of a company's business plan from the start.

Biotechnology has come of age at a time when significant changes in the U.S. regulatory process seem likely. The unprecedented and unanticipated situations of the AIDS epidemic, and the growing belief that regulatory delays disadvantage patients of many diseases, have created an urgent need to review and streamline the nation's drug approval process. At the same time, revitalized environmental concerns are leading to more stringent regulations and a reexamination of previously approved products, particularly in agriculture. Biotechnology regulation must function within these changing contexts, as well as within the evolving nature of the field itself and its underlying science. Companies must remain alert to these changes, by maintaining an active regulatory staff. Companies must go beyond mere monitoring of regulations, by actively participating in attempts to change federal and state rules, by commenting on proposed regulations when published, by maintaining contact with policy makers in the agencies, and by appropriate government relations activities to promote or prevent legislative change in important regulatory areas.

It can be argued that regulatory considerations should never take absolute precedence in developing a business plan and that concern about government regulations should never prevent a company from developing a scientifically feasible product with a known market demand. Yet it is clear that the regulatory status of a proposed product must be taken into account in strategic planning to ensure that time lines and resources are adequately estimated. And while biotechnology will undoubtedly lose its special regulatory treatment at some time in the future, it will always be true that navigating the regulatory waters will be a major factor affecting successful product development in biotechnology.

REFERENCES

Alexander, M. 1985. Ecological consequences. *Issues Sci. Technol.* Spring: 57–68.

Baum, R. 1989. Field tests of recombinant organisms shown safe. *Chem. Eng. News* 67(17):30–34.

Betz, F. 1985. Regulation of microbial and biochemical pesticides with emphasis on microbial products of biotechnology. *Chem. Times Trends* January:44–48.

Betz, F. 1990. Speech at Workshop on Microbial Pesticides, Ottawa, Ontario, March 20, 1990.

Biotechnology Newswatch. 1990a. Office of Biotechnology survives FDA's decapitation, expands role. 10(2):10.

Biotechnology Newswatch. IBA: FDA, PTO in arrears. 1990b. 10(5):1–3.

Biotechnology Newswatch. 1990c. PMA issues 1990 census of biotherapeutics in FDA pipeline. 10(8):5–8.

Budiansky, S. 1983. Regulation issue is resurrected by EPA. *Nature* 304:572.

Crawford, M. 1989. Should states regulate biotechnology? *Science* 245:466.

Davis, C. R. 1989. Biotechnology: Legislative view: The perils. *State Government News* September 1989:24–25.

Ember, L. 1989. FDA presses to speed life-saving drugs to marketplace. *Chem. Eng. News* September 25:9–13.

Fox, J. L. 1990. BSCC addresses scope of oversight. *Bio/Technology* 8(3):187.

Fritz, M. 1989. Thank you, Jeremy Rifkin. *Forbes*, October 16:268–269.

Gessellschaft fur Biotechnologische Forschung mbH. 1990. *GENTEC Update.*

Gibbs, J. N., Cooper, I. P., and Mackler, B. F. 1987. *Biotechnology and the environment: International regulation.* New York: Stockton Press.

Glass, D. J., and Owens, E. D. 1990. States noticing field releases. *Bio/Technology* 8:1064.

Industrial Biotechnology Association. 1989. Survey of state government legislation of biotechnology: Year-end review. Washington, D.C.

International Food Biotechnology Council. 1990. Biotechnologies and food: Assuring the safety of foods produced by genetic modification. *Regulatory Toxicology and Pharmacology* 12:S1–S196.

Jaffe, G. A. 1987. Inadequacies in the federal regulation of biotechnology. *Harvard Env. Law Rev.* 11:491–549.

Jones, D. 1988. Biotechnology challenges at USDA. *ATCC Q. Newslett.* 8(4):1–7.

Kelman, A., Anderson, W., Falkow, S. Federoff, N. V., and Levin, S. 1987. *Introduction of recombinant DNA–engineered organisms into the environment: Key issues.* Washington, D.C.: National Academy Press.

Knight, P. 1989. Engineered fruit and vegetable crops. *Bio/Technology* 7:1233–1237.

Krimsky, S. 1982. *Genetic alchemy: The social history of the recombinant DNA controversy.* Cambridge, Mass.: MIT Press.

Krimsky, S., Baeck, A., and Bolduc, J. 1982. *Municipal and state recombinant DNA laws: History and assessment.* Boston: Boston Neighborhood Network.

Krimsky, S., and Plough, A. 1988. *Environmental hazards: Communicating risks as a social process.* 75–129. Dover, Mass.: Auburn House.

Lindemann, J. 1990. Biotechnologies and food: A summary of major issues regarding safety assurance. *Regul. Toxicol. Pharmacol.* 12:96–104.

Mackler, B., and Lawrence, C. S. 1983. FDA will regulate bovine interferon as animal drug. *Genet. Eng. News* November/December:1.

Meeks, G. 1988. Biotechnology: There's a challenge ahead. *State Legislatures*, May/June:12–16.

Mellon, M. 1988. *Biotechnology and the environment*. Washington, D.C.: National Wildlife Federation.

Milewski, E. 1981. Evolution of the NIH guidelines. *Recomb. DNA Technol. Bull.* 4:160–165.

Moreland, S. M. 1987. Building with bacteria. *State Government News*, October:18–20.

National Research Council. 1989. *Field testing genetically modified organisms: Framework for decision*. Washington, D.C.: National Academy Press.

Office of Technology Assessment. 1988. *New developments in biotechnology—Field testing engineered organisms: Genetic and ecological issues*. Washington, D.C.: Government Printing Office.

Rich, L. A. 1988. Environmental concerns force global changes in the market. *Chem. Week* May 4:22–39.

Strauss, W. M. 1989. Biotechnology: Industry view: The promise. *State Government News*, September:24–26.

Sun, M. 1988. Preparing the ground for biotech tests. *Science* 242:503–505.

Tiedje, J. M., Colwell, R. K., Grossman, Y. L., Hodson, R. E., Lenski, R. E., Mack, R. N., and Regal, P. J. 1989. The planned introduction of genetically engineered organisms: Ecological consideration and recommendations. *Ecology* 70(2):297–315.

Van Brunt, J. 1987. Environmental release: A portrait of opinion and opposition. *Bio/Technology* 5(6)558–563.

Wisconsin Department of Natural Resources. 1987. *State agency biotechnology report and survey results: Legislative and regulatory activities*. Madison, Wis.

CHAPTER 11

BIOTECHNOLOGY AND THE U.S. GOVERNMENT: THE POT AT THE END OF THE RAINBOW?

Suzanne S. Groet

From a level of $13.5 billion in 1960, growth in federal and nonfederal support of research and development in the United States has been dramatic. This growth has been fueled in large measure by concerns of both economic competitiveness and national security. The government's continued long-term support of basic research in a multitude of areas has been critical to the development of biotechnology and to the maintenance of the United State's competitive position. With the growing awareness and demonstration of biotechnology's enabling potential in a diversity of fields, government funding of biotechnology is receiving an increasingly high priority.

RESEARCH AND DEVELOPMENT IN THE UNITED STATES

After nearly a decade during which there was no real growth in national research and development funding levels, levels in the mid-1970s increased, largely in response to the energy crisis, and grew at an average annual real-term rate of roughly 4% until 1982 [National Science Foundation 1989a]. An increased emphasis on defense-related research and development produced even more rapid annual growth (6.9% average real term) in total national research and development spending from 1982 to 1985. However, as a consequence of both federal budget constraints and decreased growth in industry support of research and development, national research and development efforts decelerated dramatically beginning in 1986, and estimated annual growth from 1985 to 1989 averaged 2% or less in real terms. In 1989, the projected national research and development expenditures of $132.4 billion constituted a 5% increase over the 1988 estimated level of $126.1 billion. After adjusting for inflation, however, this represented a real increase of only 1%.

Although the total national research and development effort increased from $13.5 billion in 1960 to over $132 billion in 1989, the government's share of total national research and development expenditures declined substantially during this period (Table 11-1). Government funding of research and development dropped from 65% in 1960 to roughly 51% by 1975. From 1985 to 1988, the government funded 48% of the total national research and development effort and slightly less in 1989 [National Science Foundation 1989a].

Of the national research and development expenditures in 1989, development efforts received 65% of the total, while applied and basic research efforts received 21% and 14%, respectively (Table 11-1). Relative to applied research and development, total funding of basic research increased at a higher annual rate throughout the second half of the 1980s. In 1989, the estimated funding level for all basic research represented a 7% increase (3% in real terms) over 1988. For applied research and development, estimated 1989 funding levels were 5% greater than 1988 (1% in real terms). During the period from 1980 to 1989, real-term annual in-

TABLE 11-1 FEDERAL AND NONFEDERAL SOURCES OF R&D FUNDS

	1960	1975	1980	1989 (estimate) in $ millions
Total	13,523	35,213	62,594	132,350
Federal	8,738	18,109	29,453	62,700
Nonfederal	4,785	17,104	33,141	69,650
Basic research	1,197	4,738	8,415	18,570
Federal	715	3,269	5,898	12,020
Nonfederal	482	1,469	2,517	6,550
Applied research	3,020	7,812	13,892	27,300
Federal	1,688	3,889	6,437	11,060
Nonfederal	1,332	3,923	7,455	16,240
Development	9,306	22,663	40,287	86,480
Federal	6,335	10,951	17,118	39,620
Nonfederal	2,971	11,712	23,169	46,860

Source: National Science Foundation 1989a.

creases in national expenditures for basic research, applied research, and development averaged 4.5%, 3.2%, and 4.3%, respectively.

FEDERAL SUPPORT OF RESEARCH AND DEVELOPMENT

As stated in the U.S. Budget, the mission of the federal government in supporting research and development is to meet its own direct needs, such that the research and development results are used primarily by the supporting agencies, as well as to assist in meeting national needs. This is particularly true of high risk basic research and other activities in which private sector investment incentive is inadequate to ensure that a strong technological foundation is in place to support national competitiveness and economic growth.

Total research and development supported by the government increased from $18 billion in 1975 to an estimated $63 billion in 1989. In 1990, the total federal research and development obligations are estimated at $69.8 billion, which consists of $67.3 billion for the conduct of research and development, and $2.4 billion for research and development facilities. This total represents a 7% increase over the estimated 1989 obligations. Figure 11-1 summarizes the estimated funding levels by government agency for 1987 through 1990 for the conduct of research and development. Of the total government spending on research and development in 1990, over 95% is supported by six major research and development agencies: the Department of Defense, Department of Energy, Department of Health and Human Services, National Aeronautics and Space Administration, National Science Foundation, and Department of Agriculture. The President's 1991 budget proposes a total research and development allocation in budget authority of roughly $71 billion, including research and development facilities.

Although the federal government is the largest source of funding for basic research in the United States, there has been a trend over the past decade for an increasing share of support for the nation's basic research to come from nongovernmental sources. During the 1980s, the federal share of the nation's total expenditures for basic research dropped from 70% in 1980 to an estimated 64% in 1989. During this period, nonfederal support of basic research increased 76% in real terms, while federal support increased by only 37%. NSF [1989a] projected an increase of only 2% in real terms in 1989 federal support of basic research as compared with a 5% increase in nonfederal support. In terms of applied research, federal support in real terms has grown annually by 3% over the 1980–1989 period. However, after adjusting for inflation, federal support for applied research actually decreased in the late 1980s.

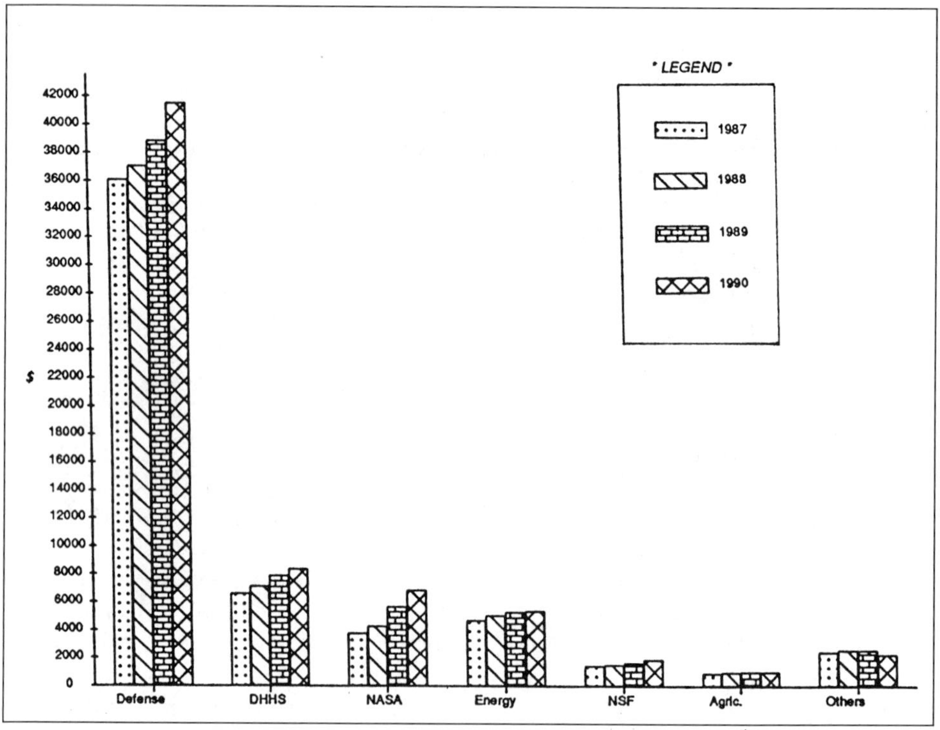

FIGURE 11-1 Federal support of research and development (millions of dollars). (*Source*: Office of Management and Budget [1989, 1990].)

An analysis by NSF [1989b] of government research and development indicates that of total 1989 federal research obligations of $19.7 billion in all fields, an estimated 29% is intramural funds for research performed by agencies of the federal government. The remainder is performed by extramural organizations, with 38% being performed by universities and colleges, 14% by industrial firms, 11% by federally funded research and development centers, 6.5% by other nonprofit organizations, and 1% by state, local, and foreign governments. Of the basic research component in 1989, over 50% is estimated to be performed by universities and colleges, 22% by intramural agencies, and only 6% by industrial firms. Intramural agencies are estimated to perform 37% of federal applied research obligations, as compared with 23% by industry and 23% by universities and colleges. In contrast, of the 1989 estimated federal obligations for development, 65% is conducted by industry.

FEDERAL SUPPORT OF BIOTECHNOLOGY

Government agencies use varying working definitions of biotechnology for the purpose of estimating expenditures; therefore, government support of biotechnology is difficult to estimate, and cross-agency support levels are difficult to compare. The Environmental Protection Agency's definition of biotechnology for example is very narrow, whereas that used by NSF is extremely broad, extending to areas in chemistry, engineering, and environmental biology. Arriving at a common working definition is particularly difficult since biotechnology is not a distinct field of science but rather a collective group of technologies that are drawn from, and applicable to, a multitude of fields. Only in the past few years has there been a concerted effort to estimate accurately spending by government agencies for research and development related to biotechnology.

In recent years, biotechnology research and development has received increased government funding priority. In 1982–1983, government funding of biotechnology was estimated at approximately $520 million per year, of which $511 million was for basic research [Office of Technology Assessment 1984]. By 1987, the estimated level had increased to $2.7 billion [Office of Technology Assessment 1988]. According to the President's 1991 budget, roughly $5 billion was invested in 1989 in biotechnology research, development, and manufacturing by U.S. industry and the federal government; of this, an estimated $3.2 billion was supported by the government (Table 11-2). In 1991, a level of $3.6 billion in budget authority has been proposed, representing a 6% increase over 1990 levels [Office of Management and Budget 1990].

Of the federal departments or agencies that provide support for biotechnology, the most significant contributor is the National Institutes of Health (NIH), which in 1987, provided over 83% of the total government funding for biotechnology. Based on unofficial Office of Management and Budget estimates of federal spending on biotechnology (Table 11-2), NIH will contribute 80% of the proposed 1991 total.

NIH, which is the government's largest nonmilitary research agency, provides the majority of government support for basic biomedical research and training. Both basic research directly relating to biotechnology, as well as broad research underlying biotechnology are supported by the NIH.

Total support for biotechnology research and development by NIH has increased from $2.3 billion in 1987 [Office of Technology Assessment 1988] to an estimated $2.8 billion in 1990. Of these totals, roughly 40% supports basic research directly involving biotechnology, while the remainder supports broad underlying research. The 1991 budget for NIH

TABLE 11-2 ESTIMATED FEDERAL SPENDING ON BIOTECHNOLOGY

	Fiscal Year		
Department/Agency	1989	1990	1991
National Institutes of Health	2,633	2,741	2,850
Alchohol, Drug Abuse, and Mental Health	107	141	183
Food and Drug Administration	15	20	25
Veterans Affairs	15	15	16
Defense	129	123	127
National Science Foundation	120	125	130
Energy	80	93	115
Agriculture	97	101	125
Agency for International Development	1	1	1
National Institute of Standards and Technology	5	5	8
Total dollars (in $ millions)	3,202	3,364	3,579

Source: Unofficial Office of Management and Budget Estimates.

includes a proposed $280 million increase over 1990, primarily for added support of basic research. In 1991, the proposed biotechnology funding level for NIH represents over 30% of its total proposed research and development budget of $7.9 billion, of which $4.5 billion is designated specifically for basic research.

One of the key missions in biotechnology research within NIH is the Human Genome Initiative, which, in cooperation with the Department of Energy supports research to map and characterize the human genome and the genomes of model organisms. This long-term investment is expected to lead to such potential benefits to society as an enhanced understanding of the genetic basis of disease, the identification of disease genes, and the development of therapies and disease prevention strategies. The 1991 proposed budget provides NIH with $108 million in budget authority for the Human Genome Initiative, representing an 80% increase over 1990 levels. A funding level of nearly $48 million is proposed in 1991 for the Department of Energy efforts relating to the Human Genome Initiative.

The 1990 NIH budget request also includes $731 million in funding for research and development efforts for the acquired immunodeficiency syndrome (AIDS) and human immunodeficiency virus (HIV). The NIH is one of the five Public Health Service agencies to be allocated funds under the National HIV Program, for which a total 1990 funding level of $921 million was proposed. Roughly 70% of NIH's AIDS- and HIV-related efforts are focused within the National Institute of Allergy and Infectious Diseases and the National Cancer Institute.

The 1991 budget also proposes specific increases in funding for agricultural biotechnology through a national research initiative. This initiative will enlarge the Department of Agriculture's competitive research grants program from $43 million to $100 million, and will increase support for animal and plant biotechnology, including mapping the genomes of important crop species. Among the long-term benefits that may result from this initiative are improved production economics, development of environmentally safe methods of controlling agricultural pests and diseases, and an increase in the quality and safety of the food supply. The 1991 proposed budget for the department includes a total of $125 million for agricultural biotechnology, representing an increase of 25% from 1990 levels.

Analyses of research and development data by budget function and by field of research and development provide additional information on government support of biotechnology and underlying sciences. Based on budget function, most of the government-funded research and development that supports the growth of biotechnology falls within the health and general science functions. Of the total 1990 research and development budget authority, these ranked second and fourth, respectively, in the top five research and development functions (defense, health, space, general science, and energy). From 1988 to 1990, health research and development accounted for approximately 12% of the total research and development budget authority, while the general science function represented nearly 4% [NSF 1989c].

Of the 1990 total research and development budget authority of $8.2 billion for health, 9% has been earmarked for development, 33% for applied research, and 58% for basic research [NSF 1989c]. The basic research component of the health function represents 42% of the total basic research supported by the government.

Within the health function, the largest component is the NIH, for which a 1990 request of $6.4 billion in budget authority, excluding AIDS and HIV funds, was proposed. This incorporates a proposed 7% increase in basic research. Nearly 65% of the total research and development request is accounted for by five institutes; the National Cancer Institute, the National Heart, Lung, and Blood Institute, the National Institute of General Medical Sciences, the National Institute of Diabetes and Digestive and Kidney Diseases, and the National Institute of Neurological Disorders and Stroke.

Research and development within the general science function,

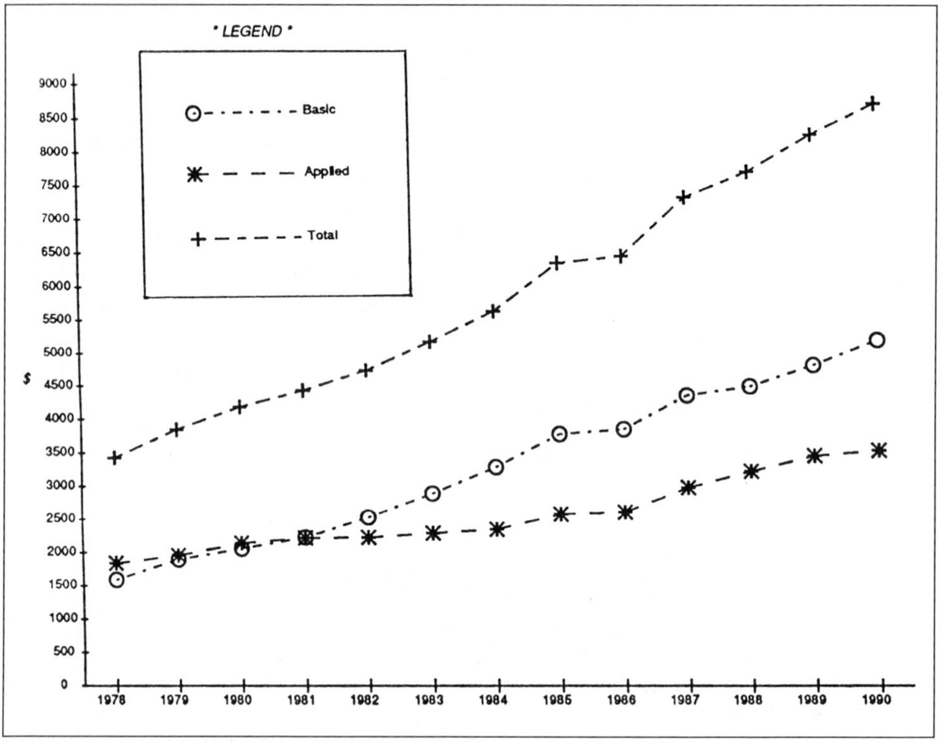

FIGURE 11-2 Life sciences research (federal obligations in millions of dollars). (*Source*: National Science Foundation, [1989d].)

which is funded by the National Science Foundation and the Department of Energy, is directed toward fundamental research in science and engineering. Of the 1990 proposed total general science budget authority, roughly 80% is devoted to research and development, of which 95% is directed to basic research. Within NSF's biological, behavioral, and social sciences directorate, 1991 obligations for cellular biosciences are estimated to increase by 10% to a total of $63 million. Molecular biosciences obligations for 1991 are estimated at $52 million.

Figure 11-2 shows the historical data for government obligations for research in the life sciences. This category includes five fields: biology (excluding environmental), environmental biology, agriculture, medicine, and life sciences not classified elsewhere. Of the total life sciences research and development obligations in 1989 and 1990, the biological and medical fields comprised over 86%. From 1978 to 1989, total federal research obligations in the life sciences more than doubled, from $3.4 billion in 1978 to an estimated $8.3 billion in 1989. During this period, the proportion allocated to basic research steadily increased from 46% of the total in 1978 to nearly 60% in 1985. Since 1985 the proportion has remained fairly constant, ranging from 58 to 60% [National Science Foundation 1989d]. Since 1978, basic research has also comprised an increasingly large proportion of the total research obligation for the broad category of science and engineering as a whole.

The federal government supports a major share of the basic research underlying or directly related to biotechnology; however, it provides only a small percentage of the capital of commercial biotechnology companies. In contrast, European governments have placed a greater emphasis on commercial biotechnology goals and on supporting industrial programs as well as collaborations between universities and industry. Like Europe, the Japanese government has also funded a lower percentage of the nation's research and development as compared with the U.S. In Japan, the government has established a series of policies and programs aimed at bringing Japan to a leadership position in biotechnology, in recognition of its key importance to Japan's future, and it has coordinated the establishment of major programs designed to foster cooperative research and development among government agencies, private companies, and universities.

According to an analysis by Burrill [1989] of the distribution of financing sources among U.S. biotechnology firms, biotechnology companies have historically relied on private equity and venture capital for the majority of their financing needs. Over the next decade public equity, debt, and strategic alliances are expected to provide the majority of financing, and only small companies will rely on private equity and venture capital. Burrill's analysis indicates that government grants have historically been most important to small companies, accounting for 8% of their financing. In 1990, this source is projected to represent 7% of the financing of small companies, but only 4% of the financing of large companies. Over the next decade, commercial sector reliance on government grants is projected to decrease overall to an average of 1%. For new entrants to biotechnology, however, government grants are likely to remain a valuable source of capital, particularly for financing high-risk feasibility programs.

SMALL BUSINESS INNOVATIVE RESEARCH PROGRAM

One government program of significant importance to early-phase biotechnology programs has been the Small Business Innovative Research (SBIR) program. This program began in 1983, after the enactment in 1982 of the Small Business Innovation Act. This Act derived from Congress's finding that, although the majority of federally funded research and development is conducted by government laboratories, large businesses, and major universities, small businesses constitute the nation's main source of technological innovations. Congress further found that small businesses are particularly cost-effective in conducting research and development and in commercializing the products of innovation.

Recognizing the potential of small businesses to contribute in large measure to the nation's technological competitiveness, Congress created the SBIR program with the stated objectives of encouraging technological innovation in the private sector, enhancing the part that small businesses play in meeting federal research and development needs, increasing commercialization by the private sector of innovations developed from federally funded research and development, and increasing the participation of minority and disadvantaged small businesses in technological innovation.

The SBIR Act specifies that each governmental agency with extramural research or research and development funds greater than $100 million must allocate a percentage of these funds to the SBIR program, with the percentage reaching 1.25% over a period of several years. The agencies that currently participate in the program are the Departments of Agriculture, Commerce, Defense, Education, Energy, Health and Human Services, Transportation, the Environmental Protection Agency, the National Aeronautics and Space Administration, the National Science Foundation, and the Nuclear Regulatory Commission.

The SBIR program consists of three phases. Phase I constitutes a feasibility study and is generally funded for up to six months and for a maximum of $50,000. The program encourages proposals on state-of-the-art, innovative, high-risk topics with potential for high commercial returns. Grantees who successfully complete Phase I and demonstrate the feasibility of their approach are eligible to apply for a Phase II award, which constitutes the bulk of the research and development effort. Phase II awards are generally granted for periods of up to two years, and, depending on the agency, are usually funded to a maximum of $500,000. Phase III is the commercialization phase of the program. Support for this phase is generally from the private sector, although federal non-SBIR funds can be obtained under certain circumstances.

A key advantage of the SBIR program is that the business ordinarily retains rights to technical data and worldwide patent rights to inventions developed with SBIR funding. The government retains a royalty-free license for its own use.

From 1983 through 1988, participating agencies have awarded nearly $1.4 billion to small businesses under the SBIR program. During this period, a total of 12,017 competitive awards were made, consisting of 9229

Phase I awards and 2,788 Phase II awards. Total SBIR awards have increased from $44.5 million in 1983 to $389.1 million in 1988.

Since 1983, roughly $150 million in SBIR funds have been invested in biotechnology (Tables 11-3 and 11-4), representing nearly 11% of all SBIR funding. A total of ten government agencies have awarded SBIRs related to biotechnology. Over 90% of this funding has been contributed by the Department of Health and Human Services, the Department of Defense, and the National Science Foundation.

SBIR funding in biotechnology has increased from a total of $9 million in 1984 to over $40 million in 1988. The biotechnology/microbiology area is among the top award groups in terms of technology areas, ranking fourth in 1988 out of a total of thirty-five technology areas. Only information processing, advanced materials, and optical/laser technology groups received higher levels of SBIR funding.

One company for which the SBIR program provided an important source of early capital for technology development leading to commercialization is IG Laboratories. This company, originally a research division of Integrated Genetics, Inc., and now a subsidiary of Genzyme Cor-

TABLE 11-3 PHASE I SBIR AWARDS RELATED TO BIOTECHNOLOGY/MICROBIOLOGY BY AGENCY

	Fiscal Year					
Department or Agency	1983	1984	1985	1986	1987	1988
Defense	224	214	180	984	442	425
Energy	150	343	150	183	287	450
National Aeronautics and Space Administration	147	0	49	100	191	188
Health and Human Services	2,273	4,311	10,250	8,523	7,046	8,381
National Science Foundation	0	0	800	680	680	880
Environmental Protection Agency	0	36	194	99	0	0
Education	0	0	117	30	0	0
Agriculture	115	163	293	321	348	194
Commerce	0	0	0	59	0	30
Interior	0	35	3	0	0	0
Total dollars (in $ thousands)	2,910	5,102	12,033	10,979	8,994	10,548

TABLE 11-4 PHASE II SBIR AWARDS RELATED TO BIOTECHNOLOGY/MICROBIOLOGY BY AGENCY

	Fiscal Year				
Department or Agency	1984	1985	1986	1987	1988
Defense	0	819	1,661	2,030	3,037
Energy	822	932	0	899	985
National Aeronautics and Space Administration	486	0	0	500	0
Health and Human Services	4,655	6,094	23,861	20,044	23,660
National Science Foundation	0	995	189	2,370	1,836
Environmental Protection Agency	0	150	299	150	0
Education	0	0	200	0	0
Agriculture	280	449	600	637	309
Total dollars (in $ thousands)	6,243	9,439	26,866	26,630	29,827

poration, pioneered the commercial development of DNA linkage analysis for the prenatal and presymptomatic diagnosis of genetic diseases. In 1982, the company received its first SBIR award for the development of a carrier test for cystic fibrosis. Multiple Phase I and Phase II awards led to the successful development of a linkage-based test for cystic fibrosis, a test and pilot program for the presymptomatic diagnosis of Huntington's disease, and technology development applicable to the development of additional genetic tests. As a consequence of its leadership position in the development of this technology, the company became the first licensed clinical laboratory in the country to offer DNA-based genetic testing services. The company has continued to participate successfully in the SBIR program and has expanded its business to offer a comprehensive range of commercial genetic testing services, which it provides to hospitals, physicians, and clinical laboratories.

In summary, the U.S. government's long-term commitment to high-level funding and support of basic research has given the United States a strong competitive edge in the basic sciences that underlie biotechnology. As the potential of biotechnology is realized in the development of commercial products and processes in many fields, the government has given biotechnology an increasingly high priority. Federal support currently exceeds $3 billion per year, most of which is directed toward basic research. Within the commercial sector, government support of research and development has represented only a small percentage of financing for biotechnology companies. The SBIR program, however, has clearly enabled many small companies to develop novel technologies into innovative products, and is likely to remain a valuable source of research and development funds for small start-up companies.

REFERENCES

Burrill, G. 1989. *Biotech 90: Into the next decade*. New York: Mary Ann Liebert.

National Science Foundation. 1989a. *National patterns of R&D resources: 1989*, Final report, NSF 89-308. Washington, D.C.

National Science Foundation. 1989b. *Federal Funds for Research and Development: Fiscal Years 1987, 1988, and 1989*, Vol. XXXVII, NSF 89-304, Washington, D.C.

National Science Foundation. 1989c. *Federal R&D funding by budget function fiscal years 1988–90*, NSF 89–306. Washington, D.C.

National Science Foundation. 1989d. *Federal Funds for Research and Development, Federal Obligations for Research by Agency and Detailed Field of Science/Engineering: Fiscal Years 1969–1990*, NSF, Washington, D.C. (Unpublished report)

Office of Management and Budget. 1989. *Special Analysis, Budget of the United States, fiscal year 1990*. Washington, D.C.: U.S. Government Printing Office.

Office of Management and Budget. 1990. *Budget of the United States, fiscal year 1991*. Washington, D.C.: U.S. Government Printing Office.

Office of Technology Assessment. 1984. *Commercial biotechnology: An international analysis*, OTA-BA-218. Washington, D.C.

Office of Technology Assessment. 1988. *New developments in biotechnology: U.S. investment in biotechnology*, special report OTA-BA-360, Washington, D.C.

CHAPTER 12

OURS, THEIRS OR BOTH? STRATEGIC PLANNING AND DEAL MAKING

James P. Sherblom

Every CEO of a biotechnology firm, at one time or another, has dreamed about setting strategic research priorities for his or her company without regard to the requirements of external funding sources. Alas, we all wake up (or most of us do) and realize that no matter how we choose to finance our company, some level of external funding for our research is critical to our continued success. A company's strategy often dictates the role and level of dependence on deal making, but in the end good deals are always critical to success. Companies that employ a product-oriented strategy have less of a reliance on external research funding than those pursuing a research-oriented strategy. Well-funded companies can afford to self-finance more of their research and hence get better terms from their partners than less well funded companies. Some companies manage to fund their programs with research and development partnerships or government grants that allow them to keep more commercial rights to resulting products. But in the end, well funded or poor, large company or small, the question becomes not if, but rather how to find appropriate research funding.

PRODUCT-ORIENTED STRATEGIES

A product-oriented strategy requires products in the near term to generate revenues to cover research costs. This seems sensible enough, but few of us start out with products or the ability to license significant products ready for introduction to the marketplace. Hence, a product-oriented strategy generally requires acquisition of a product line or company early in the company's history.

Genzyme Corporation acquired a diagnostic intermediate company, a specialty chemical manufacturer, and rights to a lymphokine product line, in three separate transactions during its first two years. These provided the core products for Genzyme's technology to build on in pursuing a product-oriented strategy.

TSI Corporation (TSI) acquired a preclinical laboratory, a biomedical testing laboratory, and a diagnostic intermediate company in its first three years. These companies provide TSI with revenues, profits, and outlets for products and services under development. The largest, Mason Research Institute, proved the ideal acquisition for entree in the pharmaceutical preclinical testing market. Mason was a leader in innovative technology and possessed a talented scientific staff. TSI's strength in genetic engineering, transgenics, and related technologies is enabling Mason to develop superior animal models to test drugs for human diseases. It was the success of these acquisitions that allowed TSI to achieve profitability in less than three years from its founding. This profitability, in turn, has allowed TSI to fund a large proportion of its research internally.

DEFINING THE ROLE FOR RESEARCH FUNDING

No decision is more important in defining a company's corporate culture than how it decides to fund its research and development budget. The select few that are fortunate enough to have operating businesses generating sufficient current profits to fund a reasonable level of research activity, such as Genzyme or TSI, usually will fund most of their programs internally. Very well funded companies, such as Amgen or Genetics Institute, can also afford to set a certain level of annual losses and fund much of their own research programs. However, even these companies, whether in response to good research programs that otherwise would not be funded or the desire to develop a critical mass of scientists in important disciplines that go beyond the company's funding capabilities, eventually turn to outside funds to augment their research programs (Fig. 12-1).

Many small biotechnology firms develop funding strategies based primarily on external research funding. For these firms a deal providing external funding represents credibility for their science, funding for their programs, and a reduction of their negative cash flow. Unfortunately, as we have seen from some recent failures of small biotechnology firms, this is generally not a viable long-term funding strategy. It often requires that

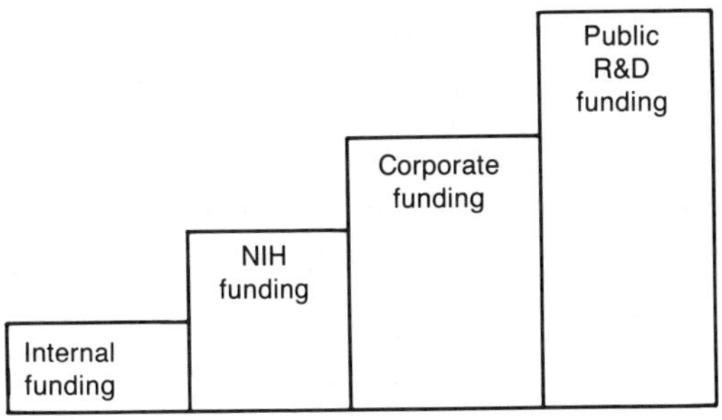

FIGURE 12-1 Steps in seeking research and development (R&D) funding.

a firm sell ever more of its technology just to keep the external sources of funds flowing. However, the decision, made early in a firm's genesis and often culturally irrevocable, to use externally funded research to augment internally funded research, or even to make it the focus of the company, drives most strategic decisions and the likely future of the enterprise.

For the small firm starting out, the logical place to start is government funding. Particularly in the area of human health, there is usually money available to fund good scientists to pursue interesting and useful medical research, from one of the National Institutes of Health (NIH). The best place to start is generally with a Small Business Innovative Research (SBIR) grant from the Institute whose mission most closely matches that of your firm. TSI's first government research funds were a $50,000 grant to produce human growth hormone in transgenic mice, $40,000 to produce genetically identical rabbits, $35,000 to produce human soluble CD4 in chicken eggs, and $49,000 to demonstrate cryopreservation of transgenic embryos.

Phase I SBIR grants can be for up to $50,000, last six months, and occasionally can lead on to Phase II grants of up to $500,000 spread over two years. There are also regular requests from the NIH for research proposals on particular topics, a list of which is published daily and weekly, and is available on request from the NIH.

When you have decided how much research you can fund from internal sources, and exhausted the possibilities from the NIH, then it is time to consider private sources of funding.

DYNAMICS OF CORPORATE PARTNERING

A perfect corporate partner, or sponsor for your research, is as much sought after and as elusive as the perfect marriage partner. First comes the bar scene, or in biotechnology the conference scene, where much longful lusting after the attributes of Merck or Sandoz generally does not progress much further than a polite exchange of business cards. Once in a great while you get lucky, perhaps scoring a series of private meetings or confidential exchanges of information, which then lead nowhere. Always expectant, facing endless rejection, the "seen and be seen" approach is as unlikely to succeed in looking for ideal corporate partners as in searching for the ideal mate.

You should also be aware that the odds are stacked against you. One major U.S. pharmaceutical company that has historically been a big funder of biotechnology firms reported the following ratios in 1989. They received 175 full-blown proposals for corporate research from biotechnology firms in 1989. Of those, they felt 125 of the proposals were in areas of interest to them and of sufficient merit to warrant detailed review. Of proposals reviewed, 50 led to negotiations and meetings with the proposed biotechnology partners. In the end, four resulted in signed deals. Three of these deals gave the pharmaceutical company more rights than the biotechnology firms had initially offered, and each was for substantially less money than the biotechnology firm felt was appropriate. As external funding by the big drug companies declines, and the number of small biotechnology firms continues to increase, the balance of power in negotiating has definitely shifted. Thus, a definite and well-conceived plan of search for a partner has never been more necessary (Fig. 12-2).

A careful search should start with a plan, and the plan should begin by defining what it is that you have (and are willing) to sell. If you give up the commercial rights to your future, you own it no more. The trick is to find something that you already own, that is of great value to a potential corporate partner, that requires your continued research involvement to achieve its promise, and that has very little intrinsic value for your company without the aid of a corporate sponsor. What you are probably seeking from a potential corporate partner is revenues, cash flow, credibility, market access, and medical expertise. Keep in mind that what the corporate partner is probably looking for is access to a product pipeline, access to new technology, and continued control over his own destiny. To the extent that you can give the corporate partner what he wants, without compromising your own objectives, you may have the basis for a beautiful relationship (Fig. 12-3).

What then are safe targets to include in a corporate partnership? You can certainly include product applications that you will not be able to develop independently. Geographic markets that go beyond your intended target market are also logical targets that are apt to have more value to a partner already in those markets than they have for you. The same is true of market segments that are outside the strategic focus of your company or ones that are too fragmented for a small firm to address cost effectively.

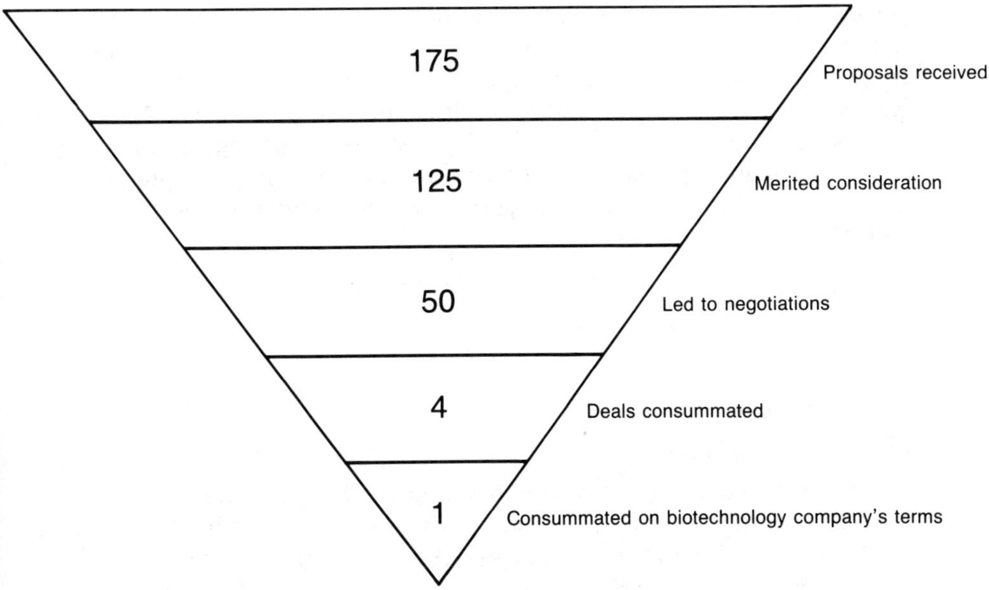

FIGURE 12-2 Research and development proposals received by a major U.S. pharmaceutical company.

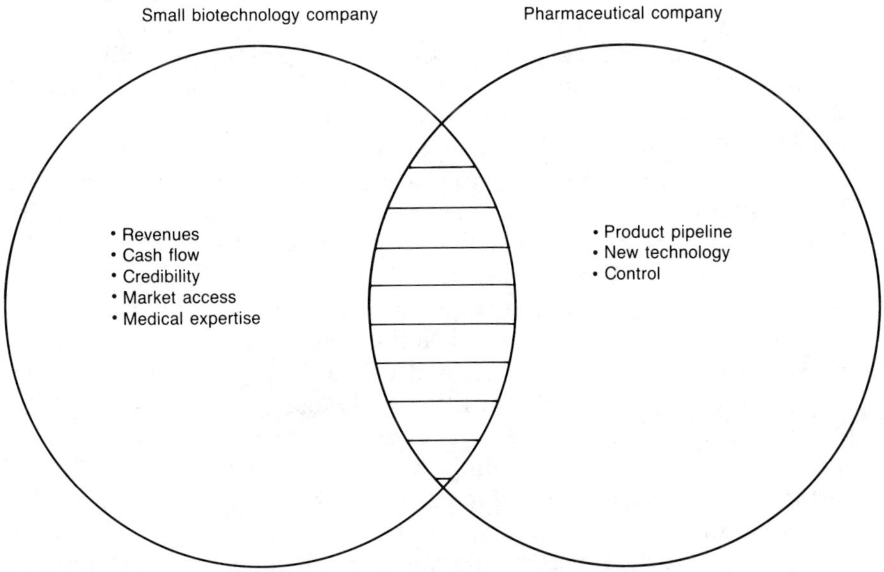

FIGURE 12-3 Goals of corporate partnering for the small biotechnology company and the pharmaceutical company.

It can make great sense to license the first product from a string of products to a corporate partner, which means the development costs of this first product funded by your partner will significantly lower the costs and risks associated with subsequent products funded by you. The trick here is to be sure that the subsequent products are as promising and real as the one you are giving away, so that you do not become a one-product company that has licensed away its future. Listed below are things "safe" to sell:

- Products too broad to develop independently
- Foreign marketing rights
- Nonstrategic products
- Highly fragmented markets
- First in a string of products

What then should not be for sale at any price? The major market for your most important, or only, product cannot be sold without implicitly selling the company at the same time. A critical skill needed to maintain your competitive advantage cannot be licensed away. Your core technology, that which is most unique and proprietary about your firm, is often of greatest value to a potential partner contemplating establishing a research effort in your area of expertise, and it is certainly what you can least afford to give away at any price. It is common sense—no matter how long or cold the winter, a farmer cannot afford to sell or eat the seed corn. Yet, it is a tension that exists in most corporate partnerships, for it is often tempting for a large company to try to acquire the core technology for itself, rather than rent access to it over the life of the partnership. The key to striking a comfortable balance is to not sell too much too soon.

When is it appropriate to seek a corporate partner? Generally later than you think. Most small companies make the mistake of beginning this critical search too early rather than too late. Why? Because every CEO, every Vice President of Business Development, can only show his prowess at deal making by announcing such a deal. Hence, often the first job of a new CEO is to try to strike a deal that will knock the industry's socks off. However, such a deal can generally be achieved only by giving away too much, too early, for far too little. If the company, or its founders or predecessors, have not spent considerable time and financial resources creating something of tremendous value, it is unreasonable to expect a partner to pay a sizable price for access. Dreams do not fetch as much as they used to.

This creates an inherent dilemma for biotechnology start-ups. The days of assembling a blue chip scientific advisory board, hiring a few scientists, building a laboratory, and then pursuing corporate partners for your research are thankfully dead. The companies recently founded on this successful early 1980s approach, are often mystified by their lack of significant funding after two or even three years of trying. Today, a company needs to have sufficient financial resources from all sources, including NIH funding, to spend two or three years developing valuable product positions, before it can reasonably expect to attract large sums of corporate funding. The only exceptions I am aware of are companies that have been able to begin life with a license to the product positions created by their founders using other people's money, generally while working at universities. As with anything else in life, there is no such thing as a free lunch.

So, you have founded your company in a unique technology niche; you have spent the requisite blood, sweat, and tears over the requisite years to create a proprietary set of potential products; you have funded this early-stage development in such a way that you still own the rights to your own future; and you have now identified a product area sufficiently advanced to warrant a corporate sponsor. How, then, do you find the right one?

Fortunately, when approached in the context of a sensible strategic plan, without looking for an undue return on your investment, finding a reasonable partner can be really quite simple. Early in a company's life, perhaps even before starting the company, a little market research should identify the dominant players in the markets best served by the company's proposed products. Such market research should be able to identify the best markets for the company to pursue on its own, as well as the markets in which it might best serve by finding a corporate partner. Such a list of potential corporate partners, which might run to twenty to twenty-five firms worldwide, can be maintained and revised as the products, research, and strategy evolve. All available information on these firms should be gathered and updated on a regular basis, including full details of their existing corporate relationships, if any. Thus research decisions during a company's critical early development years can be made with one eye on the company's business strategy and another on its appropriateness for future potential corporate partners. Following such an approach to strategic planning should result in a saleable product for your target customer in the shortest time frame. When the time is ripe, you may be surprised to find a partner actually approaching you, or you might want to send a carefully thought out proposal, or if all else fails, there are always matchmakers available for a fee.

STRUCTURING THE DEAL

Now that you have identified a group of potential partners, do you know what they would look for in a relationship? Are you sure you know what

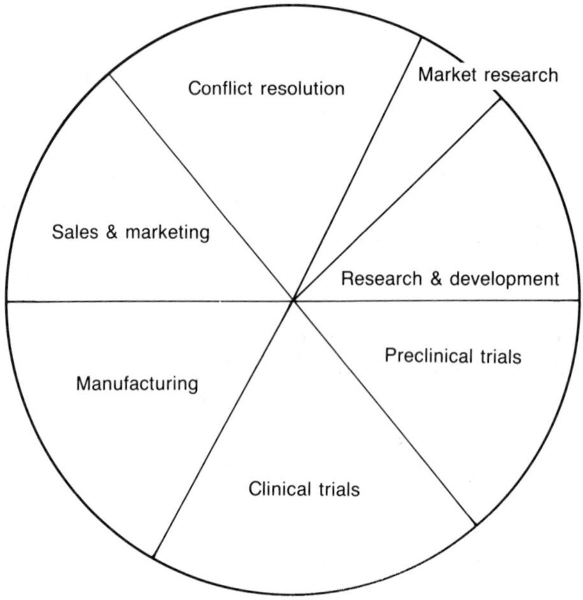

FIGURE 12-4 Slicing the pie. Who gets what?

you are looking for in a relationship? Really? Until you have both carefully defined your goals and expectations, neither of you can have any confidence that they will be satisfied. Unlike a marriage, where warm feelings often can take the place of careful discussions of the future between two people, in a corporate partnership many people are involved on each side, all with different goals and separate agendas. In a corporate relationship, while a certain amount of passion and warm feelings are helpful, careful discussions of future relations are critical. To be truly indicative of what is to come, these discussions must take place at all levels of the two companies that are likely to interact after the agreement, not just a cozy dinner between CEOs.

In structuring the deal, it is important to define exactly who has which slice of the pie—what rights and responsibilities. Who has done, or will do, the market research to decide if there is a market for the products? Who is directing the research programs? Are they jointly or unilaterally controlled? Who has responsibility for preclinical trials? Clinical trials? Manufacturing? Sales and marketing? Depending on the strengths and needs of the players, these issues have been divided in every manner imaginable (Fig. 12-4).

Perhaps even more important than who is expected to do what when, is who is responsible for redefining the work when the unexpected happens, as it invariably will. Research is by its very nature an iterative process, requiring constant reassessment depending on findings. If there is a low risk of unexpected findings requiring program reassessment, then it is probably not much of a research program. If the partnership does not include a good mechanism for conflict resolution and refocusing the work when the situation changes, it is at high risk for failure. A conflict-resolution mechanism need not be complex, but it needs to be responsive in a dynamic environment. Finding a way to reorient the project that is acceptable to both parties should always be preferable to either party abandoning the project. Achieving such a goal requires only a little foresight and plain dealing between the parties.

NEGOTIATING THE DEAL

A successful negotiation generally requires at least one, and no more than two, advocates or champions on each side. Any number of people can have an interest in the deal, any number can express opinions, but in the end the champion's job is to keep the deal on track and moving forward. Without a champion, the deal can easily get sidetracked. With too many champions, there are too many possible tracks.

Roles are generally established quite early in a negotiation. Who is in control of the pace of negotiations? Who proposes terms? Whose lawyers draft the agreement, in whose favor? Who can back out of the deal, when, and under what conditions? The parameters of the negotiation, generally set very early, often determine the outcome of many key issues (Fig. 12-5).

Who do you include in the negotiations and at what stage? The first stage is usually a general discussion among business people. The goal

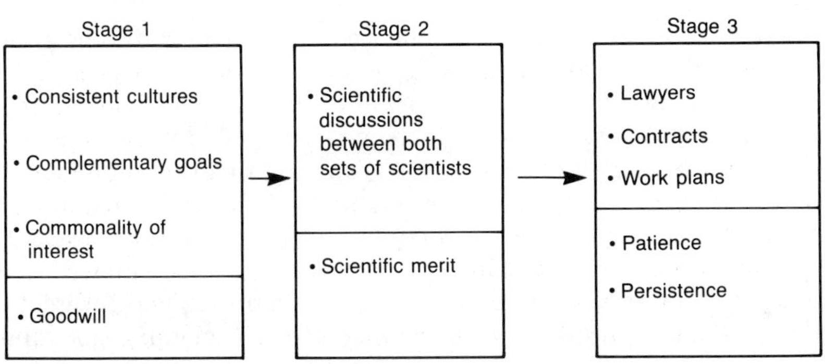

FIGURE 12-5 Stages of the deal.

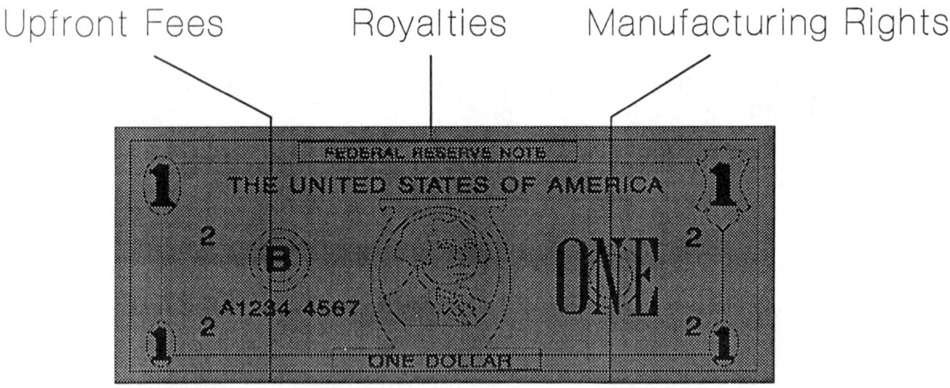

FIGURE 12-6 Form of revenues.

here is to establish whether there are consistent corporate cultures, complementary strategic goals, and a potential commonality of interest. A no-go at this stage does not cost anybody much. You go forward only if a sense of goodwill has been established. A "yes" at this stage often leads to the exchange of confidentiality agreements and the potential for beginning to spend serious money. The second stage usually involves scientific discussions including several scientists from each side. The goal here is to decide if the two groups could actually work well together, and to decide whether the proposed project has reasonable scientific merit. The third and final stage includes lawyers, contracts, work plans, tremendous patience, and persistence from the champions on each side. By the final stage, if all players have done their part well, the deal should have enough momentum in both organizations that completion of the project is easier than cancellation.

Now the key question: How much money can a small biotechnology firm reasonably ask for rights to its products? Your corporate partner is probably looking for a projected return at the 35 to 40% level, and you should expect no less. Research is risky business and needs a potentially high return to justify the investment. Your costs can include technology and expertise contributed by the company's founders, technology licensed from universities or consultants, and fully allocated research and development costs to date. Ongoing costs should be calculated as all direct costs of the project, a fair allocation of overheads and indirect costs, and a reasonable profit for opportunity costs. Generally speaking, the more certain the return, the lower the profit above real costs.

The variables you have to work with include up-front fees (generally from zero to 10% of total revenues), ongoing research funding, milestone payments (generally pure profit), and manufacturing rights or royalty on sale (generally from 1 to 10% of revenues depending on the proprietary nature and projected profitability of the products). Total revenues often range from $50,000 to $5 million or more depending on the size of the target market, complexity of research hurdles, and duration of the partnership (Fig. 12-6).

LIVING WITH THE DEAL

Making deals (like making children) is the fun part, it is bringing them up successfully that requires all the work. No negotiation, no matter how artful, can result in a preagreed solution to every possible eventuality. No contract, no matter how carefully drafted, is entirely clear on everything that was intended by the principals, when they are no longer around to interpret its clauses. Thus, negotiations and contracts are, at best, agreements at a point in time, and good relationships predictably outgrow their original contracts.

For the relationship to grow and mature well over the years requires careful management of expectations. Responsibility for the redefining of expectations as the relationship matures resides either with the deal's original champions or, more often, with their successors among the operating personnel. To maintain a commonality of interest over many years generally requires real work. Good collaborators often adopt the old truism: either we all succeed together or else all fail separately. For each party to view subsequent adjustments to the original agreement only from the point of view of their own self-interest is often to guarantee a short and stormy relationship terminating in a quick and painful divorce.

No matter how carefully you plan, how flexible you may choose to be, a relationship is always vulnerable to irreconcilable differences, and a contingency divorce plan is always to be recommended. It can perhaps even be said, that it is those who have most thoroughly contemplated the likely impact of a divorce who are best prepared to avoid it.

Central to any thought of eventual divorce is the question regarding offspring—are they yours, mine, or ours? Because you cannot guarantee commonality of interests over a long period, you should carefully decide who gets the proprietary technology in an eventual divorce, and under what terms of support. At the time a new venture is embarked on, this is a highly theoretical question, and possibly subject to easy and equitable resolution from the beginning. At the time the prospect of divorce actually looms, an easy equitable resolution is almost impossible.

HOW TO GROW OLD GRACEFULLY

To grow old together gracefully is not always the best goal of a corporate partnership. However, where it is, a substantial equity stock of one partner in the other dramatically increases the chances of success. Those who answer "yours" or "mine" to the offspring question, know that one day all corporate relationships are outgrown. Those who answer "ours" require constant planning hard work, vigilance, and a good measure of luck to succeed. May all your ventures have many offspring and if you so wish, may they grow old gracefully.

CHAPTER 13

IN THE TRENCHES: THE MARKETING AND SELLING OF BIOTECHNOLOGY

Brandon J. Price

Since the development of hybridoma and recombinant DNA techniques in the late 1970s, and continuing during the creation of hundreds of companies to exploit these techniques in the early and mid-1980s, commercial biotechnology has been driven almost exclusively by the science. This "technology-oriented" approach was suitable (and indeed most applicable) for the times. A firm must be technologically oriented when research and development make new products obsolete in a few years—and such was the case during those times for firms marketing products, as exemplified by the expanded use of monoclonal antibodies in detection of human and animal disease. The "product-to-market" events in such a firm begin with the output of product research and development, and continue in a linear fashion with production, promotion, and sales.

As biotechnology moves into the 1990s, the commercial environment is substantially different. Financing is much harder to come by than previously, and few firms, especially those engaged in bringing therapeutic products to market, have the resources simultaneously to fund full-scale clinical trials and regulatory submissions; develop, optimize, and validate large-scale manufacturing processes and facilities; and establish a sales force to bring approved products to the marketplace effectively.

It is clear, therefore, that market needs will be the driving forces of the industry during the 1990s and beyond. In its broadest sense, marketing can be defined as activities that relate an organization successfully to its environment. These activities include

- Identification of unmet or evolving needs in the marketplace;
- Coordination of the development of products and services to meet those needs;
- Distribution of the products and services to the marketplace to fulfill those needs;
- Communication to the marketplace of the ability of the products and services to meet those needs; and
- Establishment of pricing so that the company gains and maintains a level of profitability to ensure an adequate rate of return to its investors.

Since the late 1980s, when biotechnology firms began to generate significant revenues from product sales, biotechnology executives have been captured by the concept of a company's transition from being technology-driven to market-driven. As research and development is the starting point in a linear pathway for technology-driven companies, market research is the starting point for market-driven companies. This is not a viable approach. In the extreme case, a technology-driven company will produce oversophisticated products that the market does not need, while the market-driven company may produce a multitude of products, none of which is profitable. The ideal approach for biotechnology is an integrated one, whereby the market research and product research groups work together to assess the needs of the marketplace and develop the plan for producing and promoting the products.

With the above in mind, this chapter is *not* a general discussion of marketing issues confronting biotechnology firms. Neither is it a discourse on the science of marketing and its application to biotechnology. The objective of all businesses, biotechnology or otherwise, is to make money. This chapter defines the primary product areas (market segments) on which the industry is currently based and focuses on the major business and marketing concerns facing each market segment. For each segment it suggests a set of goals to achieve and strategies to employ.

MARKET SEGMENTATION

Biotechnology encompasses at least a dozen distinct market segments. The majority of revenues are currently being produced in four primary segments. Of the more than 1500 biotechnology companies in the United States, some 85 to 90% are principally involved in one of the following:

Diagnostic: Human or animal diagnostic tests used to diagnose disease or genetic deficiencies (most often monoclonal antibody or DNA-probe based).

Therapeutic: Human pharmaceutical and animal veterinary products used to treat disease or genetic deficiencies (most often recombinant DNA proteins and monoclonal antibodies).

Agricultural: Plant genetics to improve crop yields and biopesticides to fortify resistance to microbial and insect pests.

Industry suppliers: Developers of laboratory instrumentation, bioreactors, biological reagents and chemicals, and contract services such as large-scale manufacturing and biosafety testing—all germane to biotechnology.

Other applications of biotechnology are beginning to be developed in the food and energy industries, and biotechnology will undoubtedly have a major impact on environmental issues. However, these segments are not expected to produce significant revenue streams from products or services until the mid- to late 1990s.

A MARKET SNAPSHOT

More than 80% of biotechnology companies employ less than 150 people. Whereas large pharmaceutical companies can afford to be somewhat sanguine about their ability to penetrate and maintain market share, the small and medium-sized biotechnology firm cannot; they must recognize the fact that their marketing/distribution plans are critical to achieving their financial objectives. In general, the factors important in establishing a marketing approach are about the same as in other businesses. Market sizes are estimated, market penetrations are projected, distribution and sales channels are established, competitive products or services are analyzed, and pricing levels are determined. Marketing approaches have, however, been quite different among the market segments.

Diagnostic Segment

An industrywide survey by Burrill [1989] indicates that the average diagnostic firm has 32 products in the marketplace. Typically these are monoclonal antibody–based diagnostic kits that, during the mid- to late 1970s and early 1980s, replaced existing polyclonal antibody–based kits and during the late 1980s provided new types of diagnostic tests (for example, detection of antibodies to human immunodeficiency virus [HIV] in human blood). The markets for these products are geographically large and diffuse; the cost of a direct sales force to sell to medical laboratories, physicians, and hospitals worldwide is prohibitive for all but the few largest diagnostic firms. Therefore, most companies in this segment have formed strategic alliances with firms or institutions having the requisite "marketing muscle." This strategy holds true even for firms in decidedly "niche" markets. A case in point is Genica Pharmaceuticals in Worcester, Massachusetts, which specializes in detection of neuromuscular diseases. Genica offers its muscular dystrophy test to physicians in the United States through a cooperative relationship with the Muscular Dystrophy Association, which operates 230 clinics throughout the United States.

Therapeutic Segment

This market segment has the fewest products in the marketplace. It also is the segment in which whole new markets are being created. The central issues here are the enormous costs involved in developing large-scale processes for manufacture and purification of proteins, conducting clinical trials, and the long times required to complete the clinical trials and obtain product licensure from the regulatory authorities. It is interesting to note the significant overlap among development activities of most of the therapeutic firms—all of the products either being sold today or to be introduced in the next few years are the result of a rather small number of research and development programs begun in the late 1970s and early 1980s. Nonetheless, the potential market sizes for human therapeutic products range between $100 million and $1 billion. Financial success for the largest firms in this segment will consist of one or two "blockbuster" products—not unlike the traditional pharmaceutical marketplace of the past fifty years or so. Most therapeutic firms will count themselves successful if they can obtain reasonable market penetration in the larger markets, or dominate one or more "niche" markets valued in the $50 to $100 million range. To date, companies with successful therapeutic products such as Ortho Pharmaceutical Corp. (Raritan, New Jersey), Genentech/Hoffmann–LaRoche (South San Francisco, California), and Amgen (Thousand Oaks, California) have employed existing sales forces and distribution channels or developed their own to market the product directly. The large costs associated with bringing any therapeutic product to market will force most of the small to mid-sized therapeutic firms to form alliances with the large biotechnology and pharmaceutical companies early on in the clinical trial stage (Phase I/II) of product development. The few that are able to raise enough capital and persevere to product licensure will undoubtedly have to form strategic partnerships with companies having established pharmaceutical marketing capabilities and sales forces.

Agricultural Segment

The agricultural biotechnology (AgBio) segment has been much slower to mature than originally projected during the late 1970s and early 1980s. AgBio's progress has been hampered not only by the scientific challenges of developing gene transfer techniques and the small knowledge base in plant molecular biology, but by the public's concern about the release of genetically altered organisms into the environment. Nonetheless, there are over 150 firms worldwide in the AgBio industry, and product sales are expected to reach the billion dollar level by the year 2000 [Gebhart 1990]. In the early to mid-1990s, the products in this segment will be essentially replacements for products that already exist (e.g., microbial sprays replacing chemical sprays). For those products, there is a well-established distribution system already in place, and most firms in this segment are developing marketing alliances with companies already established in the marketplace.

Industrial Suppliers

Of all biotechnology market segments, the supply and service firms are farthest along the commercial road. A glance at the advertisements in any trade journal will confirm this—much of the advertising centers around the plethora of instrumentation, equipment, and supplies needed to create cell lines, perform large-scale cultivation, purify the bulk supernatant to therapeutic grade, and monitor the quality of the final product. Sales in the biotechnology instrumentation portion of this market segment alone will be close to $1.3 billion by 1993 [Gray 1989]. Of the more than 200 firms in this segment, an increasing number are offering services to the therapeutic segment of the biotechnology industry—everything from performing the technically sophisticated tests involved in qualifying cell lines as substrates for production of therapeutic-grade proteins to providing complete management services for clinical trials and regulatory submissions. A typical company in this segment will have more than 100 products (or services). The marketing approach of companies in this segment differs according to the types of products or services offered. Firms selling specialty reagents or chemicals have tended to form marketing alliances with the larger supply firms. A case in point is Promega Corporation (Madison, Wisconsin), which has an agreement with Fisher Scientific for the U.S. market. With the benefit of Fisher's large sales force, Promega has been able to penetrate its markets successfully.

On the other hand, most of the equipment and instrumentation suppliers employ their own sales forces. Many firms are large, well established, and profitable, having been suppliers to the pharmaceutical and chemical industries for many years. Newer entrants to this segment have also opted to develop their own field sales capabilities; the best known among these is Applied Biosystems (Foster City, California). Within the past five years, a number of service companies have been formed, aimed exclusively at providing quality assurance testing and clinical trial development services to the therapeutic market segment. Every one of the fifty or so companies devoted to this niche have developed in-house marketing and sales capabilities because of the limited number of clients. Most estimates place the total number of companies developing biological therapeutics at less than 300.

FORCES INFLUENCING BIOTECHNOLOGY MARKETS

There are four types of forces that affect a company's ability to serve the biotechnology marketplace and, consequently, the marketing strategies a company can employ. Two are external, which, in most cases, the company can little affect; namely, the competition and the government. What constitutes a competitive advantage in the biotechnology industry is much the same as in other industries; namely, the development of proprietary products (or services) in a timely fashion, the ability to attract and retain people of high caliber, the capability to position a company within the marketplace to the company's maximum advantage, the ability to deliver the product to the market, and the financial resources to make it all happen. These affairs are all within the control of a company; therefore, the extent to which competitors are more successful in implementing them will determine a company's success in the marketplace.

In contrast, governmental issues are a bit less certain. The unclear regulatory environment makes it difficult for companies in the therapeutic and AgBio segments to predict product introduction with any degree of confidence. In fact, what is certain is that most companies continue to underestimate the regulatory hurdles and the times required to overcome them. In dealing with new life forms and complicated manufacturing processes involving living tissue and cells, the patent situation can be equally uncertain. Litigation is proceeding on these and other related issues among companies in the therapeutic segment. Finally, the issue of third-party reimbursement will become of critical importance to companies in the diagnostic and therapeutic segments that intend to market new and innovative tests and therapies.

The third force, financial, is a mixture of external and internal, over which the company can exert some influence. The key internal issue here is the ability of a company to raise enough capital to continue to fund research and development and, in the therapeutic segment, to fund development and construction of large-scale manufacturing facilities and to complete clinical trials. Of course, the situation in financial markets, investor confidence in biotechnology, and the cost of capital are the external issues that contribute to a company's success in achieving financial objectives.

The fourth force, managerial/operational, is totally within a company's control. The issues here revolve around the company's ability to attract and retain key personnel, the continuance of a research and development capability to improve existing products and develop new ones, and control over manufacturing costs. Traditional marketing issues are key here also. There is simply no substitute for knowing your present competition and where future competition is likely to emerge and for developing product pricing strategies in keeping with market pricing and profit margin objectives.

Clearly, the business issues mentioned above are all interdependent. How these relate to the objectives and goals of the typical biotechnology firm, and ultimately the strategies employed to reach the corporate goals, are the topic of the remainder of this chapter.

MISSIONS, OBJECTIVES, GOALS, STRATEGIES, AND TACTICS

The terms *mission, objective, goals, strategies,* and *tactics* are often used interchangeably in everyday conversation, and are all too often used interchangeably by senior management in the biotechnology industry. Each of these terms should have a distinct meaning and place in the company's marketing plan. The classic means of illustrating the differences is by analogy to a college football team. One *mission* of a football team may be to develop and promote school spirit. The *objective* of the team is to achieve a desired end state—namely, to win football games. Therefore, the *goal* of the football team is to score more points than its opponent.

In industry, goals are generally expressed in terms of numbers—a magnitude to be achieved (e.g., 10% increase in sales) and a time period for achievement (e.g., by fiscal year end). Within a football game, the team will have short-term goals, such as keeping an opponent from scoring on fourth-and-one from the goal line or moving the football within range for a winning field goal.

Strategies are alternative means for achieving goals. Football teams, like companies, adopt strategies based on their strengths and weaknesses. Teams with a particularly strong running game (and a mediocre defense) prefer ball control. *Tactics* refer to the actual execution of a strategy. For a particular running play, tactics involve who will run with the ball, where he will run, and how the offensive line will block for the runner.

Some confusion in the use of these terms is due to the fact that different terms may describe the same actions taken at different levels of the company. A strategy for the marketing director may very well become the goal of a subordinate.

The purpose of the following discussion is not to examine how corporate missions or objectives are defined in the biotechnology industry, but rather to define some important goals about which companies in each segment should be thinking, and to assess which strategies may be important to achieve these goals as more and more biotechnology products reach the market during the 1990s. The astute reader will note that the goals defined below are not quantitative at all, contrary to the definition above. The quantitative parameters for each goal should be specific to each company's situation.

STRATEGIES FOR THE THERAPEUTIC MARKET SEGMENT

There is no question that, in the United States, the biotechnology industry owes its existence to the confidence of its investors. The industry as a whole is not yet profitable, despite the sales and profit figures currently reported by some of its largest members. The promise of future success is what has driven the industry during the 1980s. During the 1990s, the therapeutic segment expects to see sales increase more than fifty times current levels [Burrill 1990]. Yet at current burn rates, only about a quarter of the companies in this segment have enough cash in hand to survive more than three years, while a full 40% cannot last more than one year without additional funding (Fig. 13-1). The key objective for this segment then is financial survival. Some goals and relevant strategies are outlined in Table 13-1 and elaborated in the following discussion.

Goal 1: Extend Survival Period to Point of Revenue Generation from First Product

The obvious first strategy here is to raise additional capital. It is often stated that the cost of bringing a therapeutic to the market is in excess of $100 million. In evolving a financial strategy, it is just as important to know how the money is going to be spent as it is to know from where the money will come. The largest single expenditure from that $100 million will be in developing, constructing, and validating large-scale manufacturing and purification processes. One way to pare down that $100 million would be to seek a joint venture with a company that has relevant and demonstrated experience in large-scale manufacturing and purification of biologic products. Some of the large pharmaceutical companies

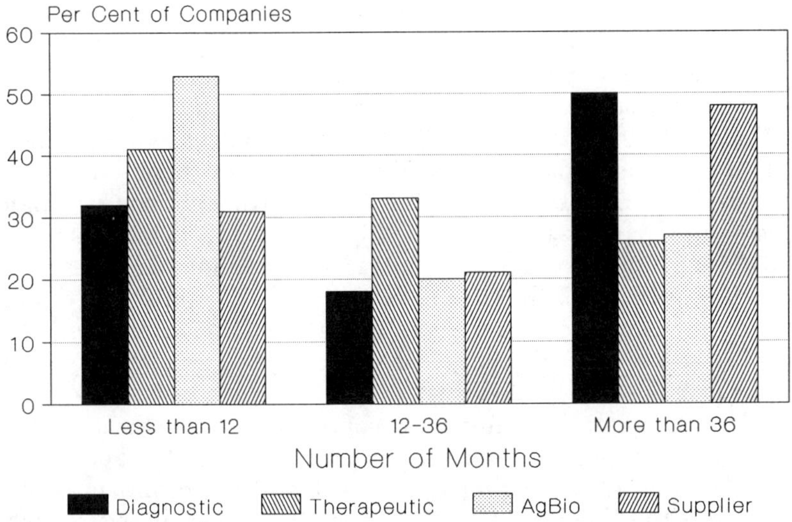

FIGURE 13-1 Distribution of survival index by market segment. Survival index (number of months) is calculated for each company as total existing (booked) cash and cash equivalents divided by monthly net burn rate. Reprinted by permission of the copyright holder (Ernst & Young) from G. Steven Burrill. 1989. *Biotech 90: Into the Next Decade*, New York: Mary Ann Liebert.

have expertise in large-scale culture of mammalian cells. There are also contract production firms with highly developed technologies for mammalian cell culture. Most companies eventually will insist on internal control of manufacturing; therefore, it is imperative that any production/ purification technologies be transferable and that the joint venture part-

TABLE 13-1 GOALS AND STRATEGIES FOR THE THERAPEUTIC MARKET SEGMENT

Goals	Strategies
Extend survival period to point of revenue generation from first product	Raise additional capital Establish a joint venture with manufacturer Shorten regulatory critical pathway
Know the real competition	Survey established pharmaceutical firms
Plan for rapid market penetration	Position the product in the market Establish sales and distribution network Determine product pricing

ner (or contract production house) has committed to transfer of that technology.

Another important strategy is to shorten the critical pathway through the regulatory bureaucracy. Since there are no formal regulations within the Code of Federal Regulations pertaining to biological therapeutics (except for vaccines), companies wishing to obtain Investigational New Drug (IND) exemption status must follow the somewhat oblique recommendations in the appropriate "Points to Consider" documents [Esber 1987a, 1987b] issued by the Center for Biologics Evaluation and Research or the published remarks of the European Economic Commission [Committee for Proprietary Medicinal Products 1989a, 1989b, 1989c]. Companies developing new biological products that can be used to treat patients with a serious or immediately life-threatening disease can make use of a "Treatment IND" exemption. Modifications of this have led to the so-called fast-track regulatory process, whereby the company and the Food and Drug Administration [FDA] participate in developing appropriate preclinical studies, Phase I safety studies, and Phase II efficacy studies. At the conclusion of Phase II, if the data are sufficient to warrant, the FDA may put a Treatment IND in place and the company can proceed to submit a Product License Application (PLA). Dr. Frank E. Young, former Commissioner of the FDA, has stated publicly [Burrill 1989] that he believes the average interval from beginning Phase I clinical trials can be reduced from the pre-1984 average of 7.5 years to as low as 3 years. The message here is that early (pre-IND) and frequent (through Phase III) consultation with the FDA can help not only to compress the time for regulatory approval, but might lead to lower overall clinical trial costs.

Goal 2: Know the Real Competition

Most biotechnology companies in the therapeutic segment consider other biotechnology companies to be their most serious competitors. This is natural, given that other biotechnology firms with similar molecular biological and cell culture expertise are most likely to be developing similar products. This is certainly borne out by the numbers of biotechnology companies working on such products as plasminogen activators, interferons, erythropoietin, and colony stimulating factors. The successful strategy here for every biotechnology firm should be to survey the *pharmaceutical* firms that may have an interest in the products under development. This interest may be due to the pharmaceutical's presence in similar markets, both diagnostic and therapeutic. Most, if not all, of the multinational pharmaceutical firms have made substantial research and development investments in biotechnology. These well established firms (with equally well established positive cash flows) have little or no need to trumpet their research and clinical findings. Nor do they wish to, as they well understand the competitive advantage of not advertising the status of product development activities. Moreover, the pharmaceutical firms are able to invest whatever resources are required to bring a product to market, including acquisition of appropriate technologies and companies. Classic examples are Bristol-Myer's acquisition of Genetic Systems in 1985, Eli Lilly's acquisition of Hybritech in 1986, and Hoffmann–La-Roche's acquisition of Genentech in 1990. The foregoing discussion on the limited financial resources available to most biotechnology firms in this segment only points up the probability that many firms with promising technologies will become the targets of the major pharmaceutical firms. The competitive landscape by the end of the 1990s will be radically different than it is today.

Goal 3: Plan for Rapid Market Penetration

Financial success is defined by a company's ability to derive revenues and profits from the sale of products. Obviously, in any market, total revenues are dependent on market share. Every biotechnology firm must develop a strategy for entry into the marketplace, must grow its share of the market (if it is not the first into the market), or must maintain market share (if it is first into the market). This plan must begin its development early in the product development cycle. To begin the plan, the company must ask itself the questions below.

How will the product be positioned within the market? Product market position refers to how customers perceive a product. This is naturally expressed in terms of a product's uses and value. For the therapeutic segment, customers are primarily physicians and medical personnel. The value of the product will be defined in both functional (practical) and monetary terms. Terms like "higher quality," "faster acting," and "less expensive" are useful in describing the product's benefits to the customer base. Of course, if the product is the first of its kind in the marketplace, market positioning becomes largely an educational task. There will be heavy reliance on data generated during clinical trials. For a completely new product in the therapeutic segment, about which thousands of physicians have to be educated, costs could easily escalate into the millions of dollars.

How will the product be distributed and sold in the United States and in foreign markets? Clearly, therapeutic products whose primary customer base is physicians will not be sold through the same distribution channels as diagnostic products or supplies. Sales forces will have to be recruited and trained—potentially an expensive proposition for a one or two product company. Companies such as Genentech, with significant cash resources, have chosen to employ their own sales force. The strategy of the majority of companies developing products in this segment will undoubtedly be to form a marketing alliance with one of the pharmaceutical firms, very likely different firms in different geographic markets. As companies grow, mature, and add to their product portfolios, increasingly the trend will be to rely on their own direct sales force in the U.S. market. If the product brings in annual revenues greater than some minimal value, say $50 million, in either Europe or Japan, some companies may opt to develop their own sales forces in those geographic regions.

What are the important factors in product pricing? There is no doubt that in this segment, pricing is and will continue to be market-driven. Most biotechnology companies rate effectiveness of existing products, what the market will bear, manufacturing costs, and cost of existing products about equally as primary factors in product pricing [Burrill 1990]. Interestingly, few biotechnology firms put any emphasis on recouping research and development costs in their pricing. Even more significantly, few firms perceive that reimbursement issues will be a factor affecting pricing. With the rising costs of medicine in general, third-party reimbursement will definitely be a cost factor during the 1990s. Companies failing to realize this are taking the risk that the perceived value of the new biopharmaceuticals will outweigh the public's resistance to increasingly high medical costs. This may be the case for truly innovative products but will probably not be for products with equivalent functionality to existing drugs.

STRATEGIES FOR THE DIAGNOSTIC MARKET SEGMENT

The diagnostic segment has been selling products for the past ten years and is a well established marketplace, with product sales providing the majority of revenues for these firms. However, about two-thirds of diagnostic companies report net losses. Relative to the therapeutic market segment, diagnostic firms are in better financial shape—half will be able to survive more than three years based on current burn rates (see Fig. 13-1). In addition, firms in this segment will continue to lead both therapeutic and AgBio segments in new product introductions for the foreseeable future. The major advantages to this market segment are lower research and development costs and a smaller regulatory burden (except for medical imaging antibodies). Barriers to entry in this market are lower; therefore, each product area tends to be more crowded and competitive.

TABLE 13-2 GOALS AND STRATEGIES FOR THE DIAGNOSTIC MARKET SEGMENT

Goals	Strategies
Develop resources for new products	Expand university contacts Create "niche" markets
Aggressively pursue foreign markets	Establish distributor base in Europe Cross-license technologies and products in Japan
Compete in major markets	Select trade shows judiciously Use telemarketing Determine selective market pricing

Overall, these firms have much lower gross margins than any of the other segments. Because most companies in this segment are not profitable, financial survival is an important issue. However, financial survival is not so much a question of raising the additional funds to support extensive product research and development or to survive through the two to three years required to obtain a product license, but rather a variety of new product development and marketing concerns related to innovativeness of product, timing in the marketplace, and penetrating the marketplace. Goals and strategies for this market segment are outlined in Table 13-2. Goal 1 pertains primarily to small, early-stage companies (less than fifty employees); Goal 2 applies equally to all companies in this segment; and Goal 3 applies more to mid-sized and larger companies, or to an early-stage company with a truly innovative product.

Goal 1: Develop Resources for New Products

On average, more than one-third of all expenses incurred by firms in this segment go toward new product research and development [Burrill 1989]. Although this is a significantly smaller percentage than either the therapeutic or AgBio segments, it is still very high given the profit margins that can be expected of products in this segment. For comparison, established pharmaceutical companies spend only about 16% of expenses on research and development. Product development resources can be external as well as internal. Indeed, the majority of firms in the diagnostic segment have established affiliations with universities, providing an excellent source of new product ideas and qualified personnel. Having an affiliation with one or two research laboratories may not be enough. The overall strategy, therefore, is to extend the base of collaborative relationships with outside resources, primarily university laboratories and government research institutes. The important point here is not so much having a proprietary technology, but being the first into a particular market with a product, and maintaining market share by continued product development and aggressive marketing. Being first into a particular market means developing working relationships with prominent researchers in that market and identifying significant technologies that can be acquired. Clearly, personal contacts are the best means of establishing such relationships. Given the scope of research in any given area, identification of promising research can perhaps be better accomplished by extensive use of computer literature banks and grant indexes. Knowledge of the most-heavily funded grants is an excellent means of identifying the leading researchers, as these grants are peer reviewed.

A secondary strategy that may be employed is to go after the niche markets ($1 to $5 million per year per product) that may be identified by the strategy discussed above. Early-stage companies are most likely to benefit from this strategy as they will be able to dominate a small market and avoid the competition in the larger markets. Controlling a niche market offers the opportunity for developing new products within that niche, as the market will generally tell a company what the next product should be. In addition, the lack of competition allows for greater profit margins.

Goal 2: Aggressively Pursue Foreign Markets

The United States market is, and will remain, the dominant market for the biotechnology industry. A surprising number of companies in this segment have sales in Western Europe (56%) and Japan (41%) [Burrill 1989]. Most of the non-U.S. sales are handled either through marketing affiliations with larger companies or through local distributors. It is important to note, though, that both the European and Japanese markets are growing in size more rapidly than the U.S. market, so their importance to competitive advantage of companies in this segment will increase during the 1990s. Many believe that by the end of the 1990s, Japan and the other Pacific Rim countries will present as large a market opportunity as Europe.

In the United States, many companies have chosen direct sales as the simplest and most cost-effective way of distributing their diagnostic products. In Europe, the strategy will be different—distributors are key to establishment of a sales base quickly and cost efficiently. Despite all the talk about 1992, European markets are and will remain culturally distinct along country borderlines. Distributors, as local representatives, understand the marketplace and will assume the responsibility for importation paperwork, inventory maintenance, customer sales, and follow-up support. Be aware, however, that distributors in Europe are generally more specialized than in the United States. Products aimed at the diagnostic market should be distributed by a specialist in that area—not in the biological supplies or research products areas.

Whatever can be said about Europe may be doubly true for the Pacific Rim countries, dominated by the Japanese market. Many of the large U.S. pharmaceutical firms maintain subsidiaries in Japan, but still have to work with their Japanese counterpart firms truly to understand the market. In many cases, Japanese firms engaged in developing diagnostic products also have active distribution channels of their own and are on the lookout for U.S. products to distribute. One good strategy here would be to seek out companies with products that either complement or extend one's own lines. By exchange of technologies or cross-licensing of products, the product lines of both companies can be enhanced, with the added advantage of reduced new product development costs. All areas of biotechnology are expected to receive heavy investment in Japan during the 1990s, so an early effort by U.S. firms to establish working relationships could pay off handsomely for the rest of the decade.

Goal 3: Compete in Major Markets

There is often a tendency for smaller companies to be hesitant about entering markets dominated by one or more of the large diagnostic firms. One can be easily intimidated by going up against a company whose direct sales force numbers in the hundreds. However, the size of a particular market may be precisely the reason for competing there. A minor share of a large market could mean substantial revenues, perhaps more than with dominance of a smaller market. Early-stage companies operate here with certain inherent advantages. For one thing, they are quite adaptable. Decision making is swift, without the layers of management and committees that often beset larger organizations. New products can often be brought from concept to market much more swiftly for the same reason. Of course, the early-stage company does not have the same marketing and sales clout that the larger firm does. One excellent technique that can be used to overcome this deficiency is the judicious selection of trade shows at which to exhibit. Everybody knows about the larger trade shows, where the established firms showcase their products in large, expensive, and well-staffed booths. These are obviously the places to be, but with a more innovative, image-enhancing booth presentation. Customers do tend to shop all the booths at any trade show.

Another technique that works is telemarketing by personnel trained not only in the proper techniques, but also armed with a thorough knowledge of the product line, including the features and benefits of each product. Admittedly, there is no substitute for a personal visit to each and every client account, but repeated phone conversations, with rapid and thorough follow-up after each, will go a long way toward enhancing a company's image and generating sales revenue.

Finally, selective market pricing on a client-by-client basis can be an effective means of gaining the attention of major potential clients, especially for firms in this segment marketing diagnostic test kits.

STRATEGIES FOR THE AgBio MARKET SEGMENT

The AgBio segment may be the least understood and most undervalued market in biotechnology. In general, this segment has focused on plant genetics to improve crop yield, prevent spoilage, and bolster resistance to microbial and insect pests. In the United States alone about $20 billion worth of chemical pesticides are sprayed on crops each year. Given public concerns about chemical pesticide spraying on apples and other fruits, it is clear that major market opportunities exist in this segment. Despite the above, AgBio does not seem to enjoy widespread confidence within the public sector and among investors. Goals and strategies for this segment out outlined in Table 13-3 and discussed in detail below.

Goal 1: Form Research and Development Strategic Alliances to Reduce Developmental Costs

Among the four biotechnology market segments, AgBio has the highest research and development costs as compared with sales—i.e., for each dollar in sales, AgBio firms spend $1.46 on research and development. This compares to $1.25 for therapeutic firms, $0.54 for diagnostic firms, and $0.11 for industrial suppliers [Burrill 1989]. The fact that this segment is less mature (with lower sales) than others contributes to this. However, for AgBio this figure has been increasing at a higher rate than the other segments for the past few years even though product sales have done likewise, suggesting that AgBio firms are accelerating product development. Given the lengthy development times and time-consuming regulatory processes that AgBio firms face, many have opted to form strategic alliances with U.S. agricultural companies. For the most part, though, these have been marketing alliances. The strategy here should be to broaden these alliances into the research and development area or, where necessary, to develop independent corporate or university alliances with strong research and development groups. Clearly, product develop-

TABLE 13-3 GOALS AND STRATEGIES FOR THE AbBio MARKET SEGMENT

Goals	Strategies
Form R&D alliances to reduce developmental costs	Broaden industrial marketing alliances into R&D Develop institutional R&D alliances
Focus more on regulatory and public awareness concerns	Establish formal government relations function Establish formal public relations function
Sell where the market is	Form R&D and marketing alliances with European agricultural chemical firms

ment costs have to be driven down for the biotechnology firms in this segment—or they will not survive.

Goal 2: Focus More Closely on Regulatory and Public Awareness Concerns

There is no doubt that the whole regulatory process has not been easy for AgBio firms. Part of this stems from the fact that the public has been uneasy about release of genetically altered organisms into the environment. Public debate on these issues has been emotionally charged, and opponents to the industry have been able to tie their arguments into environmental protection. Other issues relating to social responsibility of companies performing this research have also crept into the debate. As a consequence, years have been added to the product development cycle. The best strategy that companies can employ is to make government relations a formal and serious part of doing business. As early in the product development process as possible, the Environmental Protection Agency (EPA) should be involved. Regulatory personnel expert in the relevant EPA laws (Federal Insecticide, Fungicide, and Rodenticide Act [FIFRA]; Toxic Substances Control Act [TSCA] should conduct discussions with the EPA.

The second strategy should be to make public relations as serious a part of doing business as government relations. Appropriate environmental groups need to be consulted and informed at all levels (local, state, and national) as regulatory milestones leading to product testing in the environment are achieved. Consideration should be given to informing the press and local politicians at appropriate times. It is only through assertive and coordinated strategies such as these that opponents both on political and environmental grounds will be assured that the company can indeed be socially responsible.

Goal 3: Sell Where the Market Is

European companies should not be overlooked as alliance partners, both for research and development and for marketing purposes. The perception is that European chemical and agricultural chemical companies are more advanced in their thinking about AgBio. This can be explained by the fact that in Europe companies tend to integrate seed and chemical businesses, whereas in the United States, the seed and chemical business are separate industries. Because much of the focus in AgBio is in the seed genetics area, it may be more difficult for AgBio companies to interest U.S. agricultural chemical companies than their European counterparts. Also not to be overlooked is the fact that two-thirds of the world's market for chemical pesticides is outside the United States. A strategy of forming partnerships with European agricultural chemical firms will not only help U.S. AgBio companies to share some of the research and development burden (Goal 1 above), but could lead to penetration of the larger global marketplace.

The AgBio segment, for many reasons, has taken longer to mature than other biotechnology market segments. Agricultural business is, however, the world's largest single enterprise whose importance increases with the world's population. Despite its rocky start, AgBio will become one of the largest and most profitable segments in this industry.

STRATEGIES FOR THE INDUSTRIAL SUPPLIER MARKET SEGMENT

The industrial supply market segment is considered the most mature. Suppliers typically sell a large number of products or services, with average revenues from product sales of $25.8 million (greater than three times the industry average). Product sales account for about 95% of all corporate revenues among suppliers. Among the four segments, on average only the supplier segment is profitable, with more than half of all firms reporting a net profit. Typical income before taxes is 10% of sales. Research and development costs run at about 11% of total sales, much lower than the industry average of 42% [Burrill 1989]. The marketing issues facing supply and service firms, therefore, are typical of those found in other established industries, and they relate to recognizing the market-

TABLE 13-4 GOALS AND STRATEGIES FOR THE INDUSTRIAL SUPPLIER MARKET SEGMENT

Goals	Strategies
Establish organization image and product awareness	Position a product properly Identify and attend focused trade meetings Practice selective journal advertising Meet face-to-face with clients
Sustain client motivation and extend client base	Establish client service function In-license new products and services Expand geographic sales base Create new product/service niches

place needs and proper positioning of products or services within that marketplace.

There is an important distinction, though, between mature companies and a mature marketplace. The biotechnology industry, viewed as a marketplace, is reasonably well established and growing rapidly. There are no companies with "General Motors" dominance in the supplier market, mainly because it is quite fragmented. Some portions of this segment, such as analytical instrumentation, purification systems, and biosafety testing services, are dominated by a few major players, while others, such as fermentation systems, oligonucleotide preparation services, and cell culture media, are contested by numerous suppliers, each having minor market share. New companies are continually being formed to challenge the "established" firms, most of whom are themselves less than ten years old. Thus, there are classic marketing goals that the newer players strive toward—and some equally classic strategies available to achieve them. Most of the comments in this section are directed toward early-stage companies (i.e., companies less than five years old—which certainly applies to the vast majority of firms in this market segment!). Goals and strategies for industry suppliers are outlined in Table 13-4.

Goal 1: Establish an "Image" of the Organization, an "Awareness" of Its Products (or Services), and a "Motivation" for the Clients to Buy the Products

For early-stage companies in biotechnology (or any other industry), the image of the company is totally a reflection of its products and service reputation. Larger firms may be able to sustain a good corporate image for some time after a product's quality has degraded, and may have different images among the various markets they serve. However, image will develop a momentum of its own—and may be difficult to change once in place. The first strategy to follow is to position the product within the marketplace. As discussed under Strategies for the Therapeutic Market Segment, positioning involves a thorough knowledge of the client base as well as the strengths and weaknesses of your own product and competitive products. Importantly for the biotechnology supplier segment, it can also mean knowledge of other, potentially competitive markets. For example, Molecular Devices Corp. (Menlo Park, California) is the first to develop and market a novel instrument for detection of picogram amounts of residual DNA in biological products. Although this instrument has no competition as an automated technique, the industry standard method accepted by the FDA for testing final biological product formulations employs an entirely different technology and is offered by a number of service companies in the industrial supplier market segment.

The follow-up strategy to positioning the product is to make the client base aware of its existence, its benefits, and its features. The primary means of accomplishing this is through journal advertising, industrial trade shows, and direct sales or use of distributors. These may seem to be rather straightforward exercises—they are not. Whereas attending trade shows is probably the most cost-effective means of exposing the product directly to potential customers, deciding which trade shows to go to can be challenging. Firms that supply products such as tissue culture media and plastic ware to virtually all market segments (as well as biomedical researchers) will benefit by attending large trade organization meetings with thousands of participants. To the small supplier of services to a particular market segment, there are a number of focused scientific and trade meetings (i.e., attended by 300 people or less) that permit suppliers to set up a small booth. These meetings can be extremely effective, as a high percentage of attendees will be prospective clients.

By the same token, journal advertising will probably be more effective for the large firms in this segment selling culture media or supplies to the industry as a whole, or for the early-stage firms who wish to announce their presence and are beginning to develop an image and solicit sales leads. In total, though, there are about 1,500 biotechnology firms in the United States and less than 200 more in the rest of the world. Many companies in this market segment are selling only to a selected subset of these biotechnology firms. Therefore, as a general rule, journal advertising should augment both your trade show and direct sales efforts, and should not be the promotional centerpiece.

There is nothing better than a face-to-face meeting with clients, both prospective and current. For obvious reasons, many firms cannot cost-effectively consider this approach, as their client base is far too large and most accounts are not large enough to warrant the travel expenses. For early-stage firms selling sophisticated instrumentation or for service firms hoping to establish long-term, increasing business with a select group of clients, the direct sales approach cannot be matched for effectiveness. The most important part of that approach is a technically knowledgeable, responsive salesperson. The company's image will in large part be determined by how the sales force relates to the client base. The client's motivation to buy products or services offered by the company will be entirely affected by initial and continuing contact with the company's sales personnel.

Goal 2: Maintain the Motivation in Your Current Client Base and Extend That Base

There is one important element common to every product or service in any industry. The product or service must solve the customer's problem. This adage is so well known it has become banal in the marketplace. The health care industry, in general, suffers less from bad image in this regard than others (since when has a used-car salesman solved anybody's problem?). However, the concept of problem solving is of appropriate concern at this time, more so to the supplier market segment than others within biotechnology for the simple reason that suppliers offer far more products than any other market segment—and problems with a vendor of a critical component could very well cripple a clinical trial or a product launch. The successful company in this segment will be the one not only to offer a superior product or service, but also to work with each individual client to ensure that product or service is meeting that particular client's needs.

The first strategy here is to hire a competent, knowledgeable, and experienced client service person—someone who will learn the products, people, and functional organization of the firm. Note that this is a marketing function and differs from, for example, technical services for an instrument manufacturer. This person should be the first contact within the company for any client with post-sale problems. It is the sole function of this individual first to understand the client's problem, then to direct that problem to the appropriate individual(s) within the organization, and finally to follow up to ensure that the problem has been resolved satisfactorily. This person should have a direct line to all key managers within the organization (e.g., vice presidents of marketing, sales, operations, manufacturing, quality assurance, etc.). Satisfied clients are really a business's most important product. Satisfied clients also have a habit of spreading the word, and word-of-mouth is perhaps the most effective means of promoting a business, especially in an industry as small as biotechnology.

Strategies for expanding client bases, other than promotional, are in-licensing new products and services, expanding sales efforts to new geographic locations, and creating new market niches. Although these topics have been addressed above, they are particularly important to companies in this market segment. Companies in this segment typically spend only 11% of sales on research and development [Burrill 1989]. New products are much more likely to come from in-licensing from other supplier companies or university laboratories than from internal research and development. Expansion to non-U.S. markets has already begun in earnest in both the diagnostic and supplier market segments. Currently, about 20% of all sales for these firms come from overseas, and that percentage is expected to double during the 1990s. For biotechnology products and services, the U.S. domestic market accounts for about one-third the total world market, so expansion into these markets can be expected. The comments regarding use of foreign distributors (discussed under Strategies for the Diagnostic Market Segment) are equally relevant to firms in the supplier market segment.

Finally, companies can become successful by creating new product and service niches. The key strategy here is to recognize a market need and be able to capitalize on it. A market need can exist even in a market filled with products, simply because none of the existing firms are providing adequate solutions for the customer's problems. Examples of this are in the large-scale contract production market, where contract producers originally developed technologies more suited to multi-hundred gram production levels from suspension cells and built facilities expressly for that purpose. Because the market was not ready for these production levels in the early 1980s, these companies were unable to compete successfully for the 5 to 100 g contracts because of high facility overhead. There are now contract production firms who specialize in small-to-mid-scale culture (10 to 100 g of protein) and a variety of technologies have been developed to deal with the different kinds of cell lines employed to produce protein product (e.g., anchorage-dependent vs. anchorage-independent cells). Indeed, with the exceptions of Celltech, Ltd. in the United Kingdom and Verax, Inc. in the U.S., the large-scale production firms have all been acquired (Damon Biotech, Inc. by Abbott Laboratories, and Bio-Response, Inc. by Baxter International, Inc.), have sold their manufacturing assets (Invitron Corp. to Centocor, Inc.), or are out of the contract manufacturing business (Helix Biocore, Inc.).

KEY ISSUES FOR THE 1990s

Research, development, patents, and financing have played central roles during the 1970s and 1980s as biotechnology companies were formed and began to struggle toward the marketplace. The industry, which had been fueled primarily on promise, reached a watershed in November of 1987, with the FDA's approval of Activase (alteplase), the clot-dissolving drug from Genentech. It was a clear demonstration that a biotechnology company could indeed develop a novel genetically engineered product; establish a pharmaceutical-grade technology and facility to manufacture, purify, and formulate the product; bring the product through the trials and tribulations of clinical testing and approval; and market it quickly and in high volume across the United States. Although sales from that product have not reached expectations, it is still a blockbuster product by any measure. It is also a harbinger of things to come. Each of Amgen's approved products, Epogen and Neupogen, are expected to surpass Alteplase in sales volume. The new products pipeline is full, as hundreds of potential products enter into clinical trials and companies are working on new generations of products that will not reach the marketplace until beyond the year 2000.

As the industry grows in the 1990s, manufacturing, operations, regulatory compliance, and marketing will become the dominant themes as companies that survive the current financial crunch begin to compete in the marketplace. There will be a strain on FDA resources as greater numbers of products are submitted for licensure. A competitive environment and medical reimbursement issues will force companies to reduce manufacturing costs. Finally, though, the successful companies in biotechnology will be those that can compete effectively, in both domestic and international markets. The crucial determinant of competitive effectiveness will be how skillfully a company can create markets or establish and retain dominant market share. This is what marketing is all about.

REFERENCES

Burrill, G. S. 1989. *Biotech 90: Into the next decade*, New York: Mary Ann Liebert.

Burrill, G. S. and Lee, K. B. 1990. *Biotech 91: A changing environment*, San Francisco: Ernst & Young.

Committee for Proprietary Medicinal Products. 1989a. Notes to applicants for marketing authorizations on the pre-clinical biological safety testing of medicinal products from biotechnology (and comparable products derived from chemical synthesis. *J. Biol. Standard.* 17:203–212.

Committee for Proprietary Medicinal Products. 1989b. Notes to applicants for marketing authorizations on the production and quality control of monoclonal antibodies of murine origin intended for use in man. *J. Biol. Standard.* 17:213–222.

Committee for Proprietary Medicinal Products. 1989c. Notes to applicants for marketing authorizations on the production and quality control of medicinal products derived by recombinant DNA technology. *J. Biol. Standard.* 17:223–231.

Esber, E. C. 1987a. *Points to consider in the characterization of cell lines used to produce biologicals (1987).* Bethesda, Md.: Center for Biologics Evaluation and Research, Food and Drug Administration, Department of Health and Human Services.

Esber, E. C. 1987b. *Points to consider in the manufacture and testing of monoclonal antibody products for human use (1987).* Bethesda, Md.: Center for Biologics Evaluation and Research, Food and Drug Administration, Department of Health and Human Services.

Gebhart, F. 1990. *Genetics Engineering News* 10(1):7.

Gray, M. A. 1989. *Biotechnology equipment in the pharmaceutical industry.* Los Angeles: Strategic Directions International.

CHAPTER 14

CREATING AND MARKETING THE IMAGE OF THE BIOTECHNOLOGY COMPANY

Lisa E. Burns

Remaining competitive in today's biotechnology industry requires more than breakthrough technology and a commitment to research and development. Each year, hundreds of biotechnology companies are faced with the challenges of establishing or maintaining a corporate identity, raising capital, undertaking an initial public offering, and effectively communicating scientific accomplishments beyond the laboratory walls. While there is much debate over the variety of options available to meet these challenges, and industry experts agree that the essential ingredient for success is that of increased awareness among the scientific and financial communities and, in many cases, the national science and business media.

The role of the public relations professional, both on the corporate and agency sides, is to help the company develop a strong public image—one that reflects the true nature of the company's goals without creating unrealistic expectations among the target audiences. When used appropriately, public relations can be the most effective communications tool the company has at its disposal. Successful image building is dependent on continued corporate strategizing and ongoing communication between company management and its public relations representative(s) and with the national media and the financial community.

The following is a round-table discussion among several industry professionals on the topic of creating and marketing the image of a biotechnology company. Participants in the discussion included Lisa Burns,

President of Burns McClellan; Susan Atkins, previously Vice President of Public Affairs at Rorer, Inc., and Director of Corporate Communications at Genentech; Lyn Christenson, Senior Vice President at Hill & Knowlton, previously Vice President of Public Affairs at California Biotechnology and Manager of Communications Programs at Syntex Corporation; and Carol DeGuzman, Director of Corporate Communications at XOMA Corporation.

Lisa: The agenda here is to discuss the methods that you have found to be most effective in creating and marketing the image of a biotechnology company.

Susan: Marketing, whether it's a product or a company, has certain basic elements, and I always feel it should start with identifying the customer: who's your customer and what influences their decision? If it's the financial community, you need to determine what they need to know to make a buy decision. Then you figure out ways to get into that channel.

So that's your number one—the customer, followed by who influences that customer. And that's where we frequently come in, influencing the audiences through media and direct mail, one-on-one meetings, and presentations at all the analyst conferences.

Carol: You first have to determine what you want that image to be. You have to look at the business strategy and the scientific rationale for what the company is planning to do. Then look at other companies in our industry and determine how the company plans to proceed. Put that together and think about to whom you need to market—there's the scientific community and, obviously, a tremendous amount of marketing that needs to appeal to the financial community. For XOMA, those are the key audiences. We use the media to target the financial community and work with the scientific community through more traditional means.

Lyn: The two most important things for a company to keep in mind are how it is going to be viewed as different from the companies already in the marketplace and what will ultimately be the benefit to the patient for the product that's being developed.

Frequently, I find biotechnology companies developing common phrases [appearing in business plans and prospectuses], such as "novel therapeutics for *blank* disease area," or "We will develop a cure for *blank* disease."

These phrases don't serve the company well because, while they may eventually be true statements, they are difficult points to prove when a company is at its emerging stage. A biotechnology company needs to focus very closely on exactly what it is that will make it different from the pack.

"Positioning" a company is often phrased in terms of technology. To really be an effective marketing strategy, it [technology] should be phrased in terms of benefit to the patient: "We will produce a product which will be a new treatment for diabetes without side effects." As opposed to: "We are going to clone a particular gene or we are going to make something in a particular way or purify it in a particular way." Those are means to an end, and the marketing strategy, to be really effective in a public relations sense, needs to be stated from the point of view of the end user: "What's in it for me?"

Lisa: Let's take the scenario of a start-up company with heavy financing coming onto the scene three years after someone else has an established position in that field. How do you create a position that substantiates your research without the financial community and the media saying you're just a "me too" company? What kind of obstacles does that present and what kind of tactics would you use to avoid that?

Susan: A strategy for a company in that position would be to hire the most credible people in finance as well as in the research or the development side to do the talking, because it's going to be a hard sell. The scientific message has to be credibly delivered, because the first audience for a start-up really is the financial community.

Lyn: This certainly presents an obstacle from the investor's point of view. If a venture capitalist has already put money into a company that focuses in a particular area, it is less likely to put more capital into a second one. If the company is attempting to gain recognition for this ability in the media, that same problem will arise.

Lisa: That leads to the next question: How involved should senior management, including scientists, be involved in the marketing of the company?

Susan: One of my pet peeves on this subject occurs when senior management gets intimidated by their lawyers into not stating the positive aspects of their own company. I think they do themselves a great disservice. It is critical that senior management support the marketing and public relations efforts, because if *they* don't believe in a project, no one else in the company will.

Lyn: There is a very low level of understanding of exactly how the public relations process works. And of how powerful it can be as a result. Some of this in our industry stems from resource allocation; there is not a lot of money for things perceived to be intangible.

Carol: If your senior scientific staff is not on board with what you're doing, you can get an internal backlash that can cause different problems. The scientific people in your company, your director, vice president of research and development, need to understand what the corporate communications function is and how they can contribute to it in order that the information is not being completely oversimplified as to be inaccurately represented.

Lisa: Larry Kurtz, corporate communications director at Chiron, said that the turning point came when Chiron discovered hepatitis C virus and knew it actually had a product. "All of a sudden the scientific staff said 'Okay, now we have a product, now we have something to say. So now we would begin a real in-house corporate communications program.'

"I don't think anyone really knew what it meant—the full range of what I was going to end up doing. They thought I simply would send out press releases and call analysts. They didn't understand strategic public relations activities, the way deals and news were disclosed, how the media was handled, or the strategy for positioning Chiron to the financial community, employees, partners and so on, on a consistent basis."

In some cases, the downside for an agency can be when clients who want to use you for publicity, call up and say, "We just did this and that; send out the release." If the company has been your client for five years, that's fine, because you've been following what they have been doing for five years. But when it's, "Hey, we just did this deal," and you don't know what the history is, you haven't been in on the development process, and you may not have been thoroughly briefed on the science, it's a different ball game. We have people entering the business now even at the CEO and CFO level, who are young and may not have been through the Baxters & Abbotts and become familiar with the PR [public relations] process, who want big visibility, but, frankly haven't paid their dues to get it, and there is no real strategy.

The ideal situation is one in which management signs off early on corporate communications as a real part of the company development. It is the face we present to the world and therefore it's an integral part of the company, and you, the PR person or the agency, need to be involved.

Lyn: I think that is critical to the success of the company but it is very infrequently done. Of course, it happens in the more successful companies, some of them represented here. But very often it is something that is only brought up when the company is preparing to go public or when, as Larry said, the company has its first product. Because then there's something tangible to do.

Susan: I remember at Rorer when we were developing the corporate mission statement. [This was a company with sales in excess of a billion dollars where the chairman took a very active role.] One of the pieces of research that we came up with was that over a twenty to twenty-five year period, the stock of fifty companies that had a written corporate mission that everybody knew, performed better than companies that did not [have a mission statement].

When we are communicating verbally, we might all think we agree on the same thing, when we actually could be 180 degrees apart, and frequently are.

When you put it on paper, and often they don't even think about it until they see it on paper, they say, "Oh, my God. This is wrong; we can't say this, it doesn't reflect who we are." I think that one of the really valuable functions that PR people serve sometimes is to put it on the table. And they say, "Oh, that's how we want to be perceived. I see."

Sometimes changes are made. I think one of the most valuable services I have provided has been speech writing, because the speech writer really ends up helping to crystallize an opinion that may not be formalized until a CEO has to say those words in front of an important audience.

Lisa: Having decided that there should heavy involvement on the management side, when do you think is the ideal time for management to get involved? Does a new company need to develop a mission statement in year one?

Carol: I think a decision has to be made about how high a profile a very young company wants to have. If you don't make that initial decision, you are going to be shotgunning all over the place. You're going to spend money and expend effort in a random way that's not really going to accomplish anything.

At XOMA, before we went public, we were an intentionally low profile company, a company that focused on cancer therapeutics raising a reasonable amount of money, privately. We communicated our message basically directly to the venture capital and mezzanine financing audience, and on one kind of deal that we did on an R&D [research and development] partnership. At various times, we sat down and said, "Okay. Should we take a higher profile? Is now the time?"

Susan: It is usually a smart strategy to keep a low profile. Larry's comment was a telling one: "Until we've got a product, we don't want to say anything," and for a very good reason. I think that the climate has changed. The research story is not the big seller that it was in the 1970s and even in the 1980s. I mean, dazzle me with science—okay, but, it's not going to make your stock move the way it once did.

Good media coverage may help you with financing. I won't deny that. But it's a double-edged sword.

Lyn: My counsel to people and to companies at the early stage is to set the strategy, determine at what points during the year to two years you are going to need to be really active publicly, and be prepared to deal with those. Additionally, you do the routine PR activities from a traditional public relations standpoint, that's really plenty. The main thing to concentrate on in those early years is building scientific credibility.

Susan: That's a good point. You do need to do the routine things. You need to keep the channel open and clear to allow people in, let them know what you're doing. If your CEO gives a speech, copy the right people. Determine which members of the financial community and media may really be interested.

Lyn: Analysts and industry reporters will ask, "What can I read that's been in a journal that tells me about what you do that isn't from you, that has some objectivity?" Without that, there really is no basis for a good communications program because you have no basis for talking to the financial community. The scientific media is specialized, and there are a limited number of people who are really very influential in the media about health care, and those are the people that matter, and those are the people who will ask the question, "What's in the scientific literature?"

So development of that scientific credibility, which means spending time doing your preclinical and clinical trials, either before or after you go public is a very important step. Cal Bio went through this process. The corporate communications program was started preparatory to a fund-raising, and the contribution was significant and measurable.

Carol: To what extent in your experience has the corporate communications effort gotten involved in the direction of resources toward publications and scientific literature?

Lyn: That's an interesting question. I found at Cal Bio that it was a constant education process for the scientific staff regarding the importance of keeping the public affairs department in the loop on those papers. They really had little knowledge or understanding of why it was important to publicize. When something is published in *Science* or *Nature* or *New England Journal of Medicine,* it can be of importance to the media.

Lisa: Has it been your experience that the very same scientists that will complain about *why* we are doing this are the first people who will be in your office when the *New York Times* or the *Wall Street Journal* or *Business Week* writes about someone else, some competitive science, and *they* are not in there?

Lyn: "Good Morning America" was in the United Kingdom for a week and as part of that, their medical editors did a piece on Alexander Fleming and the discovery of penicillin. This is the kind of story that's passed around in scientific circles. Fleming is accepted by history now to have been the sole developer and discoverer of penicillin and, in truth, he wasn't. Two other British researchers at Oxford actually conducted the first human trial of penicillin. Fleming discovered penicillin, and he co-operated with the press when interest in penicillin developed. The Oxford researchers did not.

Lisa: My experience has been that scientists are trained, from high school through college, to believe that it is only really valuable to share their work with their peers—that their peers are the ones who really can add value and make appropriate judgments. For the "other" people, it makes for nice cocktail conversation, but other people ought not pass judgment on what the scientists are doing.

The reality today is that the financial community and the media pass judgment on a daily basis on the validity of the work that our companies and our client companies do. That is a concept that comes very hard to people who are not in research companies.

What happens when those same companies go public and there must be awareness among the scientists that their needs are going to be different? They will have a vested interest in making this company grow, it's their stock, it's their money.

Lyn: It is an amazing transformation. People all of a sudden are stockholders, they have some of their net worth, typically much of their net worth, tied up in their shares.

Lisa: What happens to corporate communications when the company goes public?

Carol: In XOMA's case, we began to market the company to the public financial community, which is very different from the relatively small world of private equity. We made direct contact with the sell-side analysts and worked hard to get them to come and visit the company. Over time, most major analysts have had to follow XOMA because of the company's visibility. Then, with the commitment of senior management, we were able to gain visibility on Wall Street.

The most important thing for us was that our senior management, including the chairman and the president of the company, was willing to spend the time on the telephone and in person with the financial community.

Lyn: Having looked at the industry from inside a big company, and now on the agency side, there are many companies that went public too soon from the standpoint of their ability to keep the investor interested. It becomes a communications issue, because when a company goes public they frequently do it when they say "the window is open," and when they have the opportunity to do it. And a lot of companies just jumped through the window because they thought it would be their last chance. The difficulty created in an emerging company is that you don't have enough happening in your firm to keep the investment community interested on a quarterly basis. That becomes a big communications problem, particularly if you have, like most small companies, a large retail contingent in your shareholder base.

Retail brokers tend to need feeding frequently. Institutional investors and security analysts do too, but they are much more attuned to the actual reality of progress in clinical trials. Going public can frequently make communications work very difficult, and can build expectations that can't be met. Companies take a tremendous responsibility when they go public.

Susan: And enormous hassles. I think that every company who had the choice would be private.

Carol: But it doesn't immediately change things. As time goes on you start to see how volatile all these stocks are.

Lisa: One of the things Larry (Kurtz) said was that from the moment a company goes public, they should have an internal corporate communications person. They may want to use an agency, too, but an internal person is essential. You must have someone inside who deals with management on a daily basis and whose full-time job is to communicate.

Susan: I would argue that for an agency to bill itself as an investor relations firm is just a little bit of puffery because investor relations has to happen on the inside. It is a day-to-day function. Nobody could afford an agency to truly do their investor relations. I think agencies can be advantageous to a company. Because they represent a number of companies, they have a better opportunity to go back to an analyst than a single company would, because they can talk to the analyst about another company and something else might come up and the connection might be made.

Lyn: I agree with that. There's a lot of debate in our business about where that function should report and on what level.

Lisa: That comes back to the issue of management involvement.

Lyn: Typically, to make a universal statement, it reports to finance or, less frequently, to marketing or sales. More often than not, small companies keep it under finance.

Carol: Often, the finance person will be the IR [investor relations] person and really the chief IR person.

Lyn: The CEO participates as the spokesman, but lets the corporate communication department do a lot of the media work.

Lisa: How do you think that image of the biotechnology industry, as it is communicated to the public not involved or particularly aware of the industry, affects the marketing of your companies?

Carol: Another way to look at it is, if you want to look at biotechnology versus the big drug company, does the financial community (and therefore the media) scrutinize every aspect of a clinical trial? Do the big drug companies disclose the results of their clinical trial? Are their earnings driven by what happened to those seventeen patients and who published what today?

Susan: I think that's really the story of the 1990s, the product story.

Lyn: The difference is that we're taking this new technology and advancing it much more rapidly than the large [pharmaceutical companies] are in any position to do. I think the acquisition of Genentech by Roche is a perfect example of that, of Roche saying, "Hey, you know, why don't we just let those guys do the research?"

Susan: That's one of the things that American business has recognized, because statistically you can't ignore it. The overwhelming majority of inventions come from individuals and small companies, so that in the drug industry, like any other industry, you do have to look to the smaller, swifter, and "smarter" companies.

Lisa: One of the issues recently raised is that we've been watching all this happen over the past decade. Genentech has contributed two $100 million-a-year products. There probably are not a lot of drug companies that can make that claim within that period of time, much less within their first few years of operation.

Lyn: Before biotechnology the only two companies that have made it in the pharmaceutical industry in the past 25 years are Syntex and Marion Labs. And now Marion has a significant portion of it owned by another firm, so Syntex is the only independent. Big companies usually do not talk about anything publicly until it's in Phase II clinical trials. The only way that these large companies talk about them is if their collaborators are small biotechnology companies that have to do it for material reasons. But otherwise, the pharmaceutical companies hold an analyst meeting once a year, once every two years.

Susan: When you don't have earnings and you don't have products you try to sell something else. You try to sell what you've got and what you've got is Phase I, and what you've got is discovery.

Lyn: Exactly right. You have to sell what you have.

Lisa: In my conversation with Larry, he said, "think about the fact that in this industry, unlike most others that we know, a bunch of financial analysts with Ph.D.s, most of them who have never worked in the business world, are running around to dozens of scientific meetings. They have the ability to kill or make the company, before the physicians have used the product."

Lisa: On that note, although it is not addressing analysts, how influential do you think the industry critics are in what we do? How do they affect the overall image of the industry?

Lyn: Any one of our companies or our client companies has a few shots a year, at that level of circulation, and very often we don't take advantage of them because of the conflict between what the scientist thinks should be recognized and what the reporter wants. So we do sometimes shoot ourselves in the foot in the industry by insisting on playing by science rules.

Lisa: I'll tell you why I think we have an advantage. A lot of our products are used in human beings; they are actual drugs.

Carol: There's something else that I think people don't recognize as some of the contributions of biotechnology that are in our daily lives: the pregnancy test kits that you can buy in a drugstore, which became available in about 1984, or 1985. The advances of science as they apply to someone's life are not made apparent to them.

Lisa: What are some of those advances? In other words, if you had to name a couple of milestones in the past decade in the public's acceptance of biotechnology, what would you name?

Susan: The approval of insulin, which was the first biotechnology product. Insulin is such a well-known and widely used product.

Lyn: Probably the first major AIDS product will be another milestone, whether it's a vaccine or therapeutic. Diagnostics are not in that category.

Carol: I looked at that and I don't see it as a milestone as much as I see it as lack of serious obstacles in the path of development.

Lisa: Now that we've identified some of the objectives, let's recap some of the specific tools and methods you find most successful.

Carol: I would say regular analyst contact. I'm really speaking about dealing in the financial community. Regular analyst contact from senior management is the most important tool to keep your message out there—to update your message on a regular basis, to refine and adjust your strategy as necessary. Being in New York once or twice a year is important, and participation in all the major health care seminars is equally critical because it gives the buy side an opportunity to kind of take a glance at the company. In terms of working with the media, try to keep some regular contact and talk to reporters from time to time. Don't just say you have a release, that something might be coming. Try to keep that pattern going; it just helps when you really do have something to say.

Lisa: You also have to keep your corporate materials up to date.

Carol: From a practical point of view, you absolutely have to be ready to go.

Susan: That's a very good point. Being prepared for any eventuality is something the companies do internally. Keep materials up to date when there's a new discovery, add a paragraph when there's new scientific findings, and update your financial reports.

Regular contact with media and investors is a good idea, and you can do it by phone. Depending on what state your company is in, you might just be returning phone calls all day. You try to get them to come visit your company if that's appropriate. You let them know when your CEO is going to be in their neck of the woods if that's appropriate.

Everybody can read a quarterly, but most of the time I think quarterly reports to shareholders get so tangled up in what's legally disclosable, what's material and what's not, that sometimes they're not as informative to the analyst or the financial community as you would want. A good tactic that companies can practice internally would be a letter to analysts that indexes what's coming up in their company.

Carol: Over time, you canvass the media field, so to speak, and it's no different than what you do in the analyst community: you identify or develop your allies, and you work with them and you help them. It works to everyone's advantage.

Lisa: How important is the media to the overall success of corporate communications?

Lyn: The critical part of success in an emerging company is to identify the top five or ten people on the Wall Street side or on the media side, who are essential to your being able to get your message out. Then concentrate on those people in whatever way works for your company: meetings, visits, phone conversations, packets, CEO meetings, lunches—whatever it might be. It's hard to do, but it's essential if you want the visibility that is critical to the success and survival of companies. One thing I find really useful that companies don't do very often is survey their key publics. Ask people what they think about what you're doing and get feedback.

Susan: Another tactic for smaller companies is to target the kinds of trades that will be covering their type of information, because most of them are not going to get into the *Wall Street Journal* and the *New York Times*. It's unrealistic to think that they will, and it's foolhardy to think that they need to.

The purpose of this discussion was to provide an inside look at how several of the industry leaders chose to represent their respective companies in the biotechnology industry from both a corporate and public relations agency perspective.

CHAPTER 15

WALL STREET AND BIOTECHNOLOGY: A WALK ON THE WILDER SIDE

Robert Kupor

REFLECTIONS OF THE "GO-GO" EARLY EIGHTIES

If biotechnology was born by technological advances in genetic engineering and monoclonal antibodies, Wall Street was its midwife. Creating any new company is expensive, but creating ones devoted to human health care—and especially to therapeutic drugs—is even more so. The drug industry estimates that over $200 million is spent to develop the average drug within a fully established corporation. A fledgling biotechnology company, however, must create the full infrastructure for its initial drugs (unless it licenses away rights to a drug company)—production plant, sales force, research, and administrative support. This explains why drug company executives during the early 1980s were uniformly convinced that biotechnology companies would be stillborns—asphyxiated from lack of that critical financial oxygen: cash.

But these seers failed to take into account Wall Street's eagerness to provide this cash, not only during their initial stock offerings, but during subsequent ones as well. Wall Street is an indefatigable engine when it scents profits, and it worked with a vengeance during biotechnology's gestational years of 1980 to 1986. The numbers speak for themselves. During the peak period of 1982 to 1986 alone, fifty-four biotechnology companies of all types "went public" by selling $800 million of common stock to investors. During this same period, the more mature and more successful companies additionally consummated second and even third rounds of stock offering—thirty-four rounds in all, grossing $630 million. These more mature companies also raised approximately $400 million in research and development limited partnerships to fund individual products, as well as approximately $200 million from bond issues. Simultaneously, another billion dollars of "venture capital" funds flooded into new start-up firms, which hoped to go public in the future, in nearly 300 transactions. Overall, $3 billion was funneled into the industry during this five-year period alone.

Small wonder that so much money flooded into biotechnology, because its promise glowed so bright. For the first time, any human protein could be mass produced in ultrapure, unlimited quantities. Monoclonal antibodies of exquisite specificity would likewise be available for diagnostic and "silver bullet" therapeutic use. The newfound mastery over DNA biochemistry could allow genetic analysis, human genetic engineering, and other "Brave New World" technologies. Yet all this was only the pièce de résistance, in view of all the other applications as well—revolutionary seed technologies and custom-designed crops, biological techniques for waste management that would eliminate noxious chemicals and pollutants, and improved ways to synthesize fine chemicals. The excitement was palpable, as exemplified by the public reception of Genentech's initial stock offering in 1980. Most investors recognized their limited understanding of biotechnology, but they did recognize this was a "hot deal"—this was a stock everybody had to own. Genentech shares were offered for $35 per share on that day, but frenzied trading lifted the price to $89 within several hours—creating instantaneous profits over 140% for investors wily enough to sell their mint-fresh shares before the stock closed at $71 that afternoon.

All Good Things Must End (for the Time Being)

Those halcyon days of enthusiasm could not last indefinitely, although they continued far enough into 1983 to generate a boom of major public offerings. Investors' appetites seemed unlimited for a while for anything associated with biotechnology. But, in the end, the bubble burst with alarming suddenness. Having peaked in mid 1983, stock prices began sagging by autumn and continued sliding for the next twenty months. Biogen, once a lofty $25 per share, was $4 by early 1985. Centocor collapsed from $26 to $8. Those were white-knuckle days for the fledgling industry, when drug executives' early predictions began to assume extra credibility: everyone would go bankrupt (except beloved Genentech) before any products could be launched; Wall Street would never invest another dime in such disappointments. They had failed to "deliver," and hence were permanently out of favor. Investors' commonest question during those dismal years was "Are these companies out-and-out frauds, or is *any of* this stuff real?" We argued that they were here to stay, but frankly felt like taking no chances; at our research director's urging we began diversifying into the drug area as well—"Just so you have something to do if your industry disappears."

Fortunately, forever is a long time on Wall Street, and biotechnology underwent a phoenix-like resurrection in early 1985. Several factors contributed, but none more so than an extraordinarily influential stock report with an extraordinarily simple message: virtually every biotechnology company was destined for greater profits, and sooner than anyone anticipated. Investors could get rich by investing in almost any of them, and better yet, in several. In short, buy! And they did. No one who has not worked on Wall Street can comprehend its herd mentality; attitudes shift suddenly from one extreme to another, converting skeptics into believers overnight. And new attitudes quickly generate their own autocatalytic momentum. Stock prices are to Wall Street what land battles are to warfare—they create facts. As with war, facts are documented in history by the winners. The relentless climb of stock prices through spring and summer of 1985 offered convincing proof of the report's correctness, and eager investors clambered aboard the juggernaut, fearful of being left behind. The ebullience intensified yet further with back-to-back buyouts in 1985 of Hybritech by Eli Lilly and Genetic Systems by Bristol-Myers, which both "validated the technology" and created tidy fortunes.

By 1987, however, the bloom was off, and prices began falling—for two years. In retrospect, the previous rally had seemingly lifted prices only to allow more dramatic free falls. Chiron, which was $36 per share in 1987, was $12 by late 1988. Cetus was down from $32 to $10. Once again, bankruptcy predictions arose from all sides—the companies could never survive, or would be acquired at fire-sale prices by raiders. Wall Street would certainly never invest another dime.

However, like desert camels that require only an occasional oasis, most biotechnology companies survived by dint of various collaborative product agreements with drug companies, until the mood reversed again in late 1988. Again, the transition was triggered by a stock report, this one claiming that the erythropoietin product of Amgen and Genetics Institute would post $2 billion sales within three years—i.e., would become the most successful drug in world history, by a factor of five. Both stocks soared, pulling up most other stocks as well, according to the spillover phenomenon characteristic of Wall Street—if one biotechnology company is hot, why shouldn't lightning strike others as well? So 1989 ushered in a blessed refueling opportunity for biotechnology companies which has essentially continued to the present time. Fortunately, many of them have finally attained profitability on the basis of early products, and may never (theoretically) need additional Wall Street financing.

The 1990s have therefore begun with the biotechnology industry beginning to cut its umbilical cord to the financial markets. Nevertheless, despite the industry's newfound strength and esteem, one cannot hide the fact that its decade-long track record has been an overall disappointment to Wall Street. The miracles that Wall Street initially expected have been only partially realized even today, and only after unanticipated delays and repeated refinancings that progressively diluted stock values. Throughout the 1980s, the industry acquired a reputation for promising more than it could deliver—for hype—that is only slowly receding now. Even today, relatively few investors are willing to speculate on these companies, for the stocks are too volatile, too unpredictable. One can sleep comfortably at night with an investment in AT&T or Boeing, but a portfolio of biotechnology stocks requires two Excedrins four times daily.

Why has biotechnology "failed"? Why has its fortunes fluctuated so erratically from year to year, from adored to abhorred, more so than any other industry? Is it because of the companies themselves, or simply because of their perception by analysts and investors, or simply due to mysterious biohormonal cycles?

Biotechnology Psychoanalysis

The answer, we believe, is simple: biotechnology's near-term commercial potential was considerably oversold during its inception by both the companies and the biotechnology analysts, for both intentional and unintentional reasons, to investors unable to penetrate the confusion. The subsequent volatility has reflected the interplay of multiple factors: Wall Street's recognition by late 1983 of its initial miscomprehension, followed by an ongoing succession of authentic positive and negative developments, as interpreted and misinterpreted by the companies, analysts, and investors themselves.

These elements are worth disengaging, for they are of more than historical interest—they continue to influence Wall Street's perception of biotechnology even now. Because these elements were most conspicuous during biotechnology's nascent period, we will focus primarily on this period, with the understanding that they continue in attenuated form to the present.

The Entrepreneurs

First, the companies. To begin with, biotechnology companies were overwhelmingly created by brilliant, aggressive academic scientists—as exemplified by such Nobel laureates as Biogen's Walter Gilbert and Genentech's Herbert Boyer—who lacked business experience. Unlike the situation in engineering or applied physics, in which eminent academics have long collaborated with industry, the "molecular biologists" who sired biotechnology companies were not only unfamiliar with business, but generally contemptuous of what they had long perceived as the money-grubbing and scientifically third-rate world of pharmaceutical companies. They therefore too often set up shop with unrealistic expectations and inadequate expertise. They believed that a $30 initial stock offering would suffice to fund research and development on a ten-product portfolio of drugs, which would then sail through clinical trials, Food and Drug Administration [FDA] approval, and eager physician acceptance. They had little understanding of budgeting, product development, market research, clinical trial design, or the other essentials of corporate success. Indeed, some founders seemingly could not even distinguish between commercial and academic projects, and ended up directing miniuniversities, which squandered precious years and money on fascinating but arcane research leads.

These scientist-founders did bring in small cadres of business people, of course—CFOs and COOs and so forth. But these individuals were often relatively junior themselves, or too impotent to challenge the founding scientists' wills. And another factor was sometimes at play as well—the sense of omniscience and power that so commonly arises in small, visionary enterprises, and indeed is generally so important to achieve success in the real world. No one who has not experienced life at the epicenter of a high technology start-up can readily appreciate the sense of excitement and power it creates. It is the ultimate adrenaline distillate, a narcotizing recipe—mix $30 million with an ebullient young group of academics (empowered to hire and fire), leaven with their investment bankers' adoration and their own genuine hopes of helping mankind.

Thus, much of the early industry's hyperoptimism was authentic, but not all. After all, the early biotechnology companies were quite cognizant of their dependence on Wall Street. Tens of millions of dollars were needed to start, and equal sums might be needed down the road to continue and/or expand. Accordingly, biotechnology had to be presented as bigger than life, a once-in-a-lifetime opportunity guaranteeing bonanzas to one and all. Every proposed product was presented as a little better and more proprietary than it really was, and every market as a little bigger. But above all, every product was . . . natural.

Natural! Medieval serfs probably saw little glamour in "naturalness," but the word has near-mystic resonance in the twentieth century. And for biotechnology, the talisman of naturalness was this: being natural proteins, biotechnology drugs must work, without any side effects, and therefore would sail through FDA review. Intuitively, it was so powerful a thought that few challenged it. Questions could perhaps be raised about patentability—how could anyone patent a natural protein?—but not the central dogma itself.

Of course, this dogma was erroneous, at least as a generality. It certainly holds for the minority of proteins useful for "deficiency" diseases: human growth hormone for dwarfism, insulin for diabetes, and erythropoietin for kidney failure. But most proposed protein therapeutic products are intended for industrial-dose administration into people who already possess normal endogenous levels—the interferons and interleukins. In these cases, side effects are the only guaranteed outcome, whereas years of clinical testing may be required to determine what beneficial effects such drugs might possess, if any, for what disease. Several of these "products" are still seeking a suitable disease, including once-glamorous ones such as tumor necrosis factor, epidermal growth factor, and superoxide dismutase. And one major class of these natural products—monoclonal antibodies—and not strictly natural to begin with, leading to continual problems with antigenicity. As for the FDA, no brownie points are awarded for naturalness. The FDA's boorish focus on "safety and efficacy" makes it view all drugs as black-box chemicals, to be judged on the basis of clinical value, not molecular architecture. All in all, the naturalness of biotechnology drugs never was a significant asset per se, and indeed has often been a liability. Because humans contain numerous protein-degrading enzymes in the gut and bloodstream, protein drugs must be injected, often repetitively, owing to their rapid degradation in the bloodstream. These characteristics, joined with their innately high production costs, constrain most biotechnology drugs to in-hospital use for serious illnesses, or for chronic use in rare genetic deficiencies.

The Analysts

The biotechnology analysts have been the second pivotal group in creating an understanding/misunderstanding of biotechnology. Why has this highly paid group of full-time professionals performed such an uneven job, in my opinion, of interpreting this industry to the investment world?

One reason, assuredly, is that few if any biotechnology analysts have the ideal background for their task. It is true that most have doctorates in biology, although several have financial backgrounds only. Nevertheless, virtually none has the medical training that is really required—the comprehensive knowledge of diseases and their current epidemiologies, diagnoses, and therapeutic alternatives. The outcome is, not surprisingly, highly imperfect. Genentech, for example, began claiming by 1982 that its tissue plasminogen activator (TPA) would be invaluable not only for the one million heart attack patients in the United States each year, but for the 220,000 critical cases of pulmonary embolism as well. Not until 1986 did any analyst discover, on opening a medical textbook, that embolisms kill 90% of their victims within thirty minutes—leaving a mere 22,000 patients who could conceivably be treated in time. Thus did a $450 million fantasy market for TPA collapse into a $45 million potential one. Fortunately for Genentech, the one-million-heart-attack-patient statistic proved to be inflated only fourfold.

A second reason is probably more important, however—analysts are often nowhere near the cerebrally dispassionate truth-seekers implied by their title. As denizens of Wall Street, they must play by its rules, and the foremost of these rules is "Be profitable." Because it costs large Wall Street firms $200,000 to $600,000 annually to support a biotechnology analyst, analysts must generate substantial revenues to remain employed.

So how do you make money? The quickest way is by "doing deals"—signing on biotechnology companies as corporate finance clients of one's brokerage firm in order to earn big profits on stock offerings, acquisitions, and so on. The impact of this kind of relationship on analysts' objectivity is obvious, especially in view of the powerful psychological bonding between company management and analysts that quickly supplements the purely financial one. Up to one quarter of all companies being covered by some analysts may be corporate finance clients. Because these clients often compete with other biotechnology companies, an analyst's assessment of fully half of all companies may be colored, favorably or unfavorably, by corporate finance relationships. Remarkably, investors are fully aware of these factors, yet often value the analysts' perceptions all the more, since they recognize that these analysts have favored access to company information. The problem is that the investors also assume that the analysts' bias can be screened out by perceptive questioning and listening—a feat that is indeed attainable by some, but hardly by all.

But if deals are financial home runs, every analyst must pay the daily bills with a constant stream of commission-generating recommendations to buy and sell stock. The rub is that only a dozen biotechnology companies are large enough to merit investor interest, and most are involved in only a handful of truly significant events each year. Analysts are therefore under constant pressure to discover a steady stream of "proprietary information," which can serve to encourage stock trades as well as to maintain their own "visibility" before investors. But useful information requires fast and widespread propagation, which is largely controlled by brokerage firms' sales forces. Unfortunately, sales people are generally lukewarm to biotechnology—they prefer phoning clients about IBM and General Motors, whose stocks are much more widely held than Biogen or XOMA. To make a biotechnology "call" to clients worthwhile, it must be fast, simple, and end (or begin) with a buy or sell recommendation, whether or not the information actually warrants such action. Analysts therefore confront enormous pressure to exaggerate their information's importance; an "action-oriented" call is the sine qua non of the sales forces.

A consequence of this situation is that surprisingly little relationship may exist between an analyst's reputation and actual quality. Occasionally, the relationship can even be inverse. A really influential analyst is commonly one who goes out on a limb with buy/sell recommendations that rivet attention, and then forces the stock to move in the desired direction with endless phone calls (to supplement the sales forces') and multicity "marketing trips" to clients. The time consumed in these efforts often leaves little left to do really original or comprehensive analysis. Although one might expect investors to distinguish quickly between good and mediocre analysts, the process can instead require years. Many of the predictions profferred by analysts are so long term, and can be influenced by so many intercurrent events, that many investors take years to assess analysts' track records. It must be recognized in this regard that Wall Street analysts require no credentials whatever to set up shop, other than their firms' employment letter. Investors themselves must therefore determine who is knowledgeable. But because biotechnology is particularly complex, many investors find it as difficult to determine this as it is to assess the competence of their own family physician. As with physicians, therefore, selections are often based largely on the analyst's self-confidence, attentiveness, and so on. In fact, a really influential analyst may be eagerly followed even by skeptical investors, just to come for the ride. If an influential analyst is "pounding the table" on a stock, why not buy some now and make a fast profit by selling it at higher prices next week?

This situation not only explains much of biotechnology stocks' extraordinary volatility, but another phenomenon that accounts for much of the remainder—the limited effort most analysts expend on determining the value of biotechnology companies' potential products until they are well into human testing. An outsider might expect this to be central to biotechnology analysis, especially in view of the massive information often available about biotechnology proteins even prior to cloning. After all, these proteins are natural substances that have often been studied worldwide for years (which is how companies initially select promising product candidates). But such in-depth analysis can be painstaking, and typically leads to hedged predictions of "probable success" versus "improbable," rather than definitive black-and-white answers. Sales people generally frown on such wishy-washy, complex information, and although analysts are free to write lengthy reports with learned exegesis, reports go largely unread on Wall Street. Investors are buried alive under literally hundreds of new reports weekly from major brokerage houses, and cannot read more than a fraction of them. Information is as perishable as fresh salmon on Wall Street; if it cannot be canned in a three-minute sales call, it spoils. A snappy, immediate impact piece of information is gold, on the other hand. A minor but irrefutable piece of information—e.g., that the FDA will delay a drug's review for forty-five days—will commonly generate more stock trades than a comprehensive analysis of a company's entire product pipeline. Consequently, upcoming products are generally accorded black-box status until the emergence of human clinical results, at which time stock prices abruptly soar or collapse on the "all-new" revelation.

The Investors

Investors are the third and final players in the biotechnology valuation process. Their role need not be detailed, inasmuch as they play a more passive role in the process than the other players, and already have been periodically mentioned. However, two interrelated aspects of their modus vivendi merit comment: their tendency to "trade" biotechnology stocks (rather than invest long term), and their previously mentioned reliance on fast-paced phone calls. These tendencies are the inevitable consequences of the investors' environment. Investors are paid to make profits each and every quarter for their clients, by buying and selling stock in hundreds of companies. Their orientation is necessarily short term—and necessitates reliance on rapid phone calls to assimilate the flood of fast-changing events. However understandable, the consequences are obvious: the short-term orientation clearly induces stock volatility, while the reliance on phone calls reduces their ability to assess information properly.

Having expounded the dynamics of the Wall Street–biotechnology interaction, I might conclude by noting that most aspects have gradually improved over the years. Companies have generally become more realistic and straightforward, having been sobered by experience. Analysts have gradually improved, although profitability pressures are forcing some to expand their analytical coverage to drug companies as well, thereby spreading their resources even thinner. Investors have also gained in wisdom, though some more than others. Overall, the situation has improved, but has hardly become perfect.

FUTURE DEALINGS

We must confess that this exposition of Wall Street's dynamics vis à vis the biotechnology industry's birth and growth is hardly the stuff of morality plays. Those wishing to believe that humanity is progressing toward perfection as the twenty-first century approaches will certainly find greater comfort in churches than in brokerage firms.

Does this mean that biotechnology was created largely as a snare, for the enrichment of a coterie of founders and Wall Street firms? Hardly. Virgin births are always more aesthetically pleasing than real ones, but it is worth recalling the central tenet of capitalism, which has long served the Western world so well: unfettered self-interest, if appropriately policed for wrongdoing, will miraculously generate the greatest good for the greatest number. This "invisible hand" has worked with biotechnology, too. For biotechnology *is* unquestionably one of the most hopeful technical breakthroughs in history. Biotechnology stands today at an analogous position to nineteenth century organic chemistry—with the miraculous ability to manipulate the very substance of life: proteins and DNA. The tools remain rudimentary, and the harvest to date has consisted of a handful of useful therapeutics and diagnostics. The future should yield unspeakable wonders, however—insights into the pathogenesis of all diseases, which will allow the development of novel curative drugs, as well as artificial organs and genetic manipulation. Plastics and computers are wonderful, but nothing can bring mankind closer to a millenium of well being than the mastery of human health itself.

Wall Street has contributed powerfully to this undertaking. True, it has demanded its price, but do not midwives and obstetricians as well? And do not physicians and hospitals vie equally vigorously (though more genteelly) to gain the biggest fees and market shares? From the broadest societal viewpoint, therefore, Wall Street has nothing to apologize for.

On the contrary, can it be accidental that the United States dominates biotechnology worldwide, by a wide margin? I think not. Other factors have no doubt contributed. America's stellar infrastructure of research universities and government institutes has provided a magnificent array of scientists and cutting-edge technology. America's position as the world's largest health care market facilitates rapid profitability for talented start-ups. America's esteem for risk, provides invaluable psychic underpinnings to the entrepreneurs. Even patent laws are more favorable here than abroad.

Money is the non plus ultra of the entire process, the fuel of this trailblazing flight into the future. The United States is the only country on earth with financial institutions willing to risk billions on visionary entrepreneurs. And U.S. biotechnology companies have set the pace. Biotechnology research does proceed elsewhere too, of course. Every large drug company in the world has reacted to biotechnology's siren song, and a few are developing real strengths. But how much slower would they have been without the impetus of fast-paced biotechnology companies grabbing control of this critical technology? Large drug companies are savvy but slow, no more attuned to biotechnology tap dancing than are Jurassic reptiles. Biotechnology would have developed without biotechnology companies, but far more slowly. This decade, hundreds of thousands of people who would otherwise be dead or incapacitated will owe their lives to biotechnology. Who could dispute the richness of such a return on the money invested in this industry during the 1980s? In the final analysis, this Wall Street tale has a happy ending after all.

CHAPTER 16

MERGERS AND ACQUISITIONS WITHIN THE BIOTECHNOLOGY INDUSTRY

Carol V. Hall and Harriet M. Strimpel

Since the emergence of the biotechnology industry in the late 1970s, the number of companies developing products using the technologies of recombinant DNA and monoclonal antibodies has grown to over 400. While the majority remain independent, there have been select mergers and acquisitions within this industry since the early 1980s. Table 16-1 reviews these major transactions.

While each transaction has its unique components a common motive shared by those biotechnology companies acquired by larger, established corporations, such as pharmaceutical companies, was the financial "deep pockets" of the acquirer that offered a reprieve from the constant search for new funds. The long product development cycle of human pharmaceuticals requires that biotechnology companies raise substantial amounts of cash to fund their research and development. A corporate parent can free management from this time-consuming process. Furthermore, because biotechnology companies are no strangers to working in concert with larger corporations, corporate partnering has been and will continue to be a major source of funds.

In several cases, mergers between two biotechnology companies were primarily the result of financial considerations on the part of at least one of the merger partners. In other instances, strategic additions to technology and patent bases or expansion in manufacturing and marketing potential were major motivating factors as well. These three factors and representative transactions are discussed in greater detail in the following pages.

TABLE 16-1 MAJOR BIOTECHNOLOGY MERGERS AND ACQUISITIONS

Year	Company (Acquiree/Acquirer)	Value	Type
1982	DNAX*/Schering-Plough	$30 million	Stock swap
1983	Agrigenetics*/Lubrizol	NA	NA
1984			
1985	Genetic Systems/Bristol-Myers	$294 million	Stock swap
	Hybritech/Eli Lilly	$375 million	Stock and contingent payment unit
1986	Hygeia Sciences/Tambrands	$48 million	Cash
1987			
1988	ZymoGenetics*/NOVO Nordisk A/S	NA	NA
	Advanced Genetic Sciences/DNA Plant Technology	$37 million	Stock swap
1989	Gen-Probe/Chugai Pharmaceutical	$110 million	Cash
	Integrated Genetics/Genzyme	$29 million	Stock swap
	Ingene/XOMA	$40 million	Stock swap
	Plant Genetics/Calgene	$14 million	Stock swap
	Cytotech*/Quidel*	NA	NA
	Praxis Biologics/American Cyanamid	$237 million	Stock swap
	Damon Biotech/Abbott Laboratories	$9 million	Cash
	Sungene*/Lubrizol	NA	NA
1990	Biotech Research/Cambridge Bioscience Labs	$32 million	Stock swap
	Genentech/Roche Holding (60%)	$2,100 million	Cash
	Codon*/Schering AG	NA	NA

NA = Not available
*Private company at time of transaction

Despite the large number of biotechnology companies, there has been little consolidation, and that which has taken place has occurred only recently. Consequently, it is difficult to generalize about the relative success rate of these transactions. Nevertheless, the biotechnology industry's primary assets are its employees, and the role of corporate culture has played and will continue to play a significant part in the success of each transaction. In addition, the strategic long-term focus of the acquirer has been a key factor in determining success, since its priorities determine the funding level and degree of autonomy granted to the formerly independent company.

As in other industries, the majority of biotechnology acquisitions are probably doomed to failure. The time frames for biotechnology products to reach the market are still, on average, so long that during the interim, many things can change. Furthermore, despite the great number of unknowns, biotechnology companies are not cheap. Even if a biotechnology company brings a plate full of new products to the acquisition, the price tag may never justify the transaction. In ten years will the industry look back and conclude that Roche Holding bought a bargain by paying approximately $5 billion for Genentech? The answer is likely to be no.

On the bright side, mergers between two biotechnology companies will likely fare better. If financing remains tight through the early 1990s, the top-tier biotechnology companies are in a position to pick the best of the smaller companies. Their huge cash reserves and high stock prices bring a degree of financial flexibility not seen before. Furthermore, as the success of their first products becomes easier to quantify, these top-tier companies can begin to focus on new research areas. The purchase price can be kept reasonable by stock swaps that include no premiums.

More importantly, the biotechnology companies share common backgrounds and technology bases, i.e., recombinant DNA or hybridoma technology. On the other hand, the smaller biotechnology companies squeezed for cash and under pressure from their venture capital investors to provide both a return on their investment and liquidity, may accept offers of merger that formerly would not have been considered seriously.

THE FINANCIAL MOTIVE

The biotechnology industry faces extremely long product development cycles, averaging twelve years from first discovery to market introduction, and huge developmental costs. A 1990 study from The Center for the Study of Drug Development at Tufts University, sponsored by the Pharmaceutical Manufacturers' Association, placed the cost at $231 million for the average new human therapeutic agent.

In order to pay the bills, biotechnology companies have been forced to be very creative in raising capital. Typically, a biotechnology company begins with two or three rounds of funding from venture capitalists and then proceeds to tap the private equity market through public offerings of common stock. Private institutional offerings, research and development limited partnerships, and corporate partnering have also been important sources of funds.

All this funding comes with a price tag, most often in the form of further dilution for current shareholders or in loss of various rights to

TABLE 16-2 BIOTECHNOLOGY COMPANY FINANCINGS IN 1989

Financing Strategy	Biotechnology Companies
Private offering	Over 100 completed
Initial public offering	Transgenic Sciences, Cortex Pharmaceuticals, Neurogen, ImmunoGen, ProCyte
Secondary public offering	Chiron, Genzyme
Public convertible debenture or preferred stock offering	Biogen, Amgen, Genetics Institute, XOMA, Calgene, NeoRx, Cytogen, Centocor
Private research and development limited partnership	Genentech, Genzyme
New company spin-off	Centocor/Tocor, Immunex/ Receptech

products under development. The biotechnology companies must constantly balance the need to prefund their research efforts with the cost of these funds. In the quest to raise capital, a dozen or so biotechnology companies have returned to the public markets with convertible debenture or convertible preferred stock offerings, despite the lack of earnings to pay the required interest or dividend payments. This strategy carries the risk that the necessary cash payments will erode shareholders' equity and jeopardize the company's survival if product introductions are delayed. Another new strategy of private and public biotechnology companies is the "shell" company, spun out of the parent in an initial public offering that raises the necessary funds to continue a line of research contributed by the parent company in exchange for equity in the new company. A summary of the primary financing strategies and the biotechnology companies that used them in 1989 is shown in Table 16-2.

Despite the large number of possible funding vehicles, there still have been times when raising money to fund biotechnology has been extremely difficult. Often a company finally begins to bring products into clinical testing at the same time that the dollars begin to vanish from the balance sheet at a rapid rate. An undercapitalized biotechnology company can solve its long-term financing problems by becoming part of a stronger, financially secure corporation. For the majority of the transitions in Table 16-1, financial considerations were the primary motivating factors for the acquisitions.

Three cases in point are the acquisitions described below: Damon Biotech's acquisition by Abbott Laboratories, Gen-Probe's acquisition by Chugai Pharmaceutical, and Genentech's proposed acquisition by Roche Holding. Although the three biotechnology companies share the same motive, the cases differ in other aspects. One important difference was that while Damon Biotech had little choice in accepting an acquisition proposal, this was not the case with either Gen-Probe or Genentech. Damon Biotech's acquisition by Abbott Laboratories was a last-minute reprieve that staved off liquidation proceedings, because its business strategy never bore fruit and investors refused to invest more funds. Despite the over $50 million that had been invested in Damon Biotech since its inception, Abbott Laboratories purchased the company for $9 million, approximately 1.5 times its book value.

In the second case, Gen-Probe used its alliance with Chugai Pharmaceutical primarily to infuse additional financing to strengthen its balance sheet. However, the alliance is also expected to accelerate the entry of Gen-Probe products into the Japanese market.

Finally, Genentech's proposed merger with Roche Holding freed Genentech from making the decision to cut back its basic research programs to improve its short-term earnings. Genentech has certainly not had any difficulty raising funds; indeed, shortly before the proposed merger, Genentech successfully completed a $75 million research and development limited partnership to fund work on its acquired immunodeficiency syndrome (AIDS) therapeutic, soluble CD4 receptor. The proposed merger has already achieved the effect of removing Wall Street from the list of concerns for Genentech's board of directors.

Damon Biotech/Abbott Laboratories

Damon Biotech, incorporated in 1981, was one of a few biotechnology companies offering large-scale contract production of genetically engineered pharmaceutical proteins expressed in mammalian cells. Damon was also developing internally several proprietary therapeutic proteins.

Damon Biotech began as a wholly owned subsidiary of Damon Corporation, a clinical testing laboratory. In 1983, Damon Biotech became a public company, although Damon Corporation retained a majority position. In 1988, Damon Corporation became the target of a hostile takeover attempt, and as part of the ensuing merger agreement between Damon Corporation and Nomad Partners, L.P., Damon Corporation began to seek a buyer for its Damon Biotech holding.

Unfortunately, the market for contract production of proteins never developed as expected, primarily because many biotechnology companies or their marketing partners chose to vertically integrate the manufacturing process in order to retain control of the process and any trade secrets involved in production. Significant sales revenues never materialized, and Damon Biotech's accumulated deficit grew with each passing year. In 1989, Damon Biotech had $4.2 million in revenues and a net loss of $4.5 million.

Unable to sell its stock in Damon Biotech on the public market, Damon Corporation surrendered its shares to Damon Biotech in September 1989. Since Damon Biotech's cash resources were expected to be exhausted by mid-1990, and with no other viable financing options on the horizon, Damon Biotech initiated discussions with strategic buyers. Management believed that the only other alternative was to liquidate the company.

Abbott Laboratories, a worldwide diversified health care products company that had previously contracted with Damon Biotech for scale-up production of certain cardiovascular proteins, made the only serious offer. Damon Biotech shareholders received approximately $1 per share, which was significantly above the stock price prior to the acquisition announcement. Nevertheless, Abbott Laboratories purchased the entire company for 1.5 times its book value.

Gen-Probe/Chugai Pharmaceutical

Gen-Probe, a public biotechnology company located in San Diego, California, was the first biotechnology company to be acquired by a foreign corporation. Chugai Pharmaceutical, a publicly traded Japanese corporation that markets pharmaceutical products, finalized the deal in December 1989.

Gen-Probe should be viewed as a niche biotechnology company with leading proprietary technology in the area of DNA probe diagnostics. The company was founded in 1983 to apply DNA probe technology to the development of diagnostic products. In 1985, Gen-Probe became the first company to receive Food and Drug Administration (FDA) marketing approval for a DNA probe–based diagnostic kit for the detection of mycoplasma in tissue culture samples. The company has since brought ten products to the marketplace.

For the nine-month period that ended September 30, 1989, Gen-Probe had total revenues of $7.2 million, but an earnings loss of $4.2 million. In 1989, the company sought additional financing to fund working capital needs, and possible acquisition partners were considered.

Chugai and Gen-Probe's relationship began in March 1988, when the two companies entered into a five-year, $15.5 million research and development contract. At the time of the acquisition, Chugai had paid Gen-Probe $6.2 million pursuant to the 1988 agreement.

Gen-Probe initiated discussions with Chugai in June 1989, when Gen-Probe's management informed Chugai that Gen-Probe was in negotiations with another major pharmaceutical company regarding the possible acquisition of Gen-Probe. Apparently, Chugai was pleased enough with its existing relationship with Gen-Probe to respond with an attractive acquisition proposal. Gen-Probe's management accepted the proposal, possibly with a sigh of relief. The word on Wall Street has it that Gen-Probe had offered itself to several U.S. pharmaceutical companies without receiving any offers that approached what management hoped to receive for the DNA probe company.

With the announcement of the acquisition, Thomas A. Bologna, President and Chief Financial Officer of Gen-Probe, commented to the *Wall Street Journal* that Gen-Probe would now "be able to concentrate on running the business rather than always looking for sources of financing."

Genentech/Roche Holding

Genentech, located in South San Francisco, California, was one of the first biotechnology companies established and is the undisputed industry leader. Yet it too was motivated in part by financial considerations in its decision to be acquired by Roche Holding. Roche Holding, a public company formerly known as F. Hoffmann–LaRoche & Co. Limited Company, is a worldwide pharmaceutical company with 1989 sales of approximately $6.6 billion.

Genentech's research budget has always been a high percentage of its revenues—roughly 40%. Genentech's revenues for fiscal year 1989 topped $400 million, but this figure was still below management's, and Wall Street's, expectations. Rather than cut projects to improve the income statement, Genentech chose to tap the huge source of cash in the form of Roche Holding's capital to continue the high level of research. This strategy fit well with Roche Holding's need for new products, a strong foothold in the U.S. pharmaceutical marketplace, and access to novel technologies.

THE STRATEGIC ALLIANCE MOTIVE

Biotechnology companies have actively participated in collaborative agreements with larger corporate partners since the industry began. These agreements have ranged from simple research contracts that supply short-term revenues to 50/50 joint research, development, and marketing ventures that bring revenues as well as expertise in areas that the smaller company may lack. With over 400 such arrangements on record in 1988, this works out to approximately one such arrangement per biotechnology company in 1988 alone! Acquisitions are a logical extension of a successful collaborative agreement. For example, the previously described acquisition of Gen-Probe by Chugai was a direct consequence of their prior collaborative agreement.

In the majority of transactions to date, the stronger partner was motivated by the desire to add strategically to its research base. This was the case for the 1989 merger between two biotechnology companies, XOMA and Ingene. XOMA, a top-tier biotechnology company, sought to broaden its research base in the area of molecular biology, so it acquired International Genetic Engineering (Ingene), a biotechnology company approximately one-fifth its size that possessed this expertise.

Furthermore, strategic alliances between two biotechnology companies can produce a critical mass that expands the business options available to the combined company. This was the case in the recent merger between two public agricultural biotechnology companies, Plant Genetics and Calgene.

XOMA/Ingene

Ingene, incorporated in 1984, has concentrated its research in cancer therapeutics, and its research strengths are in the area of recombinant DNA technology. Although Ingene had entered into several joint development agreements, its cancer therapeutics were still in the research phase of development. Because of its lack of clinical and regulatory expertise and because it possessed no manufacturing and marketing capability, Ingene's management made the decision in late 1988 to seek a buyer for the company. Over two dozen companies were contacted and serious discussions occurred with two, one of which was XOMA.

XOMA'S research programs are centered on the application of therapeutic monoclonal antibodies to bacterial infections, particularly gram-negative sepsis, and immune-related disorders. Founded in 1981, XOMA had completed the clinical trials for two new therapeutics and had submitted market approval applications to the Food and Drug Administration by 1990. XOMA sought to expand its research base in recombinant DNA to facilitate the development of second-generation therapeutics.

The merger between XOMA and Ingene made sense to both parties for several reasons: each was interested in the expertise that the other could provide; highly funded, XOMA could afford to pay the ongoing costs of both research groups; both companies were located in California, Ingene in Santa Monica and XOMA in Berkeley, which simplifies communication; and the combined company resulted in very little redundancy. Ingene had a lean management team, so no layoffs of key personal resulted. The research focus at Ingene has shifted from cancer to immune-related disorders, but the transition went smoothly, according to a XOMA spokesperson.

Plant Genetics/Calgene

The 1989 merger between Plant Genetics and Calgene created the largest U.S. public agricultural biotechnology company, with 1989 revenues of $30 million. Calgene's in-house research programs on rapeseed, cotton, and tomato use both traditional plant breeding and recombinant DNA technologies. The company also has a solid biochemistry group that is actively investigating the underlying mechanisms behind certain desirable traits, such as production of improved oils and insect resistance, that Calgene wishes to manipulate.

Calgene's acquisition of Plant Genetics brought Plant Genetics' alfalfa, tomato seed, and potato projects into Calgene. With five core crops in the combined company, Calgene can further leverage its existing molecular biology efforts by applying them to Plant Genetics' crops. Furthermore, Plant Genetics' strength in conventional breeding and plant regeneration complements Calgene's research strengths.

Before the merger, the two companies shared the same building in Davis, California, and as a result, the two Chief Executive Officers informally discussed the possibility of a merger many times. In early 1989, Plant Genetics was faced with the prospect of cutting back on selected long-term research projects and liquidating some of its inventory if additional financing was not secured. Since the public markets were not receptive to an agricultural biotechnology secondary offering, Plant Genetics opted instead to begin formal merger negotiations with Calgene.

An early outcome of the merger was the increased attention that Wall Street paid to the combined entity. Two months after the merger, Calgene raised $25 million in a convertible preferred stock offering that significantly strengthened the company's balance sheet. It is quite likely that neither Plant Genetics nor Calgene would have been able successfully to tap the public markets in 1989 on its own.

THE MOTIVE TO EXPAND MANUFACTURING CAPACITY AND/OR MARKETING EXPERTISE

As the new genetically engineered products enter the marketplace, manufacturing, marketing, and distribution issues will become increasingly important. Companies can choose between contracting with outside third parties or building the necessary infrastructure to vertically integrate. One shortcut is to acquire a company that has already built the necessary infrastructure.

Most biotechnology companies do not have the resources to acquire large, profitable companies; however, one such biotechnology company, Chiron, devised a creative solution to this problem, using the resources of corporate partners to enter into bidding for businesses otherwise out of its reach. Although the first attempt was not successful, it did pave the way for a future deal of a similar nature.

Biocine Company/Connaught Biosciences

One of the most unusual, though unsuccessful, acquisition attempts was Biocine's 1989 bid for Connaught Biosciences. Biocine is a 50/50 joint venture between Chiron, one of the top-tier public biotechnology companies located in Emeryville, California, and Ciba-Geigy Limited, a public, Swiss-based worldwide pharmaceutical company with sales of $14 billion in 1988.

The relationship between Chiron and Ciba-Geigy began as a research collaboration in 1984. In 1987, the relationship was expanded through the establishment of the Biocine Company, created to investigate and develop recombinant DNA-based vaccines for infectious diseases. Chiron contributed its molecular biology expertise in recombinant antigens and Ciba-Geigy contributed its novel adjuvant technology. In the past three years, the Biocine Company has entered clinical trials with promising vaccines aimed at malaria, herpes, and AIDS.

In 1989, the Biocine Company made a hostile $642 million bid for Connaught Biosciences, a Canadian vaccine company with $224 million (Canadian) sales. Biocine was interested in Connaught because of its strong marketing presence in North America.

Biocine was ultimately outbid by Institut Merieux, a majority-owned subsidiary of Rhone-Poulenc S.A. of France. However, Biocine's bid brought to light the strong relationship between Chiron and Ciba-Geigy and introduced the idea of a "joint acquisition" to the biotechnology community.

Chiron did not have the resources to pay for an acquisition of Connaught's size, but was able to convince Ciba-Geigy to pay for the majority of the acquisition up front, giving Chiron a ten-year option to purchase its own 50% share. Had this acquisition been successful, it would have catapulted Biocine, and indirectly, Chiron, into a leadership position in the manufacture and marketing of vaccines.

In 1989, Chiron used the "joint acquisition" strategy successfully in its joint venture in diagnostics with the Ortho Division of Johnson & Johnson, resulting in the purchase of E.I du Pont de Nemours & Company's $40 million blood screening division.

In this transaction, Ortho purchased all tangible assets from Du Pont, while Ortho and Chiron jointly acquired the intangible assets. This acquisition brought an existing installed instrument base and a number of diagnostic tests to the joint venture and moved the Chiron/Ortho joint venture to the number two position in the worldwide blood-screening market. Although the total purchase price was not disclosed, Chiron paid $9 million for its portion. As part of the joint venture, profits are split equally between Ortho and Chiron.

SUCCESS OR FAILURE

The Role of Corporate Culture

As in any industry, biotechnology mergers and acquisitions have not always been successful. Although it is difficult to generalize, given the small number of transactions and the fact that many of the transactions are relatively new, the role of corporate culture has figured, and will figure, heavily in the success of a merger or acquisition. For the majority of biotechnology companies, their most valuable asset is the technical staff. The management of a biotechnology company strives to maintain the high energy and creative spirit of the entrepreneurs and scientists, while at the same time providing focus and financial stability.

An acquisition can bring a cash infusion that frees the biotechnology company from its short-term financial problems and allows the scientists to focus on more long term research problems. Such was the case with DNAX Research Institute's acquisition by Schering-Plough. However, financial stability may come at the expense of the acquired's entrepreneurial culture. In the case of DNAX, the evolution from entrepreneurial culture to a more academically oriented one was successful; however, Hybritech's entrepreneurial culture was replaced with a more bureaucratic one when Eli Lilly acquired the company, and many scientists left as a result.

Moreover, while two similar biotechnology companies should, theoretically, have an easier time merging their entrepreneurial cultures, this may not necessarily be the case. The primary reason behind the failed merger between The Liposome Company and Liposome Technology was the clash between their corporate cultures.

DNAX/Schering-Plough. One of the earliest and more successful biotechnology acquisitions was Schering-Plough's purchase of DNAX Research Institute of Molecular and Cellular Biology (DNAX) in 1982. In this purchase, the two corporate cultures were kept separate and distinct, to the apparent benefit of both parties.

DNAX was founded in 1980 by three Stanford professors, Nobel laureates Paul Berg and Arthur Kornberg, and Charles Yanofsky, along with Alejandro Zaffaroni, founder of Alza Corporation. Four million dollars of initial funding came from Zaffaroni's European contacts.

By 1982, DNAX had built up a staff of approximately thirty scientists, but the next round of financing did not look promising. Research contract negotiations with Schering-Plough, ongoing at that time, turned into acquisition talks. In July 1982, Schering-Plough acquired DNAX for $30 million in a stock swap. In 1985, DNAX moved into a 50,000 square-foot facility built by Schering-Plough. Since then, DNAX's staff has grown to over 150 employees.

Schering-Plough has maintained a hands-off management style toward DNAX, even though it regards DNAX as its in-house "think tank." DNAX has always set its own research agenda and has been responsible for hiring new scientists. As a consequence, in consultation with Schering-Plough, there was an early shift away from monoclonal antibody design toward the discovery and characterization of proteins of the immune system. This research effort has made DNAX a leader in the area of cytokine research and has led to the cloning and expression of several new human cytokines.

With no financial or commercialization worries, DNAX's corporate culture has evolved from one that was entrepreneurial to one that is more academic, which has allowed the scientists the freedom to focus on long-term research problems. The success of this approach is found in the fact that until recently none of the original senior scientists had left DNAX. The two who have resigned did so to become chairmen of academic departments in their home countries.

Schering-Plough has benefited from this arrangement as well. As of early 1990, one biological discovered at DNAX is awaiting Food and Drug Administration marketing approval and another is in clinical trials. Equally important, Wall Street recognizes that Schering-Plough is a significant force in the biotechnology industry, as reflected in its stock price.

Hybritech/Eli Lilly. Eli Lilly acquired Hybritech, the dominant monoclonal antibody–based biotechnology company, for $375 million in 1986. Hybritech needed additional working capital and Eli Lilly wanted to expand its research efforts in monoclonal antibody–based diagnostics and therapeutics.

Hybritech, located in San Diego, California, was incorporated in 1981. By 1984, the company was selling diagnostics using its novel diagnostic test format, Icon® (immunoenzymatic assays). Over the years the Icon® format has been applied to an increasing range of diagnostic tests, which now accounts for over 50% of Hybritech's revenues. In 1989, Hybritech had sales of $104 million and 650 employees.

Hybritech has proven to be a partial success for Eli Lilly. On the positive side, Hybritech has grown into a strong diagnostic business, a business emphasis that Lilly would like to continue to develop. For example,

Lilly announced in early 1990 its proposed acquisition of Pacific Biotechnology, a small diagnostic business, also located in San Diego, California.

On the negative side, Hybritech has not pioneered any important new approaches in diagnostics or therapeutics since the merger. In fact, many of the scientists and management employed at Hybritech at the time of the merger have left the company to initiate or join new biotechnology start-ups.

Eli Lilly's corporate culture is primarily to blame for this significant exodus. Hybritech's entrepreneurial environment and strong competitive spirit were gone, and in their place was a hierarchical structure that did not reward creativity or provide the fast-moving environment that many of the staff had enjoyed prior to the acquisition.

The Liposome Company/Liposome Technology. In early 1989, The Liposome Company located in Princeton, New Jersey, and Liposome Technology located in Menlo Park, California, announced their intentions to merge. Both companies were started in 1981 to develop liposome-based pharmaceuticals. Because both companies were in need of cash, and Wall Street had never looked favorably on companies focusing on liposome technology, a merger would allow the companies to take advantage of complementary product lines and to realize operating efficiencies. Yet, later in 1989 the merger was called off. The primary reason given for the failed merger was that the different corporate cultures made agreement on the merger's fine points impossible. Adding to this problem was the difficulty of trying to run a bicoastal company with neither division dominating the other. These problems proved larger than the potential benefits, and the merger was canceled.

The Role of Corporate Priorities

Even if the merger or acquisition initially gets off to a good start, time and new circumstances can alter the marriage. Business subsidiaries are often the victim of change in corporate priorities and business strategy; acquired biotechnology companies are no exception. Two early biotechnology acquisitions, Genetic Systems by Bristol-Myers and Hygeia Sciences by Tambrands are examples of this phenomenon.

Genetic Systems/Bristol-Myers Squibb. Genetic Systems was purchased by Bristol-Myers in 1986 for the unheard of price of $294 million. Genetic Systems' research was split strictly into two divisions following the acquisition: Genetic Systems developed diagnostics and the Oncogen subsidiary focused on therapeutics. These two divisions reported to different management groups within Bristol-Myers.

Late in 1989, Bristol-Myers merged with Squibb, and the combined company made the strategic decision to focus more narrowly on human therapeutics. As a result, Bristol-Myers Squibb sold the Genetic Systems diagnostic division to Sanofi, Inc., the U.S. subsidiary of Paris-based Sanofi-Chemie, in early 1990. Bristol-Myers Squibb has retained the Oncogen therapeutic division.

The change in ownership did not come as a complete surprise. Genetic Systems had worked closely with Sanofi in the past. Previously, Genetic Systems had licensed what would become its primarily AIDS diagnostic test technology from Sanofi's Diagnostic Pasteur, including the marketing rights in the United States and Canada. By purchasing Genetic Systems, Sanofi recovered the lucrative North American rights to its AIDS diagnostic.

Hygeia Sciences/Tambrands. Among the early acquisitions, Hygeia Sciences was also the victim of a change in corporate direction. Hygeia Sciences, located in Newton, Massachusetts, began operations in 1981, and was one of the first biotechnology companies to focus on the application of monoclonal antibody technology to human diagnostics.

Despite growing revenues, Hygeia Sciences was never profitable and had an earning loss of almost $2 million in 1985. By the time of its initial public offering in February 1986, Hygeia Sciences had annual sales of approximately $8 million from three diagnostic products, an over-the-counter (OTC) pregnancy test and ovulation predictor test, marketed under the brand name *First Response*, and a pregnancy test for the physician's office market. The public offering netted the company $8 million and valued the company at approximately $27 million.

Hygeia Sciences had no marketing and distribution capability. Instead, prior to the initial public offering, Hygeia Sciences entered into a marketing agreement with Tambrands of Lake Success, New York, that granted Tambrands exclusive worldwide distribution rights to the two OTC diagnostic tests as well as an option to acquire the same distribution rights for future OTC products. Hygeia Sciences retained manufacturing rights for all products.

Less than a year after Hygeia Sciences' public offering, Tambrands made a friendly offer to acquire the company. Tambrand's offer was motivated by its strategic decision to diversify its own product line, thereby reducing its dependence on its one major product, *Tampax* tampons. Tambrands paid $48 million for Hygeia Sciences, a price that represented a 35% premium over the market capitalization at the time.

In 1988, Tambrand's new CEO, Martin Emmett, decided to abandon Tambrands' strategy of diversification and again focus the company's efforts exclusively on the manufacture and marketing of *Tampax* tampons.

As a result, Hygeia Sciences went on the auction block in 1989. In April 1990, Tambrands announced that Hygeia Sciences had been sold to Carter-Wallace.

Once Tambrands' overhead costs were factored in, Hygeia Sciences probably was never profitable. Insiders speculate that Tambrands received less than one times sales for Hygeia Sciences, well below the purchase price of $48 million. However, by the end of 1989, Hygeia Sciences' sales of its pregnancy and ovulation-predictor tests had grown to approximately $30 million, and the *First Response* brand was the number one seller based on the number of tests sold.

Unfortunately, the company's efforts to add to its product line have not yet been as successful. Although approval to market Hygeia Sciences' gonorrhea and group A streptococci (strep throat) diagnostics was granted for the professional health care market, the more financially significant FDA approval to market these tests to the OTC market has not yet issued.

Despite these problems, Carter-Wallace will likely benefit from the Hygeia Sciences acquisition. Although Carter-Wallace already sells an OTC pregnancy and ovulation-predictor test under the brand name *Answer*, the company does not have a large market share. By purchasing Hygeia Sciences, it bought not only market share, but a high-quality brand name as well as state-of-the-art technology. Only time will tell if this new acquisition will prove more successful for Hygeia Sciences.

BIOTECHNOLOGY COMPANY VALUATION

Mergers and acquisitions involving biotechnology companies cannot be valued using conventional techniques, such as a multiple of book value, price-to-sales ratio, or multiple of net earnings, because most of the biotechnology companies today are still mainly research boutiques with negative earnings.

Consequently, more subjective measurements must be used to arrive at a price. One technique popular on Wall Street is to project the company's revenues and earnings for several years into the future, assign a price-to-earnings multiple, and then discount this future value to arrive at a present value. This approach is fraught with obvious problems: the analyst must make a myriad of assumptions regarding future products—including market introduction dates, market sizes, and profitability—for products that are often still in the research phase of development! A biotechnology company's financial projections are subject to the same uncertainty.

For most of the recent mergers and acquisitions it will take years for management to determine if a reasonable return on its investment is possible. For example, Schering-Plough purchased DNAX eight years ago in one of the first biotechnology acquisitions, and is only now beginning to reap the rewards. In 1982, at the time of the merger with Schering-Plough, DNAX had several product ideas in the area of monoclonal antibody design but nothing more substantial. In addition, the industry was much younger, which made future financial projections almost impossible. Consequently, the purchase price of $30 million in 1982 raised many eyebrows both on Wall Street and within the pharmaceutical community.

In the ensuing eight years, DNAX proved to be valuable to Schering-Plough as a source of new biotechnology-derived therapeutics. Today, Schering has one product derived from DNAX research awaiting FDA marketing approval and another in clinical trials. Both products have the potential to generate several hundreds of millions of dollars in sales. In retrospect, Schering's management believes that the price it paid for DNAX was a bargain.

Genzyme paid $29 million for Integrated Genetics, the price set by the public market. Both Integrated Genetics and Genzyme were public companies, and Genzyme purchased Integrated Genetics in a stock swap. One year after the merger, Genzyme's management knows it bought a bargain. It accomplished this by using a technique common to leveraged buyouts, i.e., asset sales. One division was sold for $10 million. Another division, IG Laboratories, was "spun-off" in a public offering that raised $14 million, while valuing Genzyme's majority position at approximately $20 million. Genzyme not only recovered its purchase price and then some, but it kept the part of Integrated Genetics that fit with its own research focus and received a manufacturing facility as a bonus.

Integrated Genetics/Genzyme

Integrated Genetics, incorporated in 1981, specialized in the development of recombinant DNA proteins for diagnostic and therapeutic use. Promising products included cardiovascular proteins, blood cell growth factors, and endocrine-related products.

In 1986, Integrated Genetics established a joint venture, GENE-TRAK Systems, with Amoco Biotechnology Company, a subsidiary of Amoco Corporation, that encompassed all the company's activities in diagnostics. Also in 1986, the company established a commercial reference laboratory to test for genetically linked disorders.

Integrated Genetics had been a public company since 1983. Like most biotechnology companies, IG was never profitable, so additional funding was periodically needed. Integrated Genetics tried to raise $30 million in a convertible preferred stock offering in mid-1987, but was unsuccessful in this attempt. By 1988, management began to examine other options as operating losses increased.

Genzyme, incorporated in 1981 and located in Boston, Massachusetts, a half-hour drive from Integrated Genetics' Framingham, Massachusetts headquarters, is the larger of the two companies, with over 300 employees, as compared with Integrated Genetics' 140. Genzyme sells a product line of fine chemicals and biological and diagnostic products. In addition, Genzyme has built a research program in human therapeutics based on its expertise in carbohydrate chemistry and enzymology. Unlike Integrated Genetics, Genzyme has been marginally profitable since 1986.

In order to expand its research base in recombinant DNA technology, Genzyme approached Integrated Genetics with a merger proposal in early 1988, but terms agreeable to both sides could not be worked out. Integrated Genetics' pursuit of alternative financing strategies later that year was unsuccessful, so merger discussions were renewed in February 1989. The two companies agreed to merge by swapping Integrated Genetics' stock for newly issued Genzyme stock in a transaction valued at $29 million.

One pleasant surprise for Genzyme's management was the size of Integrated Genetics' pilot facilities, which it had underestimated. Instead of spending $10 million on a new pilot facility, as originally planned before the merger, Genzyme is planning to spend considerably less to adapt Integrated Genetics' existing facility.

A second means by which Integrated Genetics' ultimate purchase price was reduced came by way of Serono Laboratories, the world's largest supplier of human fertility drugs. Serono Laboratories had been Integrated Genetics' partner in the development of recombinant human fertility hormones. In order to keep this line of research, Serono bid $28 million cash for Integrated Genetics in May 1989. Because Serono was not interested in Integrated Genetics' other projects, Integrated Genetics' management elected to proceed with the Genzyme merger and then sell the fertility hormone projects to Serono for $10 million in cash.

Genzyme further intended to reduce the net purchase price for Integrated Genetics by negotiating the sale of its minority interest in GENE-TRAK Systems to the Hoechst Group of West Germany. A letter of intent was signed, and analysts expected Genzyme to receive at least $10 million in the transaction. The proposed sale never happened. Genzyme did eventually sell its position in GENE-TRAK to Amoco.

Finally, Genzyme created an independent, wholly owned subsidiary, IG Laboratories, based on Integrated Genetics' genetic testing services. Genzyme spun off a minority interest in IG Laboratories through a public offering in May 1990, which raised $14 million and valued the service laboratory at $38 million.

After these asset sales, Genzyme had effectively reduced its initial investment in Integrated Genetics to $10 million, the amount it had earmarked for a new pilot facility. Its majority interest in IG Laboratories is worth approximately $20 million. In addition, Genzyme has acquired some potentially lucrative therapeutic proteins, including a tissue plasminogen activator that Integrated Genetics' Japanese partner, Toyobo Company Ltd., may market first in Japan; strategic additions to its scientific team that will accelerate the development of new therapeutic products; and significant expertise in the manufacture of proteins using mammalian tissue culture.

Genentech/Roche Holding

The largest biotechnology acquisition to date is the proposed acquisition of Genentech by Roche Holding. In the two-stage transaction, Roche Holding will acquire 60% of Genentech's outstanding stock for $2.1 billion, then have the option to acquire the remaining shares before 1995.

Given Genentech's 1989 sales of just over $400 million, after-tax profit of $44 million, and book value of $469 million, the purchase price cannot be justified in terms of the 1989 income statement or balance sheet. Clearly, Roche Holding must believe that Genentech's product portfolio and in-house talent hold significant promise. If Roche Holding is correct, then in ten years, company executives may look back, just as Schering-Plough's executives have done, and know they bought a bargain. Yet, with such an expensive price tag, there is little room for error in their enthusiastic assessment of Genentech's potential.

It is interesting to note that in 1989, the year during which the greatest number of transactions took place, those biotechnology companies that were acquired by large pharmaceutical companies were bought at substantial premiums to the common stock price prior to the acquisitions. Yet in all cases the buyout price was still far below the stock's all-time high. Roche Holding's offer of $36 per share for half of Genentech's outstanding shares is significantly above the low twenties, where the stock was trading just prior to the announcement, but below Genentech's 1987 high of $64.75 per share. Similarly, Chugai's $6.25 offer was an almost 100% premium to Gen-Probe's stock price at the time of the acquisition announcement, but still less than the 1988 high of $8.75. Finally, American Cyanamid's $13.70 price for Praxis Biologics' shares was a whopping 175% premium over the current price of $4 per share, but still less than the 1987 high of $17.50.

Biotechnology stocks have been known for their volatility. In the three cases just cited, the companies continued to make progress, yet the stock market placed less value on those companies in 1989 than at an earlier stage in their development. As market realities become known, biotechnology stocks have adjusted accordingly. By selling below their stock highs, managements must believe that such earlier valuations were not possible in the foreseeable future.

The three mergers between biotechnology companies in 1989—Ingene and XOMA, Plant Genetics and Calgene, and Integrated Genetics and Genzyme—differed from the above-mentioned acquisitions in that they were done at essentially no premium to the market price. Investors simply swapped their shares in one biotechnology company for an equally valued number of shares in the surviving, combined entity. These transactions also differed in the relative "cost" to the acquirer. The stock market values biotechnology stocks subjectively, but there is a strong correlation between relative stock price and performance. Thus, the acquiring biotechnology company is able to use stock valued with the same criteria as the stock of the acquired company, instead of cash or stock valued with different criteria.

TYPES OF TRANSACTIONS

The two principal forms of payment for a biotechnology company have been cash and stock swaps. Table 16-1 describes various such transactions where the information is available. The two most expensive acquisitions, Hybritech's acquisition by Eli Lilly and Genentech's proposed acquisition by Roche Holding, were more complicated and contain provisions that could potentially increase the overall return to the biotechnology company shareholders.

In addition to warrants to purchase Eli Lilly common stock, Hybritech shareholders received "contingent payment units" ("Units"). These Units were expected to provide an annual cash payment to shareholders. The payment was based on a formula that included Hybritech's total revenues and gross profit. The Units cancel in 1996, or when $22 has been paid. As of five years after the acquisition, no cash has been paid to holders of the Units and none is now likely. Technically, the Units will continue to trade until their expiration in 1996, but they have virtually no value. In this case, Hybritech's post-acquisition performance did not meet the expectations required for payments to Unit holders; by including the wild-card Unit payments, Eli Lilly possibly paid less upfront than would have been otherwise required.

Genentech's acquisition is a staged acquisition in that Roche Holding will pay $2.1 billion in cash for 60% of Genentech's outstanding shares in 1990. The remaining shares will be exchanged for Genentech redeemable common stock, which Roche Holding may redeem for $38 per share in 1990, escalating to $60 per share by 1995, the date at which the redemption right expires. In addition, Roche Holding will invest roughly $500 million directly in Genentech with the purchase of its controlling interest. After Roche Holding's initial investment, the possibility remains that it will choose not to exercise its option to purchase the remaining

40%. This is certainly a possibility, particularly if Genentech's fundamentals deteriorate in the short term. Should that occur, Roche Holding will own a controlling interest in a publicly traded biotechnology company.

SUMMARY

The pace of consolidation has accelerated in the past year as biotechnology companies mature and face the prospect of paying for expensive clinical trials and developing the infrastructure to manufacture and market their biotechnology-derived products. Merger or acquisition will remain an important financing vehicle for these companies. If other sources of capital become more limiting, then merger activity is likely to increase.

There was concern that Genentech's unexpected move to be acquired by Roche Holding might be perceived as a signal that the industry was in deep trouble, but clearly that is not the case. Biogen, Amgen, Immunex, and Genetics Institute, to name just a few, are all moving important products into marketplaces around the world.

Although many of the smaller companies are strapped for cash in the current unforgiving public market, the top tier have fat cash reserves and are taking every opportunity to keep the cushion inflated. In fact, these cash reserves may provide the means to accelerate mergers between biotechnology companies. Roche Holding's acquisition of 60% of Genentech calls for Roche Holding to infuse approximately $500 million cash into Genentech, some of which may be earmarked for strategic additions to its technology base.

Despite the lack of significant merger and acquisition activity to date, many biotechnology company executives view acquisition as the final "exit strategy" for their investors. After years devoted to building basic research programs and then expanding into product development efforts, all supported by several rounds of financings including public offerings, managements expect acquirers to pay lofty sums for their companies. If other sources of capital are available, management may likely defer an acquisition to continue to build value in the hope of even larger valuations down the road. Thus, the question is not if but when a significant consolidation trend will begin.

One argument put forth against consolidation is that the many entangling relationships that biotechnology companies typically enter into in their formative years tie up valuable projects and make mergers difficult. Yet in the case of the Integrated Genetics/Genzyme merger, Integrated Genetics' existing partnership arrangements provided a source of buyers for tangential research projects. Furthermore, corporate partnering gives two corporations a chance to live together before tying the knot.

Some biotechnology companies will be well rewarded when the offer comes, but others will not be so lucky. There are simply too many biotechnology companies using the same basic technologies to address the same markets, and not all of them will end up holding proprietary technology or patents that acquirers will want to buy. For those unlucky ones, the 1990s may well become the decade of the biotechnology bust.

CHAPTER 17

JAPAN'S DILEMMA: WHERE AND WHEN?

Susan K. Clymer

Biotechnology is not an industry in the sense that we talk about steel, electronics, or automobiles. For convenience, many people refer collectively to the companies engaged in commercial biotechnology within a given national boundary as a biotechnology industry. While not an industry, biotechnology is a unique and very powerful set of tools that can be used by established industries in research, development, and manufacturing to alter the price performance of competing products. From this perspective, biotechnology will have a critical impact on international competitiveness, because it will enable companies or established industries to maintain or improve positions in core market sectors of the global economy.

Because biotechnology is not an industry per se, there is little value in debating the possibility of Japan overtaking the U.S. biotechnology industry. Instead, what we as Americans should be debating is whether or not U.S. companies will be able to compete successfully with Japanese companies in future product markets and how quickly U.S. companies will be able to translate the new biotechnologies into cost-effective industrial applications.

Though it may take Japanese companies several more years to figure out where or how to apply biotechnology to obtain maximal value added, I believe that Japanese companies will be better positioned and more successful than U.S. companies in capturing the value added in high volume product markets, particularly those with longer development horizons, such as chemicals and bioelectronics. Japan's success will be due as much to cultural predisposition as to strategic preparedness.

At present, however, many Japanese companies appear to be facing an unusual commercial dilemma—where (i.e., which product targets) and when to commit additional resources to biotechnology research in order to realize maximal commercial gain. Nowhere is this dilemma more pronounced than in the pharmaceutical field, the most advanced of all biotechnology application areas at present.

Because biotechnology is not an industry but a set of technologies that can be used by many established industries, establishing a cohesive and unified industrial policy is difficult. Unlike the computer and electronics industries, for example, which fall under the exclusive perview of the Ministry of International Trade and Industry (MITI), biotechnology crosscuts the jurisdiction of several ministries, including not only MITI but also the Ministry of Health and Welfare (Koseisho), the Ministry of Agriculture, Forestry, and Fisheries (MAFF), and the Ministry of Education, Science, and Culture (Monbusho).

Though MITI has done a splendid job of taking the lead to push the development and diffusion of the technologies throughout industrial Japan while at the same time building an infrastructure to facilitate practical application, the ultimate responsibility for generating ideas for commercial development resides in the companies.

Typical Japanese companies are risk-averse and oriented toward development and they seem stymied at present by a dearth of a priori commercial targets or product markets on which to set their sights. There are few established markets to lend strategic guidance to the companies, few perceived commercial successes to date, a perceived lack of ideas emanating from Japanese academic labs, and few new economically viable bioprocesses for making large volume products. In a sense, it seems that Japan acquired and developed a powerful set of technologies for which no user's manual exists.

Compounding the dilemma is Monbusho, which controls over half of the government's total research and development expenditure. Monbusho has done precious little in the way of increased funding or structural reform to improve Japan's basic research base—a potential source of new commercial ideas. Furthermore, there is major disagreement between companies and university scientists as to whether or not Japanese academia is capable of generating novel ideas for commercial development.

In the pharmaceutical field, for example, established pharmaceutical companies have been very slow to broaden their initial commitment to biotechnology after initially jumping in to license technology from abroad in the early to mid-1980s or after initially developing their own in-house capability. Because of unmet expectations with the first genetically engineered protein therapeutics (e.g., the interferons, interleukin-2, tissue necrosis factor, tissue plasminogen activator, and others), Japanese pharmaceutical companies are cautious about making major commitments to biotechnology, particularly to produce and develop other protein therapeutics. As long as the established pharmaceutical companies maintain a risk-averse, short-term development-minded (as opposed to research-minded) attitude toward biotechnology and persist in the belief that few valuable ideas for commercial development exist in Japanese academic laboratories, Japanese companies will continue to rely on Western countries to generate ideas and produce the scientific data necessary to justify significant increases in in-house funding of biotechnology research in the Japanese companies themselves. In Japan, companies closely monitor foreign developments and weigh the costs and benefits at every stage in order to position themselves for appropriate action when more near-term and often less risky commercial opportunities emerge.

Whereas innovative product breakthroughs are difficult to make, follow-on products come more easily and quickly. Protein engineering and design, though not yet commercial realities, will provide the ideal opportunity for Japanese companies to build on established outside scientific and clinical knowledge while obviating the need for long-term basic research. Many Japanese believe that the promise of biopharmaceuticals can only be realized through protein engineering.

As foreign research laboratories elucidate the role of specific proteins in a given disease state, Japanese companies will be well positioned to use biotechnology synergistically with well-developed in-house screening expertise to identify substances that mimic the activity of natural proteins whose role and function have been elucidated outside of Japan.

As with U.S. companies, most of the Japanese biopharmaceutical players have protein engineering groups and are developing sophisticated receptor-based assay systems. Unquestionably, the definition of a "new product" or new chemical entity and the role of patents will take on heightened international significance over the next several years as new technologies, including protein engineering, enable the practical modification of existing or promising biotherapeutics.

THE COUNTRY—MOTIVATOR AND RESOURCE TRAP

Japan cannot sustain itself in either food or energy. The country has no oil, natural gas, or iron, and only a tiny bit of coal. With a land mass less than one half that of California, only 25% of which is habitable, and a population approximately half that of the United States, it is not difficult to understand why Japanese companies strive to maximize the value of imported raw materials through highly economical use, why less risky, more cost-effective applied research is emphasized, why a development-versus research-oriented business strategy has emerged, and why technology sourcing must be international rather than domestic. To understand Japan's domestic condition is to appreciate the country's vulnerability and instinctive drive to secure raw materials and technologies that enable every last bit of value to be squeezed. In no uncertain terms, Japan's survival instinct supplies much of the raw power behind her competitive strength and compelling desire to move biotechnology into cost-effective industrial applications.

BASIC RESEARCH—THE IDEA GAP

By most U.S. and Japanese accounts, both academic and industrial, there is no technology gap between the United States and Japan. The general level of technological competence in the recombinant DNA field is pretty much equal between the United States and Japan, and because new information is disseminated rapidly via scientific journals, there is no real knowledge gap either. However, there appears to be a wide, quantitative gap between the two countries in the overall quality of basic research conducted and the ability of each country to generate ideas for commercial development through basic research. In fact, where biotechnology is concerned, there is little real evidence in Japan to suggest any national commitment in either the private sector or the government to improve or stimulate basic or fundamental research beyond its current level in order to narrow the gap.

The typical Japanese definition of basic research differs markedly from the typical American one. Whereas in the United States, basic research is oriented toward the discovery and explanation of phenomena, in Japan basic research is oriented toward early stage product research and practical application. Underscoring Japan's different sense of basic research, some companies believe that MITI is doing a much better job than Monbusho in promoting basic science.

By most Japanese accounts, overall Japanese basic research in the biomedical and agricultural fields is generally weak compared to the United States and Europe, though this is not to say that Japan does not have world-class scientists. There are just relatively fewer high caliber scientists in Japan as compared with Europe or the United States.

The hierarchical kouzasei[1] are governed by seniority. Young scientists

1. A koza is a teaching chair usually consisting of a professor, an associate, two assistant professors, and two research assistants. Between 1965 and 1989, the annual amount of

generally report that the koza system suppresses creativity and unique talent and discourages academic risk-taking, thereby promoting mediocre science. Monbusho funds are distributed and allocated based on seniority versus merit in a 2:1 ratio, which puts young researchers at a disadvantage and perpetuates the pet projects of older faculty members. Assistant professors require special permission from a professor to conduct independent research. Often, promising young scientists leave academia and move to industry, not only to obtain a salary, but also to escape academic politics and restrictions.

There are few Ph.D. programs, therefore little funding for salaries, very few new kozas created each year, and a woefully small postdoctoral system. For example, according to one source, over the past 30 years, three new kozas have been created in the Agricultural Chemistry Department of Tokyo University. Agriculture chemistry departments provide the training for a large percentage of Japan's applied microbiologists and biochemical engineers. There is a government-sponsored, two-year postdoctoral program that funds a total of 500 scientists per year.

The absence of a postdoctoral funding system in Japan not only discourages scientists from staying in academia but, coupled with the seniority-based kozas, tends to inhibit the movement of scientists from one laboratory to another and limits the cross-fertilization of ideas. Actually the seniority system can, in some cases promote the movement of scientists from one laboratory to another. In situations wherein an assistant professor is not chosen over another assistant professor to succeed the Professor, he will often take a position at a private (rural) university in order to maintain the harmony and the hierarchy of the koza.

There is limited demand for Ph.D.s in companies. Master's level students are preferred because they are not so specialized and more easily retrained to work on a variety of different projects. Therefore, there is relatively little impetus to create new programs to increase the number of Ph.D.s. Also, because there are few new kozas created, young Ph.D.s find it difficult to find academic positions.

Lack of research funds overall is problematic. The tax system discourages wealthy individuals from making contributions to science. Scientists frequently complain that Monbusho regulations impose unfair restrictions on how research funds can be used (e.g., for travel to conferences) and also inhibit the flow of private sector funding to national universities, where much of the best research is carried out.

---

funding given to each koza declined from 8.5 million yen to 3.0 million yen. The government budget for opening a new koza is 4 to 5 million yen. A laboratory bench costs about 10 million yen, and a centrifuge, freezer, cell culture hood, and incubator about 50 million yen.

In Japan, the need to strengthen the nation's basic research base is a common theme in government, academia, and industry. However, the way to go about it is the subject of considerable debate within Japan's science establishment [Sun, 1989]. Until some changes are instituted, one thing remains certain: Japan will continue to rely on the United States and Europe to generate ideas and lead the way into new commercial fields through their national commitments to basic research. Supporting this opinion are senior executives and members of boards of directors from some of Japan's leading pharmaceutical companies, one of whom openly asserted, "We need American scientists."

To compensate for the relatively weak (perceived or real) basic research base and the lack of commercial ideas coming out of academic laboratories, many Japanese companies are training their corporate scientists in the United States, funding U.S. laboratories, establishing U.S. research laboratories, and acquiring U.S. technology. Given the "openness" of the U.S. academic system and the constant need for cash in the new biotechnology firms, Japanese companies can freely piggyback the U.S. advantage and obviate the need for a national commitment to improve the basic research base, particularly as it relates to biotechnology.

Without a doubt, U.S. government funding of basic biomedical and agricultural research has established the United States as the leader in commercial biotechnology, principally because many of the research discoveries have been made by federally funded, entrepreneurial-minded scientists, whose achievements are often closely monitored by or somehow conveyed to the venture community. Thus, government funding—tied to an academic environment for creative, risk-taking pursuits—and efficient technology transfer to industry via a supportive venture community has provided the United States with its comparative advantage and present leadership position in commercial biotechnology.

Based on historical precedent, it is not unreasonable to imagine that through international diffusion of early U.S. biotechnology research discoveries, U.S. leadership will be lost to foreign companies more able to carry the United States originated technologies through succeeding rounds of improvement and application, particularly once markets have been established and strong economic incentives prevail. Though basic research is a source of industrial strength and innovation in the United States and provides comparative advantage in the emerging stages of industrial application and commercial development, U.S. competitiveness will depend as much on efficient transfer to industry, strong patents, and rapid, cost-effective industrial application as it will on *increased* funding of basic research.

U.S. policy makers need to understand that basic research funding, together with technology transfer and rapid, cost-effective industrial application are all essential if the United States is to remain competitive in commercial biotechnology. Lack of support or a breakdown in any of these component areas will undermine U.S. competitive strength. The United States wishes to maintain an open academic environment even at the expense of losing competitive advantage through technology transfer to foreign industrialized countries such as Japan. Such being the case, U.S. policy makers must consider options which not only support U.S. innovation and industrialization of biotechnology but also perhaps impose monetary assessments on foreign industrialized countries (i.e., those who do not commensurately support basic science) on their use of U.S. science

For example, levies imposed on such foreigners' use of academic science in the United States could be ploughed back into the academic research system. The additional funds would in effect help preserve the basis of U.S. innovation and competitive advantage and also in part compensate for the flow of technology out of the United States. Levies could be assessed in a variety of ways including fixed mandatory annual payments (e.g., to the National Institutes of Health), taxes on patents licensed from U.S. academic institutions, head taxes on foreign scientists conducting short-term research funded by NIH grants, a VAT tax on foreign sponsored biotechnology research conducted in U.S. academic labs, etc. Though these options may seem extreme, serious consideration in the U.S. would send a strong signal that the United States intends to maintain its competitiveness and correct asymmetries that exist in the international development of technology.

MITI—THE CATALYST

With MITI at the helm, Japan's biotechnology policy loosely fits the mold of other industrial policies. It is meant to strengthen the Japanese economy and the competitiveness of Japanese companies through more efficient use of resources and improved industrial productivity. Beginning in the late 1970s and culminating in 1981 with the formation of a national biotechnology research association, MITI and members of Japan's chemical industry (including Mitsubishi Kasei, Sumitomo Chemical, Kyowa Hakko, Mitsui Toatsu, and Asahi Chemical) forged a plan to promote biotechnology. Ultimately, the plan became part of MITI's Next Generation Basic Technology project and Vision for the 1980s.

The Next Generation Basic Technology project was designed and implemented in part to promote technologies that could help restructure Japan's depressed industries, including the chemical industry. Always concerned about Japan's resource deficiencies, MITI viewed biotechnology as a means of minimizing energy utilization and maximizing feedstock value through bioprocessing or biological manufacturing. Biotechnology also offered an avenue for diversification into higher value added products and markets such as pharmaceuticals.

Under the Next Generation Basic Technology project announced in 1981, biotechnology initially was promoted via a MITI-sponsored, national Research Association originally planned for ten years and comprised of fourteen carefully selected companies (eleven of which were chemical companies, including all those mentioned above). Within the Research Association, three individual research consortia consisted of six, three, and five companies each, were formed to conduct research in three generic applied areas, respectively: bioreactors (e.g., more efficient biological systems for producing chemicals), recombinant DNA (e.g., construction of high-level secretion expression systems), and large-scale animal cell culture (e.g., to produce pharmaceuticals and to develop serum-free tissue culture medium). In keeping with the national strategy to add value through manufacturing while reducing raw material consumption, today MITI continues to adhere to the notion that biotechnology can generate long-term cost savings in industry, slow the environmental destruction of the earth and enhance the well-being of human life. Throughout this decade, MITI is focusing on the environment: supporting companies and research and development in technology areas that promote harmony with the biosphere (e.g., carbon dioxide fixation). In 1990 MITI formally founded the Research Institute of Innovation Technology for the Earth (RITE), which will carry out research (including biotechnology) related to managing the biosphere. MITI has seized the opportunity to promote the use of biotechnology for environmental management, still reflecting the ministry's close ties with the chemical industry, which is closely associated with environmental pollution. MITI often makes extensive use of the media to bolster public support for its research goals and objectives. [*Asahi Evening News* 1990].

With a strong alliance to nonpharmaceutical companies such as the fermentation/chemical, textile, steel, and electronics companies, MITI has pushed hard to encourage and sustain private sector interest in medium-to long-term applications of biotechnology, including, for example, specialty and commodity chemical production via engineered enzymes as well as bioelectronics which also will use modified proteins.

Clearly, MITI creates attractive incentives for companies to enter and also remain in the biotechnology field. Through research consortia, for example, the purchase of expensive equipment is generously subsidized, scientists are trained, and risk, technology, and information are shared. Often the private sector looks to MITI for research subsidization to offset and justify undertaking high risk early stage (i.e., longer term) projects. It is true that MITI's financial support of high risk biotechnology projects (even as low as 10% of total funding) is often viewed by senior management as a powerful if not essential endorsement to justify risky in-house investments, and should such a project show some promising research results, it is not unusual to see as much as a twenty-fold increase in follow-on funding by the company. The goal of MITI's support of long-term research and development projects (e.g., the improvement of biological processes, the development of molecular assemblies, the practical use of protein engineering, or the exploitation of marine microorganisms) is simple: to ensure future availability of alternative technologies and resources so that Japanese companies will not lose their competitive edge.

The Research Association of 1981 promoted the development of generic technologies through specific applied projects, which spanned various industrial sectors, but since that time, other government ministries have begun to carve out their own domains for biotechnology promotion that coincide with their regulatory authority (i.e., Koseisho promotes biopharmaceutical applications while MAFF promotes applications in the food and agriculture industries).

In some respects, the promotional activities of the other ministries has diminished MITI's predominant influence over the development of biotechnology, because there are relatively fewer funds available from the Ministry of Finance for MITI's biotechnology activities. Always the opportunist however, MITI constantly seeks ways to promote biotechnology that do not compete with other ministries but that lay the groundwork for the establishment of a bioindustry over which MITI will preside to attain national goals. Through standardization, database networks, public acceptance, safety assessments, and the development of instrumentation, for example, MITI and the private sector are slowly building an infrastructure on which Japan's future bioindustry will rest.

The expressed purpose of MITI-sponsored research consortia is to develop technology. In reality, however, consortia are means by which companies subsidize the training of scientists in a given applied field. MITI ensures the diffusion of the technologies to high-caliber companies capable of developing proprietary uses for the technologies outside the confines of the consortia. Because the research and development system of Japan is vertically oriented, there is a need to institute mechanisms that promote the rapid dissemination of technologies over a broadly-defined area. (Unlike the horizontal, mobile U.S. system, wherein a great deal of information is transferred by scientists by moving from one academic laboratory to another, the Japanese system is vertical, and scientists tend to stay in one koza or in one company for most of their scientific careers.)

Though MITI's involvement in biotechnology helps to promote consensus among companies about the importance of biotechnology, it should be noted that the companies, not MITI, often generate the specific ideas for commercial application and also recommend technologies that should be promoted. Typically, specific ideas for technology promotion, for example, originate in one or two companies who then, through various means, inform MITI of the need for promotion (e.g., long-term, high risk, pivotal importance). The Protein Engineering Research Institute (PERI) is a case in point. Tonen and Mitsubishi Kasei reportedly initiated the idea.

Once the companies and MITI have agreed on the need for support, the "recruitment" process for companies begins as does public promotion of the project. In this way, a government–industry consensus is built around the importance of the technology, which in turn helps stimulate further private sector involvement and reduces the risk from an individual company perspective.

In the context of biotechnology, Japanese companies do not appear to be risk takers. In fact, there is little evidence to suggest that a large number of Japanese companies would independently fund long-term, risky biotechnology projects unless some subsidized feasibility work had already been carried out or, as was the case in early licensing deals, there is a prospect for valuable technology transfer to the company.

Thus, early on, MITI quite clearly functions as a catalyst to fuel private sector interest in specific technology areas so that proprietary, commercial spin-off projects might be undertaken. The commercial dilemma noted at the beginning of this chapter is compounded by the technology-driven, generic nature of the MITI projects, which seem to provide little guidance to companies looking for viable commercial targets.

JBA (FORMERLY BIDEC)—THE CRUCIAL LINK

In 1983, MITI along with private sector interests, established the Bioindustry Development Center (BIDEC), a unique Japanese institution that links MITI with industry and academia and through which biotechnology information is cleared and coordinated, consensus is built, policy is formulated, and a bioindustry built. Today over 300 Japanese member companies belong to BIDEC. The membership list reads like a "Who's Who" in Japanese biotechnology; its members represent industries including pharmaceutical, food and agriculture, consumer goods, automobiles, banks, textile/chemicals, petroleum, and many more. Thus, biotechnology information from almost every industry is channeled through JBA to MITI thereby providing the crucial information necessary for building a bioindustry.

MITI relies heavily on industry for guidance as to where promotional efforts should be directed. In a similar vein, industry often relies on

MITI's endorsement and support of high risk research and development areas to justify in-house investment to point the way into new fields. To maintain this type of interdependent relationship through which policy initiatives can be defined, there is a critical need for both formal and informal information exchange. JBA, functioning as MITI's secretariat, is the conduit through which such exchange often takes place and also one mechanism by which MITI influences the private sector.

Individuals from many JBA member companies characterize the JBA as a lubricant that maintains smooth communication between the companies, MITI, and academia. For newcomers to biotechnology, it is an especially convenient way to gain exposure and to collect information on commercial biotechnology. Senior JBA staff members have very good relations with the academic community (often through school ties), thus enabling MITI direct access to and input from university professors who otherwise tend to maintain an arms-length relationship with MITI, because university academics are linked to Monbusho, the source of their funds. Though they wish to remain independent of MITI's influence, they will, through JBA, interact with MITI and participate in MITI-sponsored projects administered by JBA.

JBA's executive director is an ex-MITI official, the rest of the senior staff is comprised largely of member company employees (usually scientists). The senior staff spends two to three years at JBA on leave from their companies getting to know the MITI officials, academicians, other individuals from industry, promising new research areas, and so forth. (Current JBA staff members are on leave from such companies as Kirin, Asahi Chemical, Kyowa Hakko Kogyo, Ajinomoto, JGC, Nippon Steel, Suntory and the Japan Development Bank.) There is no better biotechnology training ground or networking system for these individuals (and their companies) who, after returning to the research and development or other divisions of their respective companies, will have become part of the MITI clique of companies that participate in the formulation of biotechnology policy.

Activities and information exchange at JBA are coordinated by approximately thirty different committees comprised of the member companies and academics. The committees' activities and interests range from infrastructure development (e.g., databases, instrumentation, standardization, and public acceptance) to international harmonization (e.g., plant patents and safety guidelines) to administration of a biodegradable plastics association. As MITI's secretariat, JBA also manages MITI-entrusted research projects and carries out studies under contract from MITI.

Over the years MITI has promoted just about every possible generic applied research area of biotechnology imaginable. As a result, it is becoming increasingly difficult as time goes by for MITI to find new technology focus areas that either have not yet been promoted or that do not conflict with the expanding activities of the other ministries. Reflecting this situation, JBA's present activities emphasize its original mandate—to build an infrastructure for the development of a Japanese bioindustry and to promote international harmony in the commercial development of biotechnology.

Unlike the electronics industry, where MITI was basically the sole ministerial promoter, biotechnology is a competitive arena. Other ministries, particularly Koseisho and MAFF have taken an active interest in achieving closer ties to the private sector and in promoting and coordinating biotechnology using mechanisms similar to those established by MITI (i.e., JBA-like organizations, research consortia, and government—private sector joint ventures).

Both MAFF and Koseisho have established private sector associations to interact with companies in the industries over which they have authority and influence. Some companies complain about the costs of being compelled to belong to more than one organization. (For political reasons the companies feel that they must join the associations of the ministries through which their products will be approved, otherwise the ministry might not treat the company favorably during the approval process.) They sense a growing need for an organization to centralize and coordinate biotechnology information among all the ministries.

Because neither Koseisho nor MAFF has much experience dealing with the private sector, and both have regulatory authority, it will be interesting to see how each one balances promotion and regulation and what policies, if any, emerge from the expanding dialogue with the private sector. It will also be interesting to watch how information flow is coordinated among the various ministries. Regardless of how the individual ministries carve out their promotional turf, one thing seems clear: MITI will continue to have a pivotal role in the future development of biotechnology by exercising control over crucial components of the infrastructure on which a bioindustry is being built.

In June 1990, MAFF established the Private Sector Promotion Section, an organization geared toward establishing close relations with the private sector and supporting private sector research and development efforts. The organization will fulfill such functions as planning, coordination, and liaison for joint research and development among private enterprises and between private enterprises and universities or national and public experimental and research institutes. Though neither Koseisho nor MAFF has much experience working with the private sector, they are more eager than ever to try to coordinate and promote biotechnology in the industries over which they have jurisdiction.

In fact, MITI has provided MAFF with advice on how to go about setting up its new organization to coordinate activities in the private sector and academia. Though relations among the ministries are generally friendly, this was not always the case. Before the other ministries had established private sector promotional programs of their own, they were apparently hypersensitive about MITI's intrusion onto their turf. For ex-

TABLE 17-1 NEW JOINT VENTURES ESTABLISHED BY MINISTRY OF HEALTH AND WELFARE

Company	Investors	Research Theme	Start-up
Drug Delivery System Institute Ltd.	Asahi Chemical Industry, Ajinomoto, Eisai, Shionogi, Daiichi Pharmaceutical, Tanabe Seiyaku, Meiji Seika	Drug delivery system employing glucide as carrier	March 1988
Biosensor Laboratories Co., Ltd.	Chugai Pharmaceutical, Tosoh, Kuraray, Hamamatsu Photonics	Optoelectronics-applied biosensor for medical use	March 1988
Institute of Cytosignal Research, Inc.	Kirin Brewery, Mitsubishi Kasei, Sankyo	Separation/refining of intercellular proteins	March 1989
Vascular Graft Research Center Co., Ltd.	Sumitomo Electric Industries, Daiichi Pharmaceutical	Hybrid-type artificial blood vessels of small diameter	March 1989
Advanced Skin Research Inc.	Shiseido, Chugai Pharmaceutical, Nippon Oil & Fats	Search for skin-activating material and dosing system	March 1990
Institute for Biofunctional Research	Takeda Chemical Industries, Fujisawa Pharmaceutical, Sumitomo Heavy Industries, Tanabe Seiyaku, Yoshitomi Pharmaceutical, Wako Pure Chemical Industries, Nihon Medi-Physics	Diagnosis system for positron tomography	February 1990

ample, before Koseisho set up an investing/financing system to promote private sector biopharmaceutical research, a company involved in a MITI project to develop peptides/proteins was told by a Koseisho official that any pharmaceutical product resulting from the MITI project would not be approved. Later, Koseisho retracted its threat.

In 1987, Koseisho inaugurated a new investing/financial system (very similar to MITI's Key Technology concept established in 1985 to promote core technologies such as protein engineering, plant cell culture, biopolymer development) that is aimed at encouraging development of broad, leading-edge technology through the use of government and private sector money. The establishment of government-private sector joint venture companies (see Table 17-1) promotes "the development" of new technologies that are too big for any one company, through the interchange of numerous private sector companies. The system also encourages companies to diversify into new business fields (e.g., pharmaceuticals) via participation in the joint ventures. Very few private sector companies would hesitate to join a government-sponsored association or research project if invited to do so, particularly if the company's product approvals go through that ministry.

THE COMPANIES—TARGET SEEKING OPPORTUNISTS

In contrast to most U.S. companies, established Japanese companies are aggressively using biotechnology to diversify into new industries and product markets as well as to improve existing biological and chemical processes. For many Japanese companies, biotechnology is considered a "vitamin"—a functional supplement that can fortify the health of the corporation. As in the United States, much of the commercial interest in biotechnology emphasizes high value-added pharmaceuticals. Due to the sheer number of Japanese corporate players from a multitude of industries, however, the levels of research and development effort devoted to other fields of application (e.g., bioelectronics, plant cell and tissue culture, food and chemical processing) appear to exceed those of the United States on a quantitative basis, although it is difficult to compare actual expenditures because of inconsistencies in the definition of biotechnology.

Leading players in Japanese biotechnology generally fall into four broad categories: 1) diversified companies with extensive bioprocessing/fermentation skills, many of which already had some involvement in either the pharmaceutical (e.g., antibiotics) or agriculture fields prior to adopting the new technologies (e.g., Kirin, Kyowa Hakko, Ajinomoto, and

Toyo Jozo); 2) diversified companies from mature industries such as petroleum, chemicals/textiles, paper, and steel with limited or no previous biologic processing expertise—some twenty years ago, many of the chemical companies had been involved in the production of single-cell protein (SCP) from hydrocarbons and had thereby gained some bioprocessing expertise (e.g., Mitsubishi Kasei, Tonen, Toray, Oji Paper, and Nippon Steel); 3) established pharmaceutical companies (e.g., Takeda, Yamanouchi, and Chugai) and; 4) instrumentation and electronics companies (Hitachi, Toshiba, and NEC).

The Japanese have enjoyed a long and intimate relationship with microorganisms through the production of fermented dietary staples—for example, shoyu (soy sauce), natto (fermented soy beans), miso (fermented soy bean paste), sake, and so on. Over the years through painstaking screening, Japan has built up what is probably the world's richest library of characterized microorganisms that have potential use not only in the food industry but also in pharmaceuticals, environmental management, and chemical industries, as well as other industrial processes. Coupled with fervent support from the government and an intense sense of vulnerability from having few raw material resources of their own, the Japanese seem to have acquired an implicit, somewhat emotional resolve to seek new uses for microorganisms and exploit the properties of living systems.

Unlike most established U.S. companies involved in commercial biotechnology (with the most notable exceptions being Eli Lilly and Merck), Japanese fermentation and pharmaceutical companies as well as many others support large discovery groups whose main activity is to screen microorganisms from all over the world, not only for potent pharmaceutical materials (e.g., anticancer and antibiotic agents), but also for use in industrial processes (e.g., useful biosynthetic pathways and high tolerance to stress such as temperature, pH, pressure, solvents, and so on). Indeed, Japanese expertise in this field is highly regarded throughout the world, and after World War II, many U.S. pharmaceutical companies set up joint ventures in Japan to exploit Japanese skills.

Though the biotechnology "fever" in the pharmaceutical sector has subsided to some extent over the past five years, private sector research and development efforts continue to strengthen the industrial microbiology infrastructure of Japan. (The technological infrastructure refers to microbial screening, strain development, and bioprocessing.)

With access to well-characterized microorganisms collected worldwide, a cultural and economic propensity to use these microbes wherever feasible, and public support from MITI and the other ministries, it appears that Japan enjoys a cultural and resource advantage, as well as perhaps a political advantage in commercial biotechnology, particularly with respect to applications in food, chemicals, and perhaps environmental management. (It is difficult to compare the impact of MITI's policies with that of the U.S. government funding for basic research in the United States, especially since many companies feel politically obligated to follow MITI's lead.) What remains to be seen from a competitive standpoint is how long it will take companies to identify appropriate industrial applications for the microorganisms and how rapidly such applications will be translated into cost-effective use.

Since World War II, fermentation companies have achieved an outstanding ability to control the biosynthetic pathways that are involved in product synthesis (e.g., amino acids and antibiotics). With a strong theoretical background in molecular genetics and biosynthesis, combined with an extensive expertise in screening and strain selection and improvement (through mutation), the tools of biotechnology are expected to provide Japanese companies with considerable advantage in tailor-making microbial strains for specific industrial uses. However, in pharmaceutical applications of biotechnology, the Japanese have been quite slow and not particularly adept at independently identifying and commercializing novel biotherapeutic compounds. (The discovery of endothelian by Takeda and scientists from Tokyo and Tsukuba universities is considered by other Japanese companies to be a major milestone in the biopharmaceutical field. Some Japanese companies are analyzing the Takeda success in almost textbook fashion to try to duplicate the process of discovery for other compounds.) This appears to be due as much to a lack of discovery as it is due to a salient risk-averse attitude. With respect to applied uses in the chemical field, economics are mostly responsible for holding back commercialization, because biological systems have not yet been developed that can compete on a cost basis with many of the existing chemical ones.

Often with encouragement from MITI, not always financial, Japanese companies will maintain a small group of researchers on a small project until such time that markets develop, the economics change, or the technology is improved sufficiently for industrial use. Government support often seems to provide the impetus behind the commitment, and passive shareholders have no real role in the strategic decision making of the average Japanese corporation. Thus, many of the leading chemical and petroleum companies will maintain a small-scale project on the biological manufacture of chemicals, and should some key variables change (e.g., an oil crisis resulting in higher feedstock prices or a technological breakthrough), the companies will be ready to scale-up and take advantage of the opportunity when it arises.

In Japan, there are greater similarities in the structures and processes of the research and development organizations across industries and across sectors than is the case in societies in which industrial development was less compressed in time. As a result, it is possible to speak of "Japanese" patterns of organization in large firms without as much danger of overgeneralization as is the case with the United States and Europe [Eleanor 1988]. It is therefore not surprising to find that a majority of the diversified companies are attracted to biotechnology for many of the same reasons and that many tend to approach biotechnology research in the same manner. Reasons given for initiating biotechnology activities include

- More efficient enzymatic processes since raw materials are imported and expensive
- Diversification into higher value-added markets
  Mainline business is mature
  Costs of life-long employment are high, and new businesses are needed to absorb the fixed costs
  An unwillingness to lay off workers
- Pressure from the government
- Preparation for future resource problems and industry restructuring
- Competition from newly industrialized countries
- Establishment of high technology image to increase company's market capitalization
- Long-term gamble but not so expensive; income can go to government (taxes), shareholders (dividends), or be reinvested (biotechnology research and development)

Because biopharmaceuticals is the most advanced of the commercial biotechnology fields, the remainder of this chapter will focus on the corporate dynamics shaping commercial biopharmaceutical development and competition. The attitudes of both diversified and established pharmaceutical companies engaged in biopharmaceutical research and development provide a useful paradigm for understanding the attitudes of most companies engaged in the commercial development of biotechnology.

DIVERSIFIED COMPANIES—THE RISK TAKERS

The most aggressive players in biopharmaceutical research and development (in terms of overall in-house spending, and enthusiasm), are not the established pharmaceutical companies, as might be imagined, but more diversified companies. Many of the diversified entrants are not new to the pharmaceutical industry. For example, prior to the advent of recombinant DNA, fermentation-based companies such as Yamasa Shoyu, Ajinomoto, and others were already developing anticancer, antiviral, and antibiotic drugs, mostly because of active microbial screening programs and related nucleoside and amino acid businesses. Companies from the textile sector of the chemical industry such as Toray and Teijin were also involved in the medical field in part as a consequence of their fiber technology used to manufacture such medical devices as kidney dialysis cartridges, catheters, and apheresis apparatuses.

Quite surprisingly, few of the diversified Japanese companies appear to have a biotechnology research strategy. At first blush, it seems that the companies are wasting considerable resources by allowing their scientists to dabble in too many therapeutic research areas—building critical mass in no one area and thus seemingly jeopardizing the chances for success.

The scientists act as "probes" or sensors, monitoring the foreign literature for trends, discoveries, and advances that might provide some useful clue or insight into what areas the research group should focus on. Though it is not considered a strategy, this dabbling may well prove to be a very successful strategy to maximize opportunity and conserve resources until it becomes clear exactly where the resources should be directed.

In many diversified companies, the biotechnology research budgets peaked about 1988, after five to seven years of incremental increases, but the research and development managers believe that they will be able to maintain the present level of research funding for biopharmaceuticals for at least three more years without major breakthroughs. Because the diversified companies have established mainline businesses generating cash for biotechnology research, and because the companies are not really beholden to shareholders for quarterly earnings, Japanese companies can easily afford to adopt an unfocused, opportunistic approach.

Thus, until the Japanese companies can identify a suitable commercial target or until it becomes clear how biotechnology can best be applied for commercial gain, the "dabbling" orientation will undoubtedly continue.

In many respects, diversified Japanese companies offer an ideal environment for the development of biopharmaceuticals: plenty of cash, freedom to pursue individual research projects, and a powerful technological infrastructure to complement biotechnology. However, at present it does not appear that the companies are exploiting their inherent advantage, particularly the existing technological infrastructure.

On the whole, the diversified companies have established entrepreneurial, reasonably "free" research environments that tend to attract a unique blend of young, capable, enthusiastic scientists, some of whom have already studied in the United States and others who will be sent here for postdoctoral training by their companies. The research environment is designed to foster creativity, but much the same as in the cash-rich U.S. biotechnology companies five years ago, there appears to be a need for more focus and tighter management to ensure some level of success in the next generation of biotechnology products. Compounding the problem is the fact that many young scientists have had no prior experience in industry, and while they revel in their "academic freedom" and conduct some early-stage applied research, it is not clear that the companies are moving any closer to identifying viable product ideas. In the short term, it will be critically important for these companies to find a viable or promising biotechnology project to embrace before senior management decides to cut back on funding.

As a group, the diversified companies are more research-oriented and less risk-averse than the established pharmaceutical companies. For example, many diversified companies permit "underground" projects—individual projects of personal interest that do not have official status but that are allowed to consume as much as 20% of the scientist's time. An underground project provides an opportunity for scientists to conduct research of a purely academic nature. Because the project does not have official status, the results are not subject to the critical eye of the research and development management. Thus, if the scientist fails in this creative pursuit, there is no loss of face. Alternatively, if the project shows some promise, the scientist can choose to bring it to the attention of management and thereby gain recognition for his work. (Krestin, an anti-cancer drug discovered by Kureha Chemical Company started out as an underground project.) Without the cash flow from the mainline business, it is unlikely that such underground projects or other official high risk research would flourish.

Biotechnology is still largely expected to be used for making *new* genetically engineered protein drugs and not as a set of tools with which to conduct research. Because the companies do not generally consider biotechnology a complement to in-house capabilities (i.e., organic chemistry, and computer modeling, etc.) a substantial number of biotechnology research groups have been established as separate technological entities, not at all integrated with the rest of the company unless a scientist initiates an intracompany collaboration. Unfortunately, with this orientation, it appears that much of biotechnology's strength as a set of complementary research and development tools is being lost, unaided by the massive amounts of technology many of the diversified companies could bring to bear under better adapted, more integrative research and development organizations.

Many Japanese companies appear somewhat stymied by the apparent dearth of appropriate commercial targets in the biopharmaceutical field. From the Japanese prospective, markets are not yet well defined, there are few commercial successes to emulate, few opportunities for short-term development, and a relative paucity of ideas in Japanese academia. Still, however, management is resolute to remain in biotechnology, because if successful, the company will be infused with new life and vitality. Though a great deal of technology licensing went on in the early to mid-1980s in anticipation of a major biopharmaceutical boom, funding has leveled off and, at best, now keeps pace with overall funding increases for research and development. One major pharmaceutical company cut back on biotechnology research to the point where their P3 laboratory is now being used as a storage room!

The established pharmaceutical companies do not feel particularly threatened at present by the inevitable entry of the diversified companies into the pharmaceutical market via biotechnology. Because biotechnology products (e.g., protein therapeutics) may make up only 10% or less of the total therapeutic products of the future, the perceived competitive impact of the diversified companies is small. Second, many of the diversified companies have no means to reach the market other than by licensing their products to the established companies.

The established pharmaceutical companies are not sure how best to apply biotechnology and do not seem totally convinced that the benefits of therapeutic proteins derived from biotechnology do not yet justify the investment. Some of the largest pharmaceutical companies have excellent in-house talent apparently underutilized as a consequence of management's conservatism. As was mentioned earlier, there is a good deal of anticipation regarding protein engineering, in part because it is viewed as a refinement tool to improve existing molecules.

Though the pharmaceutical companies have evolved into more research-intensive organizations over the recent past, spending on average almost 10% of total sales on research and development, for a variety of reasons the companies do not appear to be especially research-intensive with respect to biotechnology. First, the companies are afraid to make a commitment to a single therapeutic protein for fear that it will fall short of expectations. Conventional pharmaceuticals, many identified through screening, have proven effective, and there is a lack of conviction at present about the cost-effectiveness and superiority of biotechnology-derived protein drugs. However, the efficacy and approval of erythropoietin (EPO) and the imminent therapeutic success of the colony-stimulating factors seems to have rekindled a languishing interest in biotechnology therapeutics. However, most of the major companies have had collaborations with U.S. biotechnology companies and most have been disappointed by the outcome of the projects.

Second, most companies are still very much development-oriented, especially where biotechnology is concerned—they are accustomed to screening for active compounds, not elucidating the science and established pharmaceutical companies have not yet exhibited a strong interest or commitment to funding early stage, high risk research. Within Japanese companies scientific rewards are most often based on the discovery of a new compound, not on the explanation of phenomena.

Third, because there is no other mainline business but pharmaceuticals to fund research activities and sales growth has declined because of government cuts in drug prices, established pharmaceutical companies seem more eager to channel their resources into traditional (i.e., proven) methods of drug development than into biotechnology. There is a compelling desire to launch as many slightly improved drugs as possible, since these drugs will receive higher reimbursement prices and additional value will be derived from the previous research and development.

Fourth, there are many therapeutic proteins now under international development and several different biotechnological approaches to treating the same diseases. There is no straightforward way to determine which approach might be the best except through research and development. Instead of taking a gamble on one or another approach, many of the established companies are hesitating to take any approach at all and are continuing to wait for sufficient data to accumulate elsewhere before risking an investment.

Fifth, an increasing number of established companies are beginning to look for ways to integrate biotechnology with the rest of the research and development organization (i.e., to use biotechnology as a tool to develop conventional drugs, in much the same way as western pharmaceutical companies). Similar to the diversified companies, many biotechnology research teams operate somewhat independently from the rest of the company. What seems to have escaped the general thinking of many companies is the fact that even if less than 10% of the future pharmaceutical products are biotechnology products, most if not all therapeutic products of the future will have been developed with the aid of biotechnology tools.

Sixth, with respect to licensing activities, the leading pharmaceutical companies tend to believe they have developed skills that are on a par with those of U.S. companies. Therefore, they are showing less interest in U.S. biotechnology projects.

Leading pharmaceutical companies are very conscious of their options for more intense involvement in biotechnology (e.g., collaborations, equity investments or acquisitions of U.S. biotechnology companies, establishment of U.S. research institutes), but timing is critical, and there seems to be very little sense of urgency, generally, to seize one or more of these options. From the Japanese point of view, it is a buyers' market when it comes to U.S. technology, and a wait-and-see policy, though slowly changing, still prevails. An executive in one of Japan's leading pharmaceutical companies confided to hosting two to three U.S. biotechnology companies per week last year, all of whom were seeking licensees. No licenses were signed, highlighting the general cautiousness with which many companies approach biotechnology opportunities.

Japanese companies are acutely aware that the twenty-first century is rapidly approaching. To become global powers they realize the need to bolster internationalization and increase research and development. As a group, the pharmaceutical companies agree that biotechnology will be important to global success, but is unclear when and how to exploit the opportunity maximally at minimal risk and cost.

Chugai and Yamanouchi, as well as other companies, have established solid biotechnology positions with U.S. companies through acquisition, equity, and licensing agreements (e.g., the Yamanouchi–Genetics Institute joint venture on bone growth factors and Chugai's acquisition of Gen-Probe, a diagnostic company using DNA probe technology). Though it remains to be seen which options other companies will choose, one thing is certain: international sources of technology, but more importantly, product ideas, will become a priority in the future because without the ideas, the Japanese development engines will remain idle.

CONCLUSION

Japan enjoys some cultural (i.e., demand factors) as well as resource advantages (microbial stocks) that will play a crucial role in the industrialization of biotechnology. MITI has made a commitment to build a sound infrastructure that will ensure the industrial development of biotechnology alternatives and the evolution of a competitive bioindustry. But in spite of these advantages, Japanese companies are at present still challenged by their own risk aversity, a dearth of viable product ideas, and a lack of foreign successes to emulate.

Knowing the challenges and constraints that face Japanese companies in the short run and the advantages that will assist them in the long run is a benefit to both U.S. companies and policy makers in formulating strategies to ensure future competitiveness of the United States in biotechnology.

Rather than focusing on how either the United States or Japan might win the biotechnology race, it would seem to be more prudent for policy makers and company executives to consider options that over the long run, take into account the strengths and weaknesses of both sides so that mutual benefits can be derived on a global basis. Winners will be global players and will succeed on the basis of sound strategies that maximize complementary skills and technology.

Widely recognized is the fact that Japanese companies can take a longer term view of biotechnology opportunities than can U.S. biotechnology firms, primarily because of "deep pockets," cost of capital, and patient shareholders. This is especially true of the diversified companies whose future depends on the application of new technologies to expand into new business areas.

But all too often, U.S. companies approach Japanese companies with little knowledge of the prospective partner's inherent strengths and strategic objectives. Without such knowledge it is difficult to present a rational plan that fits with and takes advantage of the partner's particular strengths or interests.

Aside from capital and access to the Japanese market, some partners might have, for example, an excellent group of molecular biologists, some trained in the United States, that will complement an early stage project. Other partners might already have highly automated, sophisticated receptor-based screening technologies which can prove to be invaluable in some projects. Many companies have rich libraries of microorganisms that might provide new drug, bioprocessing, or environmental management opportunities for U.S. firms. Additionally, many of the Japanese companies have highly skilled organic, synthetic, and computational chemists who can provide a great deal of support in the development of follow-on products.

The key point is that most Japanese companies, at least at present, lack the entrepreneurial environment to generate novel, commercially viable biotechnology ideas and look instead to U.S. companies to rectify this deficiency. However, what they lack on the idea side they more than make up for in the powerful development machinery they bring to bear in generating marketable, competitive products. These strengths must be assessed seriously and combined with U.S. strengths to create leading companies of the future. The winners of tomorrow are not going to get there alone. They will get there by careful international sourcing of technology and skills.

REFERENCES

Eleanor, W. D. 1988. R&D subsidiaries in Japan: Managing internal and external linkages in the multinational corporation. Cambridge, Mass.: MIT. (Publication number MITJSTP 88-08.)

Inoue, Yuko 1990. Japan's scientific trailblazers still toil in underfunded ill-equipped laboratories. *Jpn. Economic J.* March 17:5.

Save the Earth. *Asahi Evening News.* August 7, 1990.

Sun, M. 1989. Japan faces big task in improving basic science. *Science,* 243:1285–1287.

CHAPTER 18

EUROPHORIA: BIOPHARMACEUTICALS IN THE EMERGING EUROPEAN MARKET

Norbert Rau

Regional and global economic structures seem to be changing faster nowadays than ever before. Changes mean new challenges and new opportunities in business. Windows open wider and more frequently than in calmer times. The making of a single European market is in progress and will be bolstered by an interesting additional component by the merged German economies and the political reunification of the two countries.

Beyond the single European market as it will be formed by the EC (European Community) member states, a larger economic Europe is about to emerge with non-EC states Switzerland, Austria, Norway, and Sweden strengthening their ties to the EC and eastern nations pushing to get better access to this market.

When talking about biopharmaceutical business opportunities in a "European powerhouse," consisting of a single European market (the EC member states, including a reunified Germany) plus non-EC European countries such as Norway, Sweden, Switzerland, Austria, and some Eastern economies, it is, of course, not possible to depict exactly the nature of the opportunities and the best strategies to exploit them. Too many factors have a role that will change in the near future, although we do not know to what extent and during which period of time. There is an indisputable need for corporate strategies with respect to the developments in Europe, but we have experienced that the companies' answers to this need are far from being crystal clear. This is true for European and overseas corporations alike.

In this chapter I shall try to draw a picture of what is likely to happen to the future European biopharmaceuticals industry. Much of it is derived from my consulting experience in the biotechnology industry. And most of it is written for readers who are not very familiar with Europe.

THE SINGLE EUROPEAN MARKET IN 1993 AND BEYOND

In my eleven years of management consulting, a lot has changed within the European biotechnology environment. In the early 1980s a frequently heard word was "Eurosclerosis," which meant that Europe was over-regulated, unproductive, lagging behind in new technologies, and suffering from a brain drain—in short, a tired, lethargic old continent with easy-living, risk-averse inhabitants. "Europhobia" was a mental condition possessed by non-European businessmen and investors who had to do business in Europe, knowing of Eurosclerosis.

Later, Europhobia changed into Europhoria after the condition was treated with the Europe-1993 drug, a slow-release stimulant with few side effects such as Eurofever and, in severe cases, Euroflu. Suddenly, once-tired Europe is looking fresh and alive again, with European executives seemingly losing their interests in overseas opportunities in favor of intra-European issues. Some are even going "Eurocrazy"! And there is enough reason to be excited, because with Eastern Europe opening up, the continent's $6 trillion (U.S.) economy will be bigger than that of the United States or Japan.

The Single European Market

One of the aims of Europe in 1992 is to remove physical barriers such as custom posts, frontier controls, and restriction on the importation of certain products. Even more important is the removal of technical barriers such as different standards and regulations to which businesses have to comply in order to trade—e.g., regulations pertaining to drug formulations or food ingredients.

An example may explain the importance of this issue. In 1987, the intra-Community trade with chemicals and pharmaceuticals was ECU[1] 44.9 billion, while exports to non-Community countries amounted to ECU 32.0 billion. As a comparison, imports from non-Community countries amounted to ECU 19.7 billion. Thus, the majority of the trade with this commodity was done within the 12 EC countries, requiring them to

- Observe different labeling and packaging regulations,
- Understand approval procedures (e.g., for drugs),

[1]European Currency Units (1 ECU = $1.15 U.S. in 1987).

- Review toxicity and safety standards,
- Comprehend customs controls and paperwork at each border,
- Learn statistics collection procedures, and
- Understand indirect taxes, etc.

This required, and partly still requires, an enormous amount of paperwork, red tape, time, and money. Needless to say, removing these barriers will result in a larger, more uniform market and thus in scale effects resulting in lower costs and higher margins.

The issue of fiscal barriers is difficult. Indirect taxes vary enormously across Europe, which makes up for huge differences in the prices of certain goods. The result of harmonizing these taxes will be a dramatic change in the world of European retailing. How important it is to harmonize taxes may be shown for pharmaceutical products for which the current Value Added Taxes differ grossly from one country to the other (Table 18-1).

TABLE 18-1 LISTING OF DIFFERING RATES OF VALUE ADDED TAX (VAT) FOR DRUGS IN EC MEMBER STATES

Country	VAT for Drugs	Remark
Belgium	6%	Reduced; regular VAT is 19%
Denmark	22%	Regular VAT applies
F.R. Germany	14%	Regular VAT applies
France	5.5%	Reduced; regular VAT is 18.6%
Greece	6%	VAT system very difficult with differing rates
Great Britain	0%	Health service is entirely government determined
Ireland	0% for oral medication, 10% for non-oral medication	Both reduced; regular VAT is 23%
Italy	9%	Reduced; regular VAT is 18%
Luxembourg	3%	Reduced; regular VAT is 12%
Netherlands	6%	Reduced; regular VAT is 19%; regular VAT applies also for self-medication
Portugal	—	No VAT introduced yet
Spain	6%	Regular VAT applies

Source: RauCon [1990], based on information of Bundesverband der Pharmazeutischen Industrie, 1989.

Direct taxes will hardly be harmonized. Corporate, income, and other indirect taxes vary within the EC, and it can be foreseen that these will be used by the nations' governments as leverage in competing against each other for corporate investors. As a matter of fact, some countries, such as France, Great Britain, Germany, and Belgium, have already lowered corporate taxes or have introduced special tax schemes for distribution centers or European corporate headquarters in order to attract foreign, particularly non-European, companies.

In short, what hundreds of EC directives are aiming at, is to create

- A common market of 325 million people that shifts the balance of world trade. A united Europe will be the biggest trading bloc in the world, whereas at present the largest European market for industrial goods, West Germany, is not even half the size of the Japanese market.
- A trade zone within which goods, services, capital, and labor can move freely.
- A market without divergent norms, tariff, and regulatory restrictions.

Research undertaken by the EC suggests that the completion of the single European market will increase the gross domestic product by 5 to 7% in the medium term. This will be translated into more business and expansion of companies who take up the challenge and compete on a Europe-wide basis rather than those who remain in their local environment. Of course, there are also many doubts about whether these aims can be achieved within the time frame set:

- Can Europe's markets, split by varying cultural, social, economic, and political objectives for centuries, really be aligned?
- Can its patchwork of tariff, certification, and procurement policies be harmonized?
- Will trade barriers and conflicting standards simply cease to exist?

And, for the health care sector:

- To what extent will it be possible to harmonize grossly differing national health care systems?
- Will every person have access to the same quality of treatment and medication in any country of Europe?

The single European market issue affects health care biotechnology directly and indirectly—directly, as far as differences in regulatory situations, technology flow, and availability of qualified biotechnologists are concerned; and indirectly, as it regards factors that have an impact on the pharmaceutical industry.

Issues immediately affecting the European biotechnology health care industry are

- The EC Council directive on the contained use of genetically engineered microorganisms.
- EC regulations concerning the transport of genetically engineered organisms or of products containing recombinant DNA organisms
- EC programs for the support of research and development cooperation between companies of European countries.

Issues affecting the health care industry in general are

- Common registration for pharmaceuticals. Two alternatives (or a mixture of the two) are possible, the registration with a European Drug Administration or total reciprocity of approvals within Europe. Much effort is currently being directed at producing the legislation necessary to standardize pharmaceutical products—to ensure that quality, safety, and conditions of sale are sufficiently harmonized for a free market. It seems, however, that this process will extend well beyond 1992.
- Standardization of Good Laboratory Practices and Good Manufacturing Practices. In 1992 the application of approved GLP/GMP will probably be obligatory.
- Patent issues: closer harmonization of national rules on patent law for better patent protection. Among EC politicians there is a growing awareness of a need to extend patent protection to a period that would grant the same protection that companies in the United States and Japan now enjoy.
- Pricing and health cost reimbursement regulation—it would probably be too optimistic to expect this any time soon.
- Labeling and packaging of products and the information provided about them.

Issues affecting the entire industry are

- Harmonization of administrative requirements pertaining to all matters of logistics and transport.
- Harmonization of information flow (standards in data communication).

A number of important harmonization goals will probably affect the health care industry only marginally, such as standardized public procurement procedures or harmonization of banking and financial services.

An Even Greater European Market to Come

Although some see the single European market as an ultimate goal, most regard it only as a stopover on the way to a Greater Europe. Other European countries will strengthen their ties with the European Economic Community in order to benefit from the trade with this market.

EC + EFTA. First to benefit from this trade are the members of the European Free Trade Association (EFTA)—Austria, Finland, Iceland, Norway, Sweden, and Switzerland—as well as Liechtenstein. Added up, the EFTA member countries have a population of 32 million. Austria is currently negotiating conditions to become a member of the European Commission.

Today, the EFTA member states are the EC's most important trade partner. Two thirds of all EFTA exports go to EC countries and Switzerland (after the United States) is the second most important market for EC products and the third most important EC supplier. Already in 1972 Switzerland had signed a Free Trade Agreement with the EC that has gradually eliminated customs duties on industrial goods involved in mutual trade. Switzerland's proposal to include the other EFTA countries in the agreement as well were turned down by the EC. Later, in 1984, EC and EFTA created the concept of the European Economic Space and in 1989 decided to commence formal negotiations on this subject in 1990. The formal negotiations are aimed at the simultaneous conclusion of a multilateral and comprehensive treaty between the EC and the individual EFTA member states. In effect, the single European market will be extended to EFTA territory on January 1, 1993—namely, the free movement of goods, services, capital and persons. A number of flanking policies such as competition and corporate and environmental laws will be included, but no policies on customs union, taxes, or agricultural policy.

The reason I am stressing the proposed EFTA agreement is the fact that these countries are important trade partners and that some EFTA member states are important biopharmaceuticals manufacturing countries, namely Switzerland, Austria, Sweden, and, to a lesser extent, Finland. Hoffmann–LaRoche, through numerous acquisitions, equity stakes, partnerings, and agreements is one of the most important genetic engineering giants in the world, Sandoz is polishing up its rDNA projects, Vienna has become a center of human health care and molecular biological research, with companies such as Boehringer Ingelheim and Genentech deeply involved, while Pharmacia and Kabi represent two of the most important Swedish groups with biotechnology businesses. Closer ties to the EC will open up important markets to them.

EC + EFTA + Eastern European countries. But the real long-term boost for Europe is expected to come from the changes of the political and economic systems in the Eastern European countries. Eastern European countries are finally rejecting central planning and state ownership in favor of free markets and private property. Some suspect that Czechoslovakia, Hungary, Poland, and other Eastern European countries could enjoy economic miracle growth similar to that of West Germany after World War II.

Most of these countries have significant research programs in biotechnology but currently lack the ability to bring their inventions to the market. There are many biotechnology opportunities in industrial modernization of the chemical and health care industries and in environmental management.

One of my longest stays ever in a traffic jam lasted for one hour twenty minutes. This was on the Autobahn near Leipzig, East Germany, just before the Economic and Monetary Union went into effect on July 1, 1990. Looking around, I noticed that eight to nine of ten cars were from West Germany, France, Switzerland, and other European countries. A similar thing could have happened to you if you wanted to book a hotel room, tried to reserve a table in a good restaurant, or tried to place a cross-border telephone call. Businessmen from all over the world were flooding the country during this period and exploring opportunities.

This did not apply only to what formerly was called East Germany, it was much the same in Hungary, Czechoslovakia, and Poland. These countries together have a population of over 80 million. Although a great deal of the early money being spent by Western companies goes into building the basic infrastructures necessary to operate, companies see tremendous opportunities. According to the United Nations Economic Commission for Europe, joint venture registrations soared throughout Europe in the second half of 1989. The figure topped 3300 by year-end, including 1500 Soviet joint ventures (of which only 50 were with U.S. companies), versus a meager 165 joint ventures in early 1988.

Of all former Eastern Bloc countries, the Eastern part of Germany has a head start in the recovery of its economy. Prior to the reunification, the East Germans have enjoyed the greatest prosperity in Eastern Europe, with a per capita income of 75% of the EC average, as compared with 70% in Czechoslovakia and 40% in Poland. Investors worldwide are enthusiastic at the prospects that Germany's reunification will add to Europe's growth spree. Some 100,000 new firms were expected to be founded in East Germany in 1990, and 53% of all West German firms planned some activity in East Germany. However, much investment will still have to go into basic infrastructures supporting a functional economy: the cost of modernizing East German roads and railways alone will run to $40 billion (U.S.) over the next 20 years: the cost of cleaning up the contaminated environment to $250 billion.

The East German Parliament ("Volkskammer") decided to join West Germany politically on October 3, 1990. Just one year before, on the occasion of the 40-year anniversary of the "German Democratic Republic," exleader Erich Honnecker proclaimed that the Berlin Wall would persist for the next one hundred years. With political and economic systems and conditions changing radically almost monthly it is extremely difficult to foresee economic developments on a reasonably sound basis. It is, however, safe to say that after a transition period a system will be in effect that is identical to the present West German political and economic system and the German pharmaceuticals market will have increased by some 26% within about one year!

Czechoslovakia will probably be the next country to recover economically. What is not widely known is the fact that unlike its neighbors, Czechoslovakia had a buoyant economy and a strong industrial base inbetween the two World Wars. In Czechoslovakia, there is still a lot of entrepreneurial ability that the other countries are lacking. A drawback may be the unhealthy dependance on heavy industries and the constitution of the country, which does not allow private industry. Changing this constitution will not be simple and will take a long time.

Nevertheless, analysts believe that traditionally strong areas such as chemicals, textiles, shoes, glassware and porcelain, bicycles, cars, and aircraft could appeal to foreign investors. Curiously, the first Czechoslovakian joint venture to be set up in 1987 with a foreign firm, was a biotechnology enterprise, Tessek. Its partner is Senetek of Aarhus (Denmark). Tessek is an internationally known leader in chromatography equipment applied in chemistry and medicine, in the pharmaceutical and food industries, and in environmental testing.

Hungary pioneered Eastern Europe's first attempts at reforms. Joint ventures were legalized in 1972, management and pricing decisions were decentralized in 1984, and majority foreign ownership came into force in 1988. However, authorities still have discretionary power to interpret the new laws, which makes putting up a joint venture difficult. Also, harsh foreign exchange rules hamper international business to a great extent.

Research prospects are good in Hungary. The government spends about 2.5% of the gross national product on research, and Hungary accounts for almost half of Eastern Europe's filings with the European Patent Office. A spin-off of these research efforts was Vepex Biotechnika. It was launched in 1984 by the Biological Research Center of the Hungarian Academy of Sciences. With $5 million (U.S.) in equity capital, Vepex is doing basic biotechnological research and acts at the same time as a technology transfer and licensing agent. It has developed a more efficient champagne fermentation method and is developing heat-tolerant enzymes for use in detergents.

Many Differences Will Remain

At first sight, it looks as if the single European market will lose its heterogeneity, but many factors that contribute to this heterogeneity will remain the same after 1992. Take the cultural diversity, including management styles. It makes quite a difference whether you try to enter into a strategic alliance in Denmark or in Italy. And even within Italy there is enough cultural diversity between the north and the south to cause some headaches. André Laurent [1986] obtained the following figures when asking executives whether they believe it is important to be able to give precise answers to questions asked by their subordinates. Answering with yes were executives in

United States	16%
Sweden	13%
The Netherlands	18%
Denmark	27%
Great Britain	30%
Federal Republic of Germany	40%
Switzerland	40%
Belgium	49%
France	59%
Japan	77%

Considering the results of these interviews, one can easily imagine that the cultural differences of management within Europe are enormous.

Other differences relate directly to marketing and advertising. In order to understand the essential differences you need only observe the kind of messages that advertisements in British, French, or German magazines imply. A British reader may find a French-type advertisement rather indecent, while a German will probably not quite understand the typical English humor behind an advertisement in a British journal. Also, if you plan to base an ad campaign in daily newspapers you should first find out how important newspapers are in the countries under consideration. According to Axel Springer Verlag, the average daily newspaper circulation is 269 per 1000 inhabitants in the United States and 575 in Japan, it is 524 in Sweden, 421 in Great Britain, 408 in Germany, but only 191 in France and 82 in Italy.

Then, of course, there is no common language. However, roughly half

TABLE 18-2 FOREIGN LANGUAGES SPOKEN BY EUROPEAN BUSINESS PEOPLE

	All	United Kingdom	France	Germany	Italy	Spain	Nether-lands	Denmark
English	53%	100%	28%	32%	23%	21%	78%	36%
French	34%	8%	100%	15%	21%	19%	22%	3%
German	29%	4%	6%	100%	4%	3%	41%	13%
Italian	16%	1%	3%	4%	100%	3%	2%	0%
Spanish	12%	2%	3%	3%	2%	100%	3%	1%
Dutch	7%	1%	1%	2%	0%	0%	100%	0%
Danish	8%	0%	0%	2%	0%	0%	1%	100%
Any foreign	46%	12%	39%	50%	35%	43%	100%	40%
No. of languages per speaker	1–3	1–3	1	1 or 2	1–3	1 or 2	1–5	1–3

Source: Ogilvy and Mather [1989]. (Personal communication to RauCon GmbH.)

of all European business people speak one or more foreign languages, except for British business people, of whom only 12% speak a foreign language, and the Dutch, who are polylingual by tradition (Table 18-2). In the light of the common language it seems understandable that for American investors the country of preference is the United Kingdom, where they made 80% (by value) of their EC acquisitions during the 12 months from October 1988 through September 1989. When it comes to doing business in a multilingual environment, the British or American managers of these acquisitions have a clear disadvantage as compared with their colleagues in other European countries.

There is no European currency yet, although the ECU may replace national currencies in some future day. There are differing legal and fiscal operating environments and different industrial relation patterns. Consumer tastes and attitudes vary enormously across Europe, for reasons related to history, culture, physical characteristics, and political structures, as does public perception of high technology and genetic engineering.

SOME STRATEGIC CONSIDERATIONS FOR BIOBUSINESS IN EUROPE

For biopharmaceutical companies almost everything from research and development to marketing will be affected by the provisions of the single European market. Take research and development, for instance. There is evidence that European research institutions will coordinate (leading to less duplication) research efforts in order to counterbalance the United States and Japan's dominance in research. For a biotechnology company, therefore, it will be necessary to widen input from national to European research institutions. The same applies for graduate recruitment schemes that now must include all European institutions, because people will be able to move freely and work wherever they want, without the need of a special permit.

A special aspect would be research and development in sensitive areas. For example, it is not allowed to do embryo research in Germany, whereas it is permitted in Great Britain. Similar differences in national legislation may exist in other areas, and companies will now be more flexible in moving laboratories and scientists, especially as university degrees will be mutually recognized. In this respect it is important to note that mutual recognition of medical qualifications has been in effect since 1975, allowing doctors to practice in other member states.

Product variety and product line length must be also reconsidered in the light of new competition. Manufacturing costs may decrease because of scale effects and cheaper transportation. Europe 1992 may open up better and cheaper opportunities for raw material, intermediates, labor, and know-how. Inventory holding costs may decrease if distribution policy takes advantage of the new opportunities.

Finally, marketing will probably experience the greatest impact because of changes in market structure, distribution, pricing, and marketing organization. Media deregulation will lead to larger media consumption and therefore advertising growth rates, which should buy more readers for your advertisement dollars, pounds, francs, or marks.

Now I would like to stress some viewpoints that are of particular importance for the biopharmaceuticals industry.

The Difficult Choice of Corporate Location

All European countries, regions, and major cities have one thing in common: they are located exactly in the center of Europe, if you believe their advertising. Their ads will always exhibit a (more or less distorted) map of Europe with concentric rings around the respective location or a starlike arrangement of distance lines connecting to major places and suggesting corporate heaven: corporate and individual taxations are most attractive (by the national bureau of internal revenue's standards), delightful tax incentives exist, union activity is virtually nil (although there is a public transportation strike every time you visit the place), the natural environment is plentiful, and the quality of life is just superb.

In the past three years I have conducted various location criteria surveys and site location studies in Europe and North America for a number of European authorities and industrial groups. In the course of these surveys I have interviewed managers at more than 500 biotechnology companies. For them, the seven most important location criteria for biotechnology research and development and production facilities were

- Availability of qualified personnel (skills, and work ethics),
- Communication and infrastructure (telecommunication, roads, and airports, etc.),
- Market proximity,
- Local exchange of experience, cooperation, and synergies,
- Local financial climate (including taxation and subsidies),
- Presence of biotechnology and other research institutes, and
- Business-oriented community and local government.

All other factors, such as raw materials, utility costs, labor costs, and regulatory issues, ranked far behind, although I would not neglect the latter.

Qualified personnel are available where pertinent research is done, i.e., where universities and life sciences research institutes are found. Generally, the nations known for high productivity also have highly educated labor forces. European universities are nowadays moving toward more European approaches, offering multilingual courses, language training, expanding the notion of equivalent degrees, and cooperating with other universities. An example of the latter was the signing of a convention in December 1989 of the Universities of Strasbourg and Mulhouse (France), Basel (Switzerland), and Freiburg and Karlsruhe (Germany), which included, for example, a single curriculum for the study of biotechnology, mutual acceptance of degrees, and free right to transfer from campus to campus. Needless to say, this trend toward a more uniform academic educational system makes it a lot easier for companies to recruit qualified scientists internationally.

Remunerations for scientists and executives vary enormously within Europe. P-E Inbucon Ltd. found that the average total remuneration for a head of finance of a small manufacturing company, aged 45, married, and with two children ranged from £34,910 (Brit.) in Ireland to £41,656 in Great Britain to £46,515 in France to £48,274 in West Germany to £57,358 in Italy! You should, however, consider the differing income taxes and different costs of living before you choose the country where you want to become an executive.

There are also considerable differences in the indemnifications that a company has to pay to an employee in the case of his dismissal or of the closure of the entire facility. The latter can become extremely expensive in Belgium, where it can reach ECU 187,000 (roughly $400,000 U.S.) for a single employee who has served the company for 20 years.

Or take into account that you have to buy (or lease) a company car fleet for your executives, marketing people, and representatives. According to the European Commission the following differences exist:

Country	*Index*
Luxembourg	100
West Germany	105
Belgium	109
France	124
Great Britain	129
Netherlands	134
Ireland	165
Denmark	207

These differences are attributable in only a minor way to different value added tax rates, i.e., the differences are in basic prices and other, nonharmonizable taxes, and will persist after 1992. For example, if you have to buy twenty company cars that would cost ECU 15,000 each in Germany, the decision to locate in the Netherlands would cost you some ECU 54,000 or more than $100,000 (U.S.).

As regards telecommunication, Europeans look forward to being totally interlinked by ISDN (Integrated Systems Digital Network) in the first half of the 1990s, with the exception of, perhaps, the most remote and unpopulated areas. Most of the bigger cities and urban areas are already connected by fiber-optics networks that serve ISDN. (ISDN is a standard that defines the implementation of communications, voice, data, and video over a single cable.) Once this service becomes available everywhere in Europe, the issue of locating a firm where the best telecommunication facilities, teleports, and so forth are, will become obsolete.

A company or a branch (e.g., marketing) that depends on telecommunication should also compare the different costs of telecommunication. These can differ widely among the European countries. A transatlantic telephone call from Rome, for instance, costs about 50% more than a call from Belgium. Likewise, a telex from Italy costs about 70% more than one from Germany.

The same applies for energy costs. National gas and electricity authorities across the European Community are jealously guarded fiefdoms, which is one of the reasons for grossly varying energy prices. Any company whose production costs are significantly determined by energy prices should seriously consider the data in Table 18-3.

Corporate tax is still a major issue based on location. The rates of corporate tax and withholding taxes currently in existence in the EC are

TABLE 18-3 INDUSTRIAL ENERGY PRICES (ELECTRICITY, GAS)

Country	ECUs per Gigajoule of Gas	ECUs per Kilowatt-Hour of Electricity
West Germany	3.95	0.07
Belgium	3.19	0.05
France	2.85	0.04
Spain	2.71	0.06
Netherlands	2.69	0.05
Italy	2.49	0.05

Source: RauCon [1990], based on EC data.

shown in Table 18-4. The effective rate payable may, however, be substantially lower than the statutory rate if depreciation rates are taken into account. In addition, tax holidays, accelerated depreciation, and tax exemptions in certain cases vary from country to country, making it necessary for an investor to compare the taxation systems meticulously.

It is not always wise to choose a location where the lowest tax rate is paid. Taxes are used, at least in part, to finance infrastructure projects, research and development, environmental measures, and public services, all of which are important for high technology industries. If the taxes are low, the tax revenues for the government are low, too. This can only mean low expenditures for important things like infrastructure and research or, alternatively, higher public debt, which will ultimately result in tax increases.

These are only some examples that show how difficult it is and even will be after 1992 to choose the proper location for a company in Europe. In the following, I shall address some issues that are more biotechnology-specific.

Why Global Technology Input Is a Must for European-based Companies

Europe has seven of the world's top ten chemical companies and six of the top ten drug makers, but all were slow off the mark in biotechnology despite excellent basic research. European companies have been smart enough, however, to pursue a global technology input strategy, and this is typical not only for biotechnology. In high technology in general, an increasing statelessness of large industrial groups can be observed, which is particularly evident in Europe. (There is a difference between "multina-

TABLE 18-4 CORPORATE TAXATION IN EUROPE

Country	Corporate Tax	Capital Gains Tax	Withholding Tax on Dividends	Withholding Tax on Interest
Belgium	43%	21.5%	25%	25%
Denmark	50%	50%	30%	0%
France	39% (undistributed)	15–42% 42% (distributed)	25%	45%
Germany	50% (undistributed) 36% (distributed)	50% 36% (distributed)	25% 25%	0% 0%
Greece	46%	30%	42–50%	46%
Ireland	43%	36–60%	0%	35%
Italy	36% 16.2% (local tax)	36%	32.4% 16.2% (local tax)	12.5, 15, and 30%
Luxembourg	34%	34%	15%	0%
Netherlands	35%	35%	25%	0%
Portugal	36.5%	36.5%	20% or 25%	20% or 25%
Spain	35%	35%	20%	20%
United Kingdom	35%	35%	0%	25%

Source: Bernard Jeffcote (1989). (Courtesy of Ernst & Young, London [1989].)

tionality" of a corporation, which always includes a home nationality, and "statelessness," which does not). Whereas American concerns seem to be late in catching up with this trend, European world corporations, worried by the emergence of regional trading blocs, move factories and laboratories around the world.

The reason for this movement is quite obvious if one looks at the sales of these corporations outside their respective home countries. Whereas American chemical groups do only roughly half their sales abroad (DuPont, 44%; Merck, 47%; Johnson & Johnson, 50%; Dow Chemical, 54%; in 1988–1989), Europeans earn two-thirds or more of their turnover in foreign countries (BASF, 65%; Bayer, 65.4%; Hoechst, 76%; Degussa, 78%; Boehringer Ingelheim, 78%; ICI (UK), 78%; and SmithKline Beecham (UK), 89%). Swiss-based Hoffmann–LaRoche and Sandoz do 96% of their sales outside Switzerland, which is not really amazing considering the size of their home market. It is not surprising, therefore, that these corporations rely on a free flow of goods, capital, manpower and technology in each direction. Except for a few latecomers, world corporations scour the globe for leading scientific ideas in biotechnology and every other field.

Many of the research findings that have led to today's molecular biology–derived products, originated from Europe. However, the United States has provided the ideal climate for the commercialization of these findings. As there is a general tendency of international concerns to make technology and research and development–oriented strategic investments in places of the world where an input from technology leaders can be expected, it seems obvious that investments in American biotechnology companies are dictated by corporate strategic thinking rather than by anything else.

Patenting—Or What Has EPO in Common with Biotechnology?

For any company involved in the biopharmaceuticals business in Europe, the patent situation is a major issue. And investors from the United States should carefully watch the developments in this respect. There are enough differences between American and European patent rights to cause biotechnology inventors and investors some headaches. This has become apparent for the first time when Harvard researchers, who had patented a transgenic mouse with an inclination to develop cancer, tried to do the same in Europe. The Munich-based European Patent Office, whose acronym "EPO" seems to be about the only thing it has in common with biotechnology, said no. They argued that the technology in creating the mouse could be patented, but not the mouse itself or its offspring, which for EPO is an animal breed and thus not patentable.

This requires some explanation. The current patent laws of most member states were adopted in the late 1970s as a result of the 1973 European Patent Convention, i.e., their origins date from a time when the extent of biotechnology inventions could barely be imagined. The European Patent Convention excludes from patent protection "plant or animal varieties or essentially biological processes for the production of plants and animals," not applying to microbiological processes or products. (The phrase "essentially biological process" refers to a traditional breeding process, not to a microbiological process.)

To make things even more complicated, one just has to look into the nature of the EPO. This institution establishes common patent protection for all of the EC member states, except Ireland and Portugal, as well as the non-EC states of Austria, Liechtenstein, Sweden, and Switzerland. At the same time, the patent courts of individual nations retain the right to defy EPO guidelines—a right that is often used to support national industries. European pharmaceutical manufacturers, for instance, complain bitterly about Greece, Portugal, and Spain, who allow their drug producers to copy products from Germany, France, and Britain.

The differing patent laws and their interpretation in the European countries, in the United States, and in Japan have been the cause for many lawsuits and political interventions. In the early 1980s, U.S. officials often protested the copying of microchip designs by European firms, while the United States was equally often accused of allowing their companies accused of theft or copying to carry on selling, even if they lost in court.

Obviously, there are enough reasons to improve things. The European Commission has now proposed a directive on the legal protection of biotechnological inventions that would allow inventors to benefit from patent protection as effective as that in the United States and Japan. If successful, universal EC protection could become a reality by January 1, 1993.

The proposed directive states that the subject matter of an invention should not be considered unpatentable for the reason only that it is composed of living matter and that only plant and animal varieties, i.e., in the genetically fixed and stable form of a variety, should be excluded from patentability. Inventions concerning plant or animal cells, cell lines, tissues, and larger parts may be patented. It also states that microorganisms and biological classifications other than plant or animal varieties shall be considered patentable matter. Patentable subject matter, according to this directive, shall be any human intervention that consists of more than selecting any available biological material and letting it perform an inherent biological function under natural conditions. Also, the proposed directive seems to aim at the patentability of naturally occurring matter notwithstanding that the substance existed as an unseparated part of other material prior to the recognition of its existence and industrial applicability.

Another patent issue affects the entire pharmaceutical industry, not just the manufacturers of biopharmaceuticals: the erosion of patent protection. The average length of time required for developing a drug in Germany has grown from four years in the mid-1960s to about twelve years in the early 1980s. Because patent protection is only for a period of twenty years, the time when revenues are earned from the product may shrink to eight years or less. The normal duration of a pharmaceutical patent thus hardly allows the patent holder sufficient time to recoup the research and development costs incurred.

The pharmaceutical industry proposes an extended patent protection with the EC planning to introduce an "actual commercial exploitation period" of sixteen years, i.e., the time after product approval. (This period is seventeen years in the United States and fifteen years in Japan; the French government plans to introduce an actual exploitation period of seventeen years.) To this end, the Commission has produced a preliminary draft proposal for a Council Regulation on the creation of restoration certificates for pharmaceutical patents. It will enable the duration of pharmaceutical patents granted by an EC member state to be extended on application by the holder of the patent in question, provided a valid authorization has been granted for the marketing of the products that are the subject of the patents.

European Genetic Engineering Regulation Finally Coming to Terms

Of all European countries only Denmark and West Germany had genetic technology laws at the time this chapter was written. In France, Belgium, Great Britain, the Netherlands and Ireland, the contained use of genetically engineered microorganisms was regulated by specific guidelines. In Italy, it was included in a general regulatory framework for industrial operations and was handled on a case-by-case basis. In other countries, such as Spain, Portugal, and Greece, no regulation existed at all.

In Denmark, the "Environment and Gene Technology Act" of 1986 was born of a regulatory crisis when Nordisk Gentofte intended to produce recombinant human growth hormone from *Escherichia coli* and Novo Industri planned to produce human insulin in yeast. (Since then Novo Industri and Nordisk Gentofte have combined to form Novo Nordisk.) The law was drafted quickly and came into force in June 1986. In late 1989, Danish law makers eased certain provisions of this law after some skillful public relations.

Genetic engineering regulation in Germany has come a long way. Although a "Code of Practice to Ensure Protection Against Dangers Arising from Recombinant Nucleic Acids Constructed *in vitro*" was introduced in 1978, politicians felt that with genetic engineering moving quickly from basic research to industrial use, it had to be regulated by a law. A Parliamentary Commission was formed to study the "prospects and risks of genetic engineering." This resulted in a lengthy but excellent report, which the Bundestag began considering and discussing in 1988, thereby shaping the Genetic Technology Law. Several drafts have emerged, with considerable revisions each time. The development culminated in parliamentary sessions of the Lower House (Bundestag) and the Upper House (Bundesrat) in October 1989, with a spectacular filing of more than 250 objections against the current draft. Finally, in March and May 1990, the final draft passed both Houses and the law was put into effect in July 1990.

In Germany, the contained use of the most commonly used genetically engineered microorganisms (which belong to the lowest risk category) still requires authorization. However, it does not involve disclosure of particulars and public hearing procedures. In France, after some changes in the regulatory requirements, merely notifying the authorities suffices.

However, much is going to change at the end of 1991 because the Council of the European Communities adopted a directive on the contained use of genetically modified microorganisms on April 23, 1990 (EC Directive 90/219/EEC). The member states of the EC are obliged to bring into force laws, regulations, and administrative provisions not later than October 23, 1991. This will mean that a number of European countries that have not had appropriate regulations will have to introduce new laws or amend existing ones in order to comply with the EC requirements. As the council adopted the directive with reference to Article 130s of the treaty establishing the European Community, member states will have to adopt the directive as a minimum requirement, leaving it to them to introduce additional regulatory provisions on a national level.

The directive distinguishes between two types of operation:

- Type A—any operation used for teaching, research, development, or nonindustrial or noncommercial purposes and that which is of a small scale (e.g., 10 liters culture volume or less);
- Type B—any operation other than Type A.

Furthermore, it classifies genetically modified microorganisms as follows:

- Group I—microorganisms that are nonpathogenic, have no adventitious agent, and have a proven and extended history of safe use or built-in biological barriers, which, without interfering with optimal growth in the reactor or fermentor, confer limited survivability and replicability, without adverse consequences in the environment.
- Group II—other than those in Group I.

Any installation to be used for the first time for operations involving the contained use of genetically modified microorganisms requires the submission of a notification containing names of responsible persons, address of the installation, a description of the nature of the work to be undertaken, and a summary of the risk assessment. Separate notifications must be made for the first use of genetically modified microorganisms. The administrative requirements necessary for obtaining approval are outlined in Tables 18-5 and 18-6.

TABLE 18-5 ADMINISTRATIVE REQUIREMENTS FOR rDNA OPERATIONS ACCORDING TO THE EC DIRECTIVE

	Users of genetically modified microorganisms classified in	
Microorganism	*Type A Operation*	*Type B Operation*
Group I	Keep records on the work carried out (made available to the competent authority on request)	Before commencing the contained use, must submit a notification
Group II	Before commencing the contained use, must submit a notification	Before commencing the contained use, must submit a notification

Source: RauCon [1990].

Public hearing is not mandatory for any of the above operations. The directive leaves it to the member states to provide that groups of the public may be consulted on any aspect of the proposed contained use if considered appropriate.

Confidentiality is ensured, at least in theory, by the provision that the notifier may indicate which information in the notifications submitted should be treated confidentially in order not to harm his competitive position by inappropriate disclosure. However, verifiable justification must be given in such cases and it is at the will of the competent authority to decide which information will be kept confidential. Some information

TABLE 18-6 PROCEDURE OF APPROVAL OF OPERATION

Microorganism	Type A Operation	Type B Operation
Group I	No approval necessary	May proceed sixty days after submission of the notification or earlier with the agreement of the authority
Group II	May proceed 60 days after submission of the notification or earlier with the agreement of the authority	May not proceed without the consent of the authority, which shall decide ninety days at the latest after submission

Source: RauCon [1990].

may not be kept confidential at all, in particular the description of the genetically modified microorganisms, name and address of the notifier, purpose of the contained use, location of use, and safety and emergency response measures.

The directive further provides for measures to be taken in the event of an accident and requires the member states to consult with other member states liable to be affected in the event of an accident in the drawing up and implementation of emergency plans. Member states are also required each year to submit to the Commission a summary report of all the contained uses classified in Group II in Type B operations, including the description, proposed uses, and risks of the genetically modified microorganisms.

The Drag with Public Perception and Environmental Groups—A Communication Skills Issue

For a while, the poor public perception of biotechnology in Germany was seen by the international biotechnology community as one of the most inconceivable social phenomena in this business. In the meantime, the public perception issue has become dominant not only in Germany and Europe, but also in the United States, as recent statements of industry leaders and administrators have shown. Ronald E. Cape [1990], Chairman of Cetus Corp., complains:

> *We have delivered on what we promised. We have delivered technology and products. Some of us have already delivered returns to investors. But the public still doesn't value us. We still aren't trusted.*

In a letter of April 25, 1990, to the members of the Association of Biotechnology Companies, Hon. Robert A. Roe, chairman of the U.S. House of Representatives Committee on Science, Space, and Technology wrote:

> *To the public, Superfund sites and nuclear waste dumps are the results of decisions the public was not involved in or even informed about. If the public has little influence in determining the direction and uses of biotechnology it will not go to the trouble to educate itself about the science of biotechnology, and it will respond to biotechnology's power out of fear rather than reason.*

The Organisation for Economic Co-Operation and Development (OECD) [1989] found that new biotechnology is distinguished from other major technologies of the twentieth century by the fact that its impacts on the quality of life (its human and social consequences) are arriving earlier and may go deeper than macroeconomic impacts measured by productivity, investment or gross domestic product growth. The OECD also states that in contrast to the scientific assessment of DNA safety, there has been little or no convergence of public attitudes toward biotechnology.

It is evident that academia and the biotechnology industry have to develop skills to communicate with the public. They have to communicate their intentions, their developments, their safety measures, and the benefit for mankind that biotechnology-derived diagnostics, drugs, vaccines, crops, and cleanup methods represent. Public relations work does not replace proper communication with the public, but it is a prerequisite for it.

Over the past couple of years, I have been telling industry that their communication strategy needs improvement and must go beyond mere public relations work. I am absolutely convinced that Ronald Cape is right when he states that public policy skills are the most important "commercial development" skills for biotechnologists to possess in the 1990s.

In general, the poor perception of biotechnology in Europe may be attributed to a number of factors:

A commonly growing disbelief in science and technology. Science and technology had become a substitute religion for many during the period between the end of World War II and the early 1980s. In the 1970s and 1980s, an intellectual counterculture nourished by leftist governments in Germany, Great Britain, France, and Italy created an aversion to technology and industry, and problems of economic growth, Chernobyl, Bhopal, Superfund sites, and so on have shaken the confidence in the omnipotence of science.

A fear that genetic research may reveal that our functioning, behavior, will, and destiny is determined by genes to a much greater extent than we previously thought. This would mean the death of human dignity and uniqueness—a vision unacceptable to many. In Germany, a touch of romantic transfiguration of nature may add to this.

A number of ethical issues closely related to this fear. How far are we allowed to go in genome analysis, gene therapy, and prenatal screening? People feel uneasy at the thought of being classified in groups as genetically fit and unfit. In Germany, the trauma experienced with euthanasia contributes to this uneasy feeling.

A fear of loss of control over releases. Accidental releases from contained uses of rDNA microorganisms and deliberate releases of microbes, plants, and animals may get out of control and cause irreversible damages.

The last point is very closely related to a broad environmental movement observed everywhere in the world but that may be expressed more in Europe, and especially in Germany, than elsewhere. A recent opinion poll found that in West Germany, 82% of the consumers take environmental considerations into account when shopping, versus 67% in the Netherlands, 55% in Great Britain, and 50% in France.

But the degree of "environmentalism" is certainly leveling off among the European countries. This can be seen from the development of the "Green Parties" in Europe. The German "Green Party" is presently entering a stage of consolidation on a low level and even losing support (in fact, they lost all of their seats in German parliamentary elections in 1990), whereas other European countries such as Great Britain, France, and Belgium may be good for surprises in the near future as the European Parliament elections in June 1989 have shown.

In any case, the times when a chemical group could erect a manufacturing facility wherever it pleased are over forever. Even in Ireland, a country that had been advocating industrial investments more than any other European country, people are not willing to pay the price for "jobs at all costs" as chemical giants such as Sandoz and Dow Chemical had to experience. In Germany, the Greens and the "Gen-ethic Network" are battling the biotechnology industry, although their position has become weaker after the German Genetic Technology Law came into force in July 1990. In Italy and Portugal, very peculiar constellations exist: Portugal has a coalition of environmentalists and communists (Coligação Democrático Unitária), while in Italy, Communists have a particular interest in environmental issues, probably because they provide an instrument to fight industry and capital. Every European country has its environmental groups and all of them fight genetic engineering—some more, others less, successfully.

Needless to say, this situation needs a public relations and communications policy from the side of the biotechnology companies that is customized to the peculiarities of differing cultural heritages and ways of political thinking. The U.S. type of public relations, which is usually accompanying corporate quarterly reports in order to keep Wall Street happy, is by no means adequate. Dealing with the heterogeneous European public requires communications skills that go far beyond that to address the various ethnic groups in a sophisticated manner. Fortunately, these can be grouped together into segments with certain degrees of cultural and/or linguistic similarity (Fig. 18-1).

Cultural and Linguistic Variety Calls for Market Segmentation

Communication and marketing in a Greater Europe will require a matrix-type refinement of communication and marketing strategies (see Fig. 18-2). The microsegmentation of markets by consumer habits, lifestyle

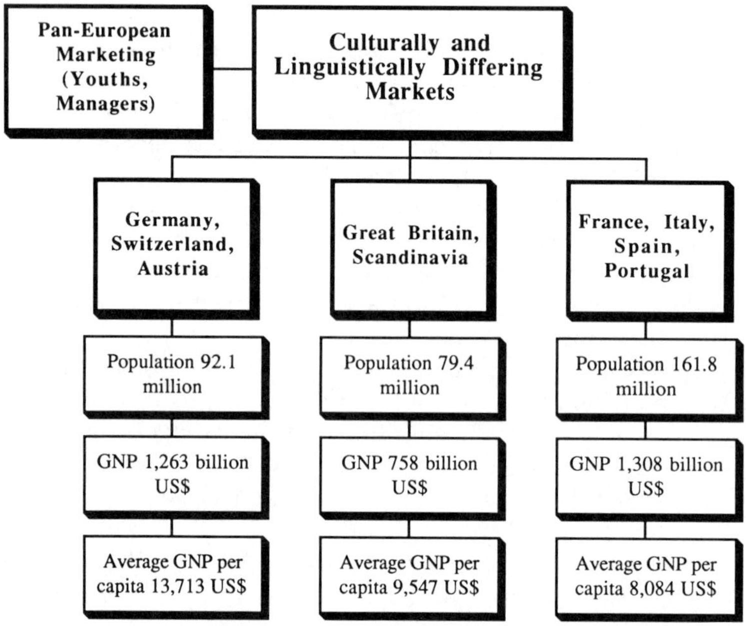

FIGURE 18-1 Relative demographics and wealth of Greater Europe.

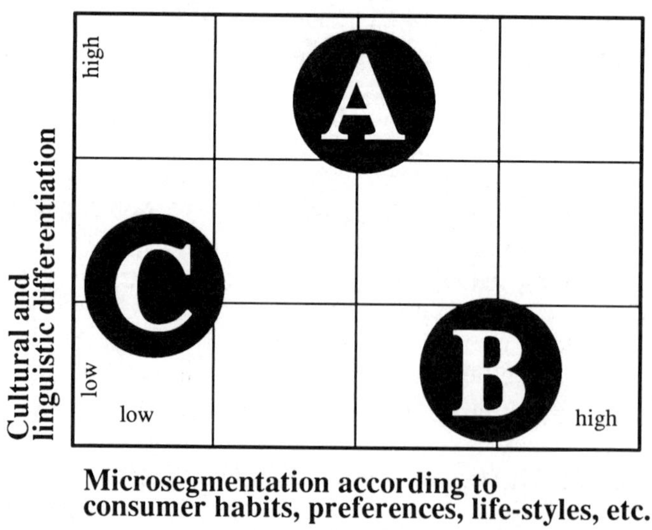

FIGURE 18-2 A need to refine communication and marketing in Greater Europe will require a matrix-type conceptual framework. For example, Type B marketing would correspond to a microsegment with no cultural and only minor linguistic differentiation (pan-European marketing), e.g., marketing of a cancer drug to specialized treatment centers.

preferences, and other characteristics will be compounded by cultural and linguistic market segmentation. For example, a male impotency drug will have to be marketed in a microsegment formed by aging men or those whose sexual performance is hampered by side effects of drugs (e.g., diabetes treatment). At the same time the choice of methods and ways to promote this drug will have to differentiate subtly between the more liberal north, the conservative Anglo-Saxon area, the life-enjoying French, and the Roman Catholic south of Europe, where a certain machismo cannot be denied. Communication and marketing of this type would probably correspond to Type A in the marketing matrix.

Type B marketing would correspond to a microsegment with no cultural and only minor linguistic differentiation (pan-European marketing), e.g., the marketing of a cancer drug to specialized treatment centers.

Finally, Type C could be a pain reliever with a low degree of cultural, but medium degree of linguistic, differentiation in marketing.

To make things easier and to attain some reasonable scale effects in communication and marketing, one could group major countries into culturally and linguistically differing markets as proposed in Figure 18-1. In Germany, Switzerland, and Austria you should certainly use German language in all of your advertisements, brochures, etc. This is in spite of the fact that many Germans, Swiss, and Austrians are able to read English. However, they want to be addressed in their own language. In France, Italy, Spain, and Portugal, it will be more difficult to find people who read English or each other's languages. Although they have a reasonably common cultural heritage, each nationality will have to be addressed in its own language, which diminishes the attractiveness of this huge market. Also Great Britain and Scandinavia have similar cultural traits and Scandinavians do understand English, although they would prefer to be addressed in their own language.

SOME CLOSING REMARKS

I have tried to create an understanding of the market opportunities emerging from a Greater Europe by increasing confusion over its heterogeneity despite a "Europe 1992" and over a number of uncertainties regarding the developments in the Eastern parts of Europe. If I have provoked the assumption that business in Europe is business as usual, which can be planned from a desk in New York, Tokyo, or Sydney, I have failed. If I have been able to convince the reader that business in Europe requires a maximum of continual on-site attention over the next few years, an input of strategic information exceeding the usual level, and expertly designed business strategies, I have been successful.

In short, my recommendations are

- Keep a close and continuous watch on what happens during the next five years. Windows will open fast and unexpectedly.
- Increase your direct information input from Europe and on European matters. If you read it in your trade magazine, it may be too late. Opportunities will present themselves for those who stay online and react fast.
- Do not leave the planning and designing of your European business strategies with somebody who is not an insider, you need an expert on European biopharmaceuticals business.
- Always keep in mind that in spite of the harmonization of laws and standards, the differences that will remain will be the determining factors for the success of your business. Diligent consideration of these differences will pay off.
- Always be aware of fast and unexpected changes in Europe, but at the same time do not underestimate the determination of European business managers and politicians to create an economic powerhouse that will be on equal terms with the major economic centers of the world.

REFERENCES

Cape, R. E. 1990. What is biotechnology's public role after we have delivered what we promised? *Genet. Eng. News* 10(2):4.

Jeffcote, Bernard. 1989. Imperfect harmony. *Corporate Location Europe.* October 1989:18–19.

Laurent, A. 1986. The cross-cultural puzzle of international human resource management. *Human Resource Management* 25(1):91–102.

Organisation for Economic Co-Operation and Development. 1989. Biotechnology—economic and wider impacts. OECO, Paris.

RauCon Bioinformatics & Consulting GmbH. 1990. Health care biotechnology in the Federal Republic of Germany. Dielheim, Germany: Rau, Norbert.

CHAPTER 19

BIOTECHNOLOGY MEGATRENDS: LESSONS FROM THE PAST—VISIONS OF THE FUTURE

Roger E. Shamel

In his book, *Megatrends* and the recent sequel *Megatrends 2000*, John Naisbitt [1982, 1990] sets forth "a handful of new directions" for the 1980s and 1990s. In the updated version, "biotechnology" is identified as one of the ten overarching trends that will influence our lives in the 1990s.

What is the significance of biotechnology in terms of a decade-long or century-long historical perspective? What are its historical roots? Where does it stand today? Where is biotechnology headed in the twenty-first century? These are some of the broad questions that I will cover in this chapter. My approach will be briefly to sketch out the key milestones that led to the birth of biotechnology, to give a capsule summary of the current industry status, to list the lessons of the 1980s, and, finally, to take you on an imaginary journey to the year 2020 for a vision of the longer-term potential impact of biotechnology on this generation and the next. Finally, the chapter concludes with a plea for every reader to help fight the scientific illiteracy that threatens to limit unnecessarily our society's ability to harvest the fruits of biotechnology.

When asked to write the final chapter in this book, I was both honored and humbled. Certainly, in my consulting work at Consulting Resources, my colleagues and I have had many occasions to observe, contemplate and write about trends in biotechnology. But translating that experience into something worthy of capping off a book written and edited by such an august group was another matter. After considerable research, writing, and rewriting, I offer this attempt to present a mega-review and forecast of biotechnology. Before moving beyond the introduction, I wish to thank the editor and the publisher for their patience, and also to thank the many individuals whose ideas have contributed to mine, especially my colleague, Joseph Chow.

ORIGINS AND OUTLOOK: THE MEGAVIEW

"Biotechnology," in the traditional sense, has been with us since antiquity. As illustrated in Figure 19-1, man has used or manipulated biological systems to produce alcoholic beverages, leavened bread, cheese, and other products since the beginning of recorded history. Selective breeding over the centuries gave us our domesticated plants and animals—all without ever slicing or splicing a gene in the laboratory.

Although the origin of the "new biotechnology" (defined here as the use of scientifically engineered biological systems to produce useful products) could be traced to the elucidation of DNA's helical structure by Watson and Crick in 1953, it was not until the development of recombinant DNA techniques in 1973 that we really began to enter a new era. From 1973, it took only eight years, until 1981, for the first major products of the new biotechnology to reach the marketplace. The most well-known of these were the monoclonal antibody-based diagnostics, which have been developed and widely marketed as home pregnancy test kits. From this relatively simple beginning, diagnostic and therapeutic biotechnology products in the U.S. market alone have product sales which have grown from zero in 1980 to an estimated $1.4 billion in 1990.

Although it is difficult to make solid predictions on the basis of such a short history, our current belief at Consulting Resources is that biotechnology sales will continue to expand at a compound real growth rate exceeding 20% per year through the remainder of the century. We thus expect U.S. biotechnology industry sales to exceed $12 billion (in 1990 dollars) by the year 2000 (Table 19-1). More than half of all biotechnology industry revenues will continue to be generated by human therapeutics during the 1990s. Human diagnostics, the "granddaddies" of the industry, will grow at a mere 18% per year and contribute an additional $2.2 billion to the century-end total. Sales of agricultural products, which are just about to get off the starting block, should begin to catch up with human diagnostics sales by the year 2000. All other products, including those involving specialties and contaminant monitoring, will contribute another $1.1 billion to sales by the end of the century.

Beyond 2000, we expect biotechnology sales to continue to expand at double-digit, albeit slower, growth rates through the first several decades of the twenty-first century. In other words, from a "megatrend" point of view, biotechnology is still a relatively young industry.

Antiquity	Alcohol, cheese, crop and animal breeding, fertilizers
1869	DNA isolated
1928	Penicillin discovered
1953	Structure of DNA postulated
1961	Genetic code deciphered
1973	rDNA Experiments
1975	Hybridomas created
1976	Genentech
1981	MAb Diagnostics
1982	Humulin; animal vaccine
1983	Foreign expression of plant gene
1985	Transition from R&D to market diagnostics, therapeutics, plant genetics
1986	Hepatitis B vaccine; human growth hormone; interferon-alfa
1987	Field test of "Frostban," tPA
1988	"Lipolase," field testing of genetically engineered crops
1989	Erythropoietin
1990	Agricultural and food products, fine chemicals, specialty chemicals, commodity chemicals?
2000	Energy, biochips

FIGURE 19-1 Biotechnology: historical perspective. (*Source*: Consulting Resources Corporation.)

Because of overzealous, if not misguided, publicity, the industry has had several false starts. At this point, when the industry is perhaps fifteen years old, one must remember that many industries require between 50 and 100 years to move through the four stages of the industry life cycle before they reach maturity. As illustrated in Figure 19-2, we believe that biotechnology, in 1990, is just entering the "growth" phase of the life-cycle. The relative current life-cycle positions of the U.S. pharmaceutical and chemical industries are also shown for comparison.

TABLE 19-1 U.S. BIOTECHNOLOGY PRODUCT SALES FORECASTS FOR THE 1990s (MILLIONS OF 1990 DOLLARS)

Key Sectors	Base Year 1990	Forecast Years 1995	Forecast Years 2000	1990/2000 Growth (%/Yr.)
Human therapeutics	840	2600	6700	23
Human diagnostics	430	1050	2250	18
Agriculture	55	420	2200	45
Specialties	60	210	900	31
Contaminant monitoring	5	50	200	45
Total	1390	4330	12250	24

Source: Consulting Resources Corporation.

We will return to a more vivid description of "biotechnology future" later. First, however, let us examine what has emerged from the decade of the 1980s in terms of fundamental developments and key products.

THE CURRENT SITUATION: A 1990 ASSESSMENT

What have been the major milestones of the 1980s? What slate of products has the biotechnology industry delivered to date?

The 1980s represented the first decade of commercialization for bio-

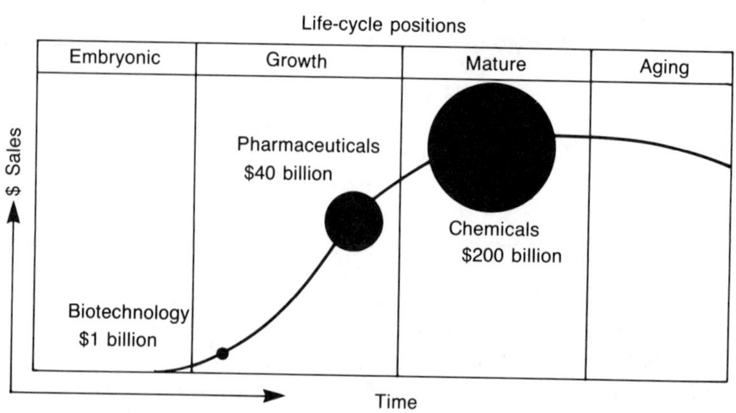

FIGURE 19-2 Relative size and maturity of key industries: 1990. (*Source:* Consulting Resources Corporation.)

technology. As the decade opened, Amgen was founded, Genentech sold its first public stock offering, and the U.S. Supreme Court ruled that life forms could be patented. During the remainder of the decade, the following milestones were reached:

1981	First monoclonal antibody diagnostic tests sold Gene synthesizing machines produced First human oncogene discovered
1982	First recombinant DNA (rDNA) product (human insulin) sold
1983	Amgen, Biogen, and Chiron go public First expression of a foreign gene in plants
1984	DNA "fingerprinting" developed First biotechnology vaccine (hepatitis B) engineered
1985	Second rDNA product (human growth hormone) sold First major acquisition (Hybritech by Eli Lilly)
1986	First genetically engineered plant (tobacco) field tested First biotechnology vaccine (hepatitis B) sold
1987	First genetically altered bacterium field tested Third rDNA product (tissue plasminogen activator) sold
1988	First patent granted for genetically altered animal First major acquisition involving two biotechnology companies (Advanced Genetics Services by DNA Plant Technology)
1989	Erythropoietin approved to boost red blood cells Second wave of biotechnology/biotechnology acquisitions (Plant Genetics by Calgene, Integrated Genetics by Genzyme, etc.)
1990	Genentech acquired

To cap off the decade, in February 1990, Roche Holding Ltd. paid $2.1 billion for 60% of Genentech, thus absorbing the industry's "flagship company," and obscuring the true meaning of "success" in the minds of many biotechnology company executives.

Carrot-and-Stick Theory of Product Commercialization

The "carrot-and-stick" theory can serve as a useful tool for understanding the profile of products either on the market or under development. The concept of "market need" as a driving force for product commercialization is not new, although it has recently become very much in vogue to talk about the need to be "market-driven." Every successful product, past, present, or future, is fundamentally satisfying a market need. Of course, labor, technology, and capital have important enabling roles along the way, but, on the bottom line, the rewards come to those who most successfully satisfy the market need.

So much for the "carrot." My definition of the "stick" is slightly different from its normal use in the "carrot-and-stick" context. Instead of being limited to the image of a painful blow, it is any hurdle that slows or stops one's progress toward successful commercialization of a product. A few examples would be regulation, red tape, being outmaneuvered by the competition, and market rejection due to faulty market research.

One of the best examples of a product that has led its proponents along a tortuous path between carrot and stick is bovine somatotropin (BST). Although many companies correctly identified a market need (more efficient and increased production of a staple food—milk), some misjudged the negative public reaction that accompanied tampering with Mother Nature's pure, unadulterated milk. The irony here is that BST is a harmless, natural protein that is already present in milk.

As a practical application of the carrot-and-stick theory, one can develop, through brainstorming or other approaches, matrices of driving forces and/or commercialization hurdles to help predict the relative rate and direction of product development. One such matrix is shown in Figure 19-3, which compares estimated technological availability and estimated market need. Viewing this matrix, one would correctly conclude that human diagnostics and therapeutics would be the most successful early products of biotechnology.

Product Status

Human therapeutics and diagnostics. There are now seven types of biotechnology-based human therapeutic products on the U.S. market, as shown in Figure 19-4. We estimate total sales derived from those products to be about $840 million for U.S. companies in 1990.

Erythropoietin (EPO) is the most recent addition to the list, which also includes human insulin, human growth hormones, alfa-interferons, OKT-3 (a monoclonal antibody that prevents kidney transplant rejection), hepatitis B vaccine, and tissue plasminogen activator (tPA). Over 100 other new therapeutics are at various stages of regulatory approval, and several dozen of those are expected to reach the market in the early 1990s. On the horizon are a number of monoclonal antibody–based therapeutic products, more interferons, and several colony-stimulating factors (CSFs).

At least two important potential AIDS (acquired immunodeficiency syndrome) biotherapeutics are currently in clinical trials; CD4 and

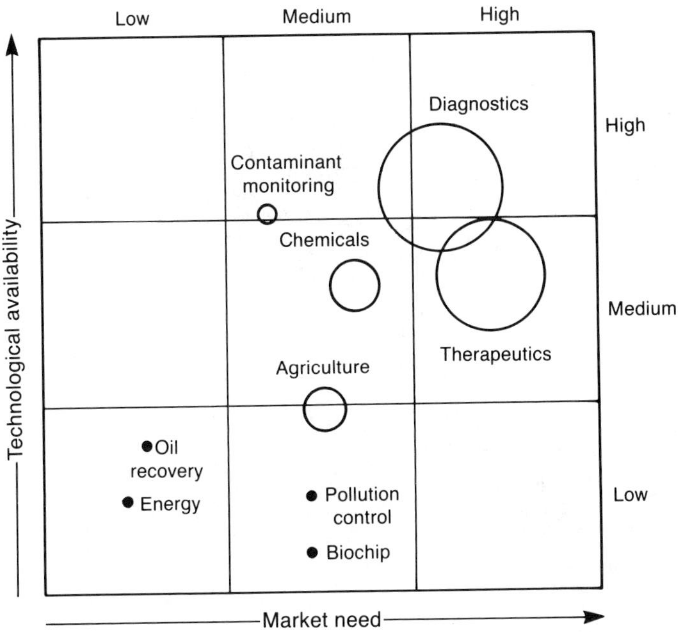

FIGURE 19-3 Technological availability/market need profile. Circle size indicates 1990 market size. (*Source*: Consulting Resources Corporation.)

GLQ223, a purified form of a plant protein (trichosanthin) derived from the root of Chinese cucumbers.

Human diagnostic products have blossomed rapidly in recent years, with more than 300 products on the market (mostly home pregnancy test kits) and new ones continually being introduced. Total 1990 sales for the human diagnostics segment are estimated at about $430 million.

Agricultural biotechnology. Although products currently under development in the agricultural sector will offer real benefits to both farmers and consumers, some, such as the BST mentioned earlier, will also generate heated controversy. Realizing the potential for misunderstanding, many companies are undertaking or planning substantial consumer education efforts to counter unwarranted fears.

Sales of biotechnology products in the agricultural segment will be approximately $55 million in 1990. Products currently on the market include biopesticides, such as a *Bacillus thuringiensis*–based pesticide to combat the Colorado potato beetle; animal health care products, such as scours and pseudorabies vaccines; and a variety of agricultural diagnos-

Pre	Clinical trials I	II	III	PLA/NDA	On Market	Year of entry
					hINSULIN	83
					hGH	85,87
					INF-α	86
					OKT-3	86
					HEP-B VAC	86
					tPA	87
					EPO	89
				IL-2		91
				INF-γ		91
				F-VIII		91
				G-CSF		91
				G7-CSF		91
				Septic shock MAb		91
			INF-β			91
			EGF			91
			GCR			92
			SOD			92
		ANF				93
		TNF				94
		M-CSF				94
		CD4				94
	IL-4					94
	AIDS vaccine					95+
	IL-1					95+
	IL-3					95+
	PF-4					95+
	IGF-1					95+
	Argatroban					95+
	Relaxin					95+
	PDGF					95+
IL-6						95+
IL-7						95+
BCGF						95+
Lung surfactant						95+
FGF						95+
IGF-2						95+
MSF						95+
PROT-C						95+
MIS						95+
TGF-β						95+
Adipsin						95+

FIGURE 19-4 Therapeutic Product Status—1990. ANF = atrial natriuretic factor; BCGF = B-cell growth factor; CD4 = soluble T4 receptor; G-CSF = granulocyte colony-stimulating factor; M-CSF = macrophage colony-stimulating factor; GM-CSF = granulocyte macrophage colony-stimulating factor; EGF = epidermal growth factor; EPO = erythropoietin; FGF = fibroblast growth factor; F-VIII = Factor VIII; GCR = glucocerebrosidase; INF-γ = interferon gamma; HEP-B VAC = hepatitis B vaccine; hGH = human growth hormone; hINSULIN = human insulin; IGF-1, 2 = insulin-like growth factor 1,2; IL-1,2,etc. = interleukin-1, 2, etc.; INF-α = interferon alpha; INF-β = interferon beta; MSF = megakaryocyte stimulating factor; MIS = mullerian inhibiting substance; OKT-3 = Ortho kidney transplant-3 monoclonal antibody; PDGF = platelet-derived growth factor; PF-4 = platelet factor 4; PROT-C = protein C; SOD = superoxide dismutase; TGF-β = transforming growth factor beta; TNF = tumor necrosis factor; tPA = tissue plasminogen activator. (*Source*: Consulting Resources Corporation.)

tics. Commercialization will begin this decade for many of the more important potential agricultural products (such as genetically engineered crops, for which a number of field trials already are under way. Pesticide- and herbicide—resistant tobacco plants, and insect-resistant tomato and corn plants have also been tested and will likely come to market in the 1990s.

Specialties. Estimated 1990 sales in this broadly-defined category are about $60 million. Products include biotechnology-derived specialty chemicals, pollution control products, and microbial mining products. Enzymes and various food-related products, such as vitamins, are also included. However, most current biotechnological processes for specialties do not provide distinct advantages, and substantial research and development investment is required before better bioprocesses are developed.

Contaminant monitoring. Until now, this segment has been a small but dynamic part of the biotechnology industry, with sales for 1990 estimated to be $5 million. The potential for growth in this segment will depend heavily on acceptance by government, industry, and commercial testing laboratories and the level of public concern over environmental contamination.

By using immunoassay technology presently found in home pregnancy test kits, simple tests have been developed that enable quick, sensitive, and relatively inexpensive consumer-operable checks for contaminants in fruits, vegetables, and drinking water. Currently, these tests are being evaluated by government and industry and are not mass-marketed to consumers.

Mergers and Acquisitions

In concluding this section, a few comments are in order regarding the current status of biotechnology merger and acquisition activities. The trend toward mergers and acquisitions has long been predicted by industry watchers. Most of the recent activity has involved mergers between biotechnology companies, although several have involved acquisitions by the blue-chip giants. This rash of weddings within the biotechnology company family began in late 1988 with the merger of DNA Plant Technology and Advanced Genetic Sciences, two well-known plant biotechnology companies.

As the industry moves from the embryonic to the growth stage, we expect consolidation to continue, with an average of about twenty to twenty-five (mostly small) companies disappearing each year through mergers, acquisitions, or failures over the next decade.

LESSONS FROM THE 1980s

What are the key lessons of the 1980s for the biotechnology industry? Of the dozens that could be cited, I have developed a list of "seven virtues" and "seven sins," which ought to be etched into every biotechnology executive's bathroom mirror by now.

First, the seven virtues:

1. *Develop a sound, "street-smart," business plan.* Most potential start-ups never even get started because of the shortcomings of the business plan. In the eyes of the venture capitalist or other potential investor, a plan that does not take into account the inevitable problems of slow regulatory approval, gradual market acceptance, and aggressive competition is a "no funder."

2. *Hire good people, especially a great CEO.* Sound, innovative technology filling a market need and good financial backing are necessary but not sufficient to the success of a venture. By far the most critical factor in determining which ventures "make it" and which do not is the dedication, talent, and experience of the management team members. Get the best.

3. *Get products on the market early.* Most start-up biotechnology companies fail because they do not get early product revenues. Even the most patient investors are not willing to wait more than a few years to see cash flowing in from products. Although it can be done, starting with a human therapeutic product is clearly a high-risk strategy.

4. *Work closely with appropriate regulatory bodies.* We live in an age of regulation. Most start-ups assume an adversarial relationship with state, local, and federal authorities. This is a mistake. It is much better to assume a partnership role, to supply information on your plans and, more importantly, to find out exactly what must be done to get approvals. As always, the interpersonal element can make a critical difference.

5. *Play by the rules.* No unauthorized roof-top, laboratory-basement, or field experiments. Not only do you risk fines, delays, and a damaged reputation, but you also provide fuel for the fires of public fear—possibly setting back the entire industry. Above all, biotechnology companies must be good corporate citizens in order to build and maintain public confidence and support.

6. *Deliver what you promise.* As is the case in all business, but especially importantly in biotechnology (where the credibility of product commercialization timetables have come under special scrutiny), you must set and meet realistic commercialization goals.

7. *Focus on what you do best.* In the early 1980s, many biotechnology companies attempted to pursue too many avenues of product development. These meant, in some cases, simultaneously working on human therapeutics and diagnostics, agriculture, chemicals, pollution control and other product development options. By the latter part of the decade, focusing on one or two related product areas had quite logically and correctly become the rule.

Next, we turn our attention to the seven sins, which more than one member of the emerging biotechnology community committed in the 1980s:

1. *Do not rely merely on the "elegance" of the technology.* This is the most common mistake made by those who have spent their lives in the academic world, where it sometimes seems that "elegance is everything." As hinted before, what sells is a product that meets a market need at an attractive price, no matter how inelegant the technology.

2. *Do not build unrealistic expectations.* A question: How do you introduce the largest selling new drug in history yet have it be viewed as a failure? Answer: Promise the world that its sales will be several times larger than what is actually achieved. This is the lesson of Genentech and the "billion dollar" t-PA.

3. *"Do not put all of your eggs in one basket."* Why do these aphorisms come in so handy? Because the lessons are old ones: it is only the technology that is new. Genex focused nearly its entire effort on developing and supplying G.D. Searle with the raw materials for NutraSweet's aspartame sweetener. Then Searle surprised Genex by cancelling its contract and began to manufacture aspartame itself. Genex barely survived and has never regained its former position of prominence.

4. *Do not perform unauthorized experiments.* Whether it is a well-intentioned attempt to accelerate the fight against Dutch elm disease at Montana State University at Bozeman, or a similarly motivated effort to more quickly develop the technology to save crops from frost damage in California, unauthorized experiments are a bad idea—bad for the company, and bad for the industry as a whole. A graphic illustration of the consequences of the Advanced Genetic Sciences experience in Oakland, California, is shown in Figure 19-5.

5. *Do not underestimate commercialization hurdles.* For whatever reason (probably because most new companies are started by people who are, by their very nature, optimistic), it always seems to take longer than expected to get your product on the market. So be sure to build in a contingency factor or safety net of some kind. Problems can unexpectedly arise in many areas: the laboratory work can fail, the regulatory approval can be delayed, and, as Monsanto has learned from its commercialization of bovine somatotropin, an uninformed public can easily turn against you when you start trying to improve on cow's milk.

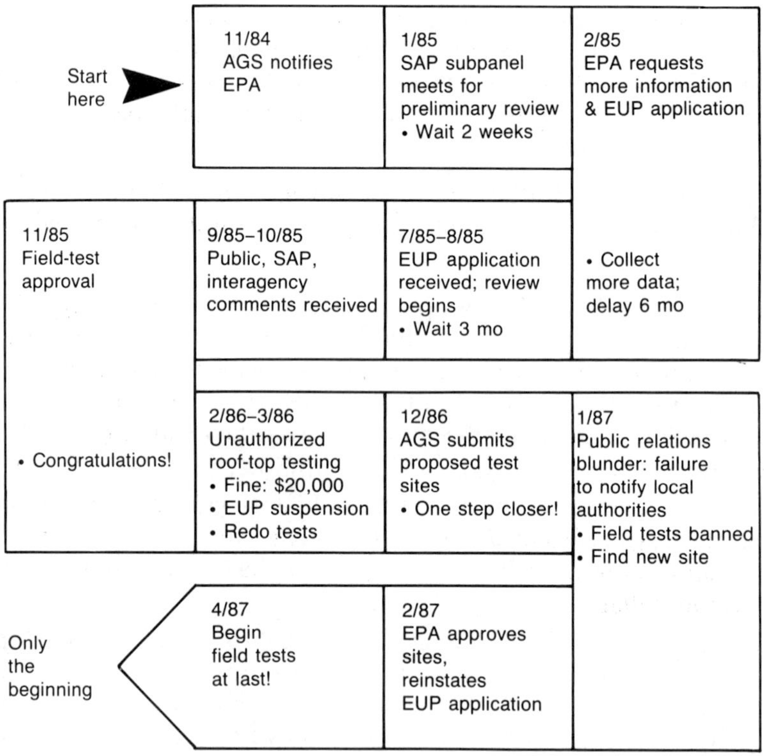

FIGURE 19-5 Advanced Genetic Sciences (AGS) case study. EPA = Environmental Protection Agency; SAP = Scientific Advisory Panel; EUP = Experimental Use Permit. (*Source*: Consulting Resources Corporation.)

6. *Do not buy a "pig in a poke."* Again, the aphorisms say it best. The fat, juicy "pig" that you think you are buying and that you hope to serve up for dinner, may turn out to be a "dog" when you finally get home and open the poke (or bag). On more than one occasion we have done some research for a prospective buyer of a biotechnology company only to find, on closer examination, that the potential acquiree was a different animal than was expected.

7. *Do not run out of money.* This, of course, is the most important of all things to avoid. It may not be over "until the fat lady sings," but when you cannot pay your bills, you know the end is near.

I would like to summarize and conclude this section by directing your attention to what we believe are the seven most critical challenges in biotechnology, depicted in Figure 19-6.

1. *Financial*—balance short-term cash needs with expected longer-term payoff
2. *Human resources*—attract, hold, and motivate the best managers, engineers, and scientists
3. *Research and development*—fashion more complete understanding and greater control over the human immune system and plant physiology
4. *Regulatory*—accelerate the speed and lower the cost of bringing safe, effective, and valuable products to market
5. *Production/engineering*—develop effective, low-cost technology for scale-up and purification
6. *Marketing*—control and direct investor/market hopes and fears; obtain cost-effective distribution through appropriate avenues
7. *International*—establish strategic alliances; optimization and coordination on a global scale

FIGURE 19-6 The seven most critical challenges in biotechnology. (*Source:* Consulting Resources Corporation.)

VISIONS OF THE FUTURE: 20/20 FORESIGHT

What is the outlook for biotechnology? What will be the major trends? Is it poised for explosive new growth or simply looking at steady, evolutionary progress? What direction will it take in the 1990s and beyond?

We believe that the biotechnology industry is still in its infancy. In this section, I hope to stimulate your thinking regarding the potential for biotechnology to affect our health, our food, our plants and animals, our environment, and other critical parts of our lives in the twenty-first century.

Picture, for a moment, a world totally free of disease. A world where there are no troublesome insects, where perfect, high-yield crops can be grown without pesticides and without nitrogen fertilizer. Picture a world where birth defects are eliminated, where computers really can think like humans. Finally, picture a world where waste disposal and energy availability are no longer problems, where pollution can be reduced or eliminated.

These are some of the broad visions that a biotechnology future may deliver sometime during the next century. Let us turn our attention now to a closer look at some future products, industry developments, and industry trends.

Products

In our view, the biotechnology industry is moving into a rapid growth phase. The flow of products in the 1980s has been only a trickle. Nearly everyone agrees that the rate of product introduction will increase dramatically during the 1990s.

Most biotechnology products in the early 1990s will remain for the pharmaceutical market, where the technology is further developed than in other sectors and where profit expectations are highest. However, research in food and agriculture is also expected to lead to a host of new products, as illustrated in Figure 19-7.

What about developments in the second half of the decade? If one focuses on only the two primary technologies—hybridomas and rDNA—the range of possibilities is much broader than commonly realized. As illustrated in Figure 19-8, hybridoma technology, which has been commercialized significantly to date mainly in the area of *in vitro* human diagnostics, holds the promise to offer at least a dozen different product categories. Many of these product concepts are just beginning to emerge from the laboratory, yet some of them will be commonplace by the next century. For example, catalytic antibodies, or "abzymes," will enable us to produce specialty polymers with performance characteristics that are both valuable and unattainable by any economical means today.

A similar promise of unfulfilled potential is provided by rDNA. As illustrated in Figure 19-9, another dozen potential product categories await development here. The most important products commercialized to date from this family are the therapeutic proteins, such as human growth hormone and tPA. Yet to be developed, but ultimately equally important, are the rationally designed drugs that will be commonplace by the year 2020. The product possibilities in agriculture have been at least touched on above, as have some of our visions relating to foods, materials, chemicals, energy, and pollution control.

One vision of the year 2020 in medicine will serve to illustrate what we expect to be clearly on the horizon, if not achieved at that time—truly customized medicine. As illustrated in Figure 19-10, the evolution of drug development has just recently begun to move slowly from the "classical" trial-and-error screening method, to the use of rDNA techniques to produce therapeutic proteins. In the S curve illustrated here, drug development will evolve through protein engineering, to new expression systems, to therapeutic peptides, and ultimately to what I will call, for lack of a better term, the "custom gene machine."

How would a custom gene machine potentially have an impact on medicine? Picture this: sometime during the 2020s, a patient will check into the hospital with a rare, often fatal, disease for which no mass pro-

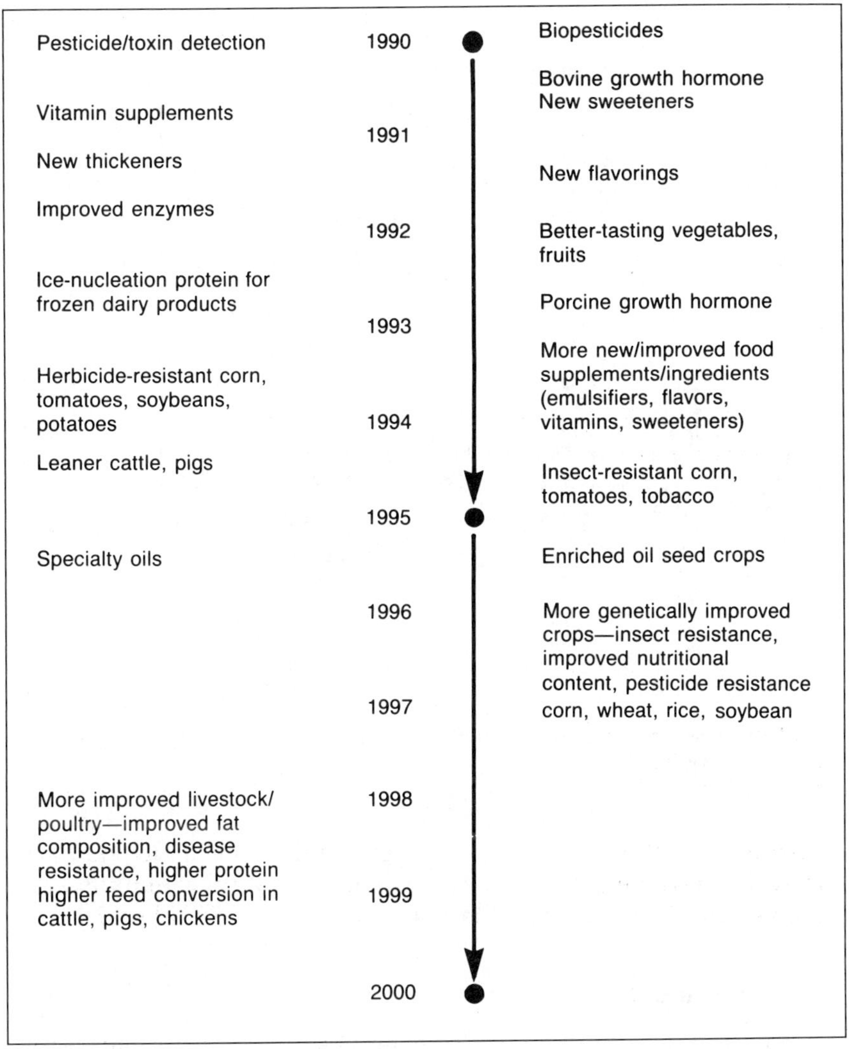

FIGURE 19-7 Biotechnology developments affecting foods and agriculture. (*Source:* Consulting Resources Corporation.)

duced drug has been developed. Rather than telling the next of kin that little can be done except to try "enjoying" the time that remains, the doctor will begin running some diagnostic tests. Most of these will have been developed through biotechnology. Then, with the aid of a computer, he or she will proceed to instruct a nearby custom gene synthesis machine to produce a needed protein, peptide, or other biochemical to counteract the

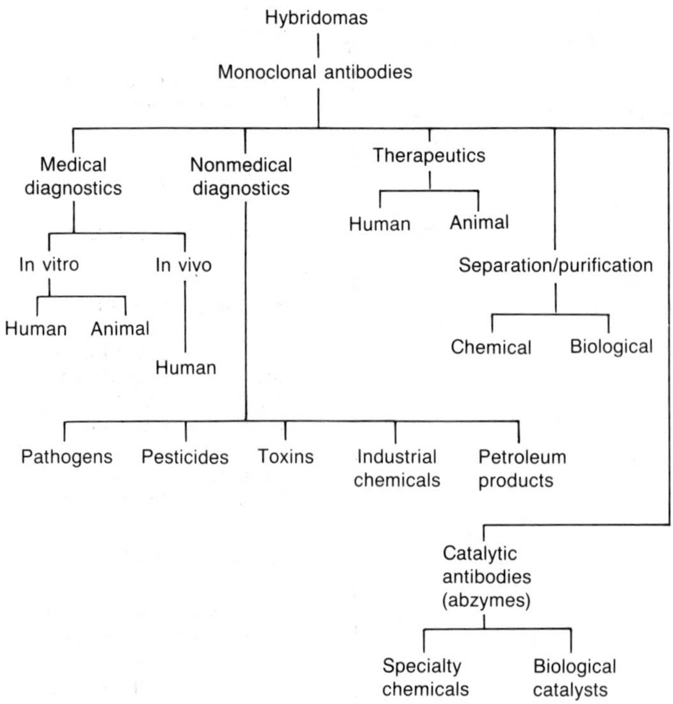

FIGURE 19-8 Products flowing from hybridoma technology. (*Source*: Consulting Resources Corporation.)

disease. At that point we will have come full circle, to the point at which medicine revolves more closely around treating the individual, rather than synthesizing drugs on the basis of what can be sold to the masses.

The Industry

In the 1990s, as the industry enters its growth phase, marketing will steal the limelight from research and development in terms of relative strategic importance. Management style and corporate strategy will need to shift as cost control and cost cutting become key. Biotechnology companies will become more formal in terms of organizational structure and business practices. The critical challenge will be to manage the shifting financial, marketing, regulatory, and production priorities depicted in Figure 19-11.

Industry consolidation will continue, if not accelerate. Already, fewer new companies are being started. Mergers and acquisitions will almost certainly become more frequent. In ten years we expect the number of

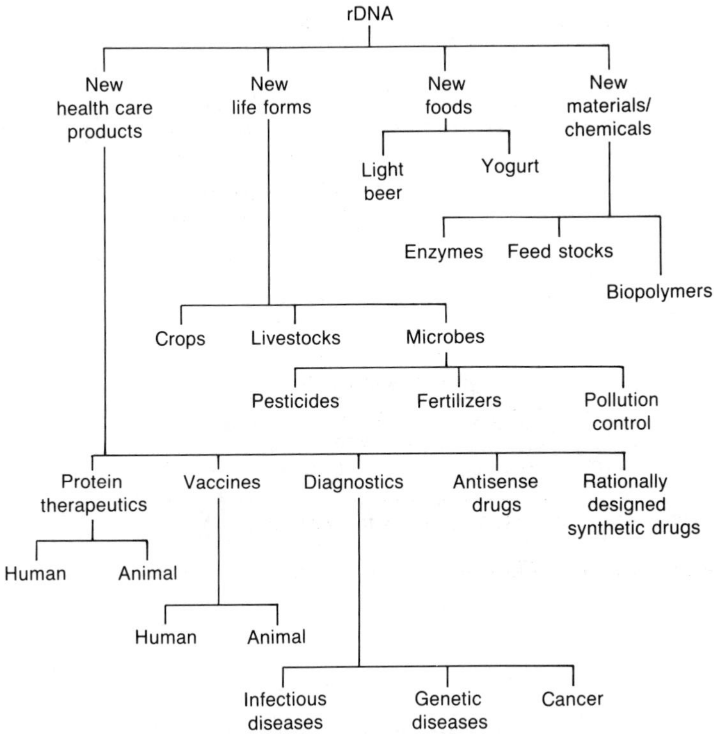

FIGURE 19-9 Products flowing from rDNA technology. (*Source*: Consulting Resources Corporation.)

independent biotechnology companies to be about half what it is today. More and more, the small companies will disappear and the large companies will flourish.

The coming of 1992 in Europe will bring an acceleration of the existing trend toward universal socialized medicine. It will also bring more uniform and stricter regulatory standards, the imposition of good manufacturing standards throughout the European economic community, and, quite likely, stronger competition on a world-wide basis. The impact of restructuring in the Soviet-bloc countries can only be positive, in terms of market growth.

Overarching Trends

As we draw both this chapter and the book to a close, one might ask: What are the "megatrends" that will be influencing biotechnology in the

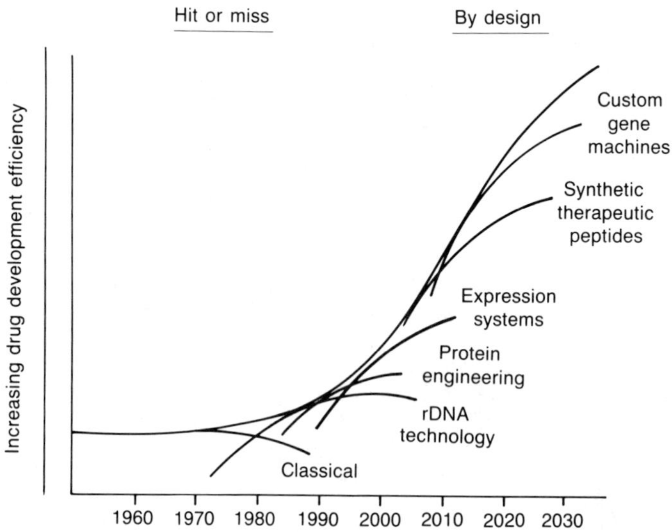

FIGURE 19-10 Evolution of drug development. (*Source*: Consulting Resources Corporation.)

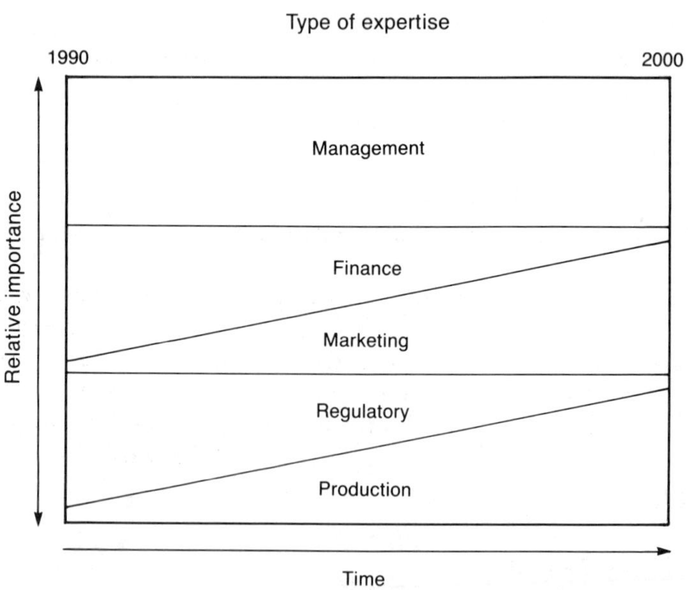

FIGURE 19-11 Shifting critical success factors. (*Source*: Consulting Resources Corporation.)

1990s? There are probably as many ideas for this as there are authors writing about biotechnology. In my opinion the key trends are the following:

1. We will move from fighting symptoms to fighting disease-causing factors.
2. We will move from treating disease to preserving wellness.
3. We will move from random testing to rational synthesis for new drugs.
4. We will move from mass medicine to customized medicine.
5. We will move from a chemical and physical basis to more of a biological basis for technology.
6. We will move from selective breeding to genetic manipulation for improving animals and plants.
7. We will move from chemical reactions to biochemical reactions to produce more products.

Certainly there are more than these seven, but these will at least serve as signposts for some important developments as we head towards the twenty-first century.

A PLEA FOR MORE EMPHASIS ON SCIENCE IN EDUCATION

If the United States, and, in a broader context, mankind, is to benefit fully from the potential of biotechnology, it is extremely important that the development of biotechnology not be unnecessarily restrained by the fears bred of ignorance. Fears of "killer tomatoes," or "rampant new strains of insects" are given far more attention than they deserve. When fear is mixed with ignorance, the inevitable outcome is retrenchment. The sad result could be that human lives, as well as many attractive commercial opportunities could be lost because the beneficial potential of biotechnology is unnecessarily restrained. As pointed out by Naisbitt in *Megatrends 2000,* "Technology is not inherently evil. It is neutral. How we use it is key." I would like to add that we can only use it to its full potential if we, and the next generation, understand it.

REFERENCES

Naisbitt, John. 1982. *Megatrends.* New York, NY: Warner Books, Inc.

Naisbitt, John, & Aburdene, Patricia. 1990. *Megatrends 2000.* New York: William Morrow & Co, Inc.

INDEX